Human Retroviruses

Human Retroviruses

EDITED BY

Bryan R. Cullen

*Howard Hughes Medical Institute, Section of
Genetics and Departments of Microbiology and
Medicine, Duke University Medical Center,
Durham, NC 27710, USA*

IRL PRESS
at
OXFORD UNIVERSITY PRESS
Oxford New York Tokyo

Oxford University Press, Walton Street, Oxford OX2 6DP

*Oxford   New York   Toronto
Delhi   Bombay   Calcutta   Madras   Karachi
Kuala Lumpur   Singapore   Hong Kong   Tokyo
Nairobi   Dar es Salaam   Cape Town
Melbourne   Auckland   Madrid
and associated companies in
Berlin   Ibadan*

Oxford is a trade mark of Oxford University Press

*Published in the United States
by Oxford University Press Inc., New York*

© *Oxford University Press, 1993*

*All rights reserved. No part of this publication may be
reproduced, stored in a retrieval system, or transmitted, in any
form or by any means, without the prior permission in writing of Oxford
University Press. Within the UK, exceptions are allowed in respect of any
fair dealing for the purpose of research or private study, or criticism or
review, as permitted under the Copyright, Designs and Patents Act, 1988, or
in the case of reprographic reproduction in accordance with the terms of
licences issued by the Copyright Licensing Agency. Enquiries concerning
reproduction outside those terms and in other countries should be sent to
the Rights Department, Oxford University Press, at the address above.*

*This book is sold subject to the condition that it shall not,
by way of trade or otherwise, be lent, re-sold, hired out, or otherwise
circulated without the publisher's prior consent in any form of binding
or cover other than that in which it is published and without a similar
condition including this condition being imposed
on the subsequent purchaser.*

A catalogue record for this book is available from the British Library

*Library of Congress Cataloging in Publication Data
Human retroviruses / edited by Bryan R. Cullen.
p.   cm.
Includes bibliographical references and index.
1. Retroviruses.   I. Cullen, Bryan.
[DNLM: 1. Gene Expression Regulation.   2. HTLV Infections.
3. HTLV Viruses.   4. Retroviridae.   QW 166 H91837]
QR414.5.H8   1993   616'.0194 — dc20   92-48274
ISBN 0-19-963383-5 (hbk.)
ISBN 0-19-963382-7 (pbk.)*

*Typeset by
Footnote Graphics, Warminster, Wilts.
Printed in Great Britain by
The Bath Press*

Preface

Although retroviral research has a long and productive history, this field has clearly been galvanized over the last decade by the discovery of human retroviruses and their emergence as a major public health concern. In this volume, I have attempted to highlight those areas of current human retroviral research that will be of most interest to molecular biologists. Appropriately, the volume focuses particularly on HIV-1, the best understood and most pathogenic human retrovirus. Irvin Chen and colleagues begin by reviewing the molecular basis for cell tropism in HIV-1, an area of considerable importance to the regulation of HIV-1 pathogenesis *in vivo*. Gene regulation in human lentiviruses is discussed by Gary Nabel, focusing on the role of cellular transcription factors, and by Matija Peterlin, discussing transcriptional *trans*-activation by the viral Tat protein. The complexity of post-transcriptional regulation of HIV-1 gene expression is revealed in the chapter by Tris Parslow. The role of the various other auxiliary proteins encoded by the primate lentiviruses is discussed by Ron Desrosiers from the perspective of the simian model of AIDS. Finally, chapters by John Brady and colleagues and by Rolf Flügel serve to review the molecular biology and pathogenic potential of the human T-cell leukemia and foamy virus families, retroviruses that are, from the molecular biological perspective, equally as fascinating as the more pathogenic human lentiviruses. My thanks go to each of the authors for their willingness to pursue this project and for their patience with, and attentiveness to, my editorial attempts to minimize repetition and maximize logical flow. I can only hope that this volume will prove to be of interest both to researchers in the field and, most importantly, to younger scientists deciding on their future research careers.

Durham, NC, USA Bryan R. Cullen
November 1992

Contents

List of contributors	xiii
Abbreviations	xv

1 An introduction to human retroviruses — 1
BRYAN R. CULLEN

1. Introduction	1
2. The retroviral replication cycle	2
2.1 Receptor binding and entry	2
2.2 Reverse transcription and integration	4
2.3 Regulation of retroviral gene expression	5
2.4 Virion assembly and release	8
3. Retroviral taxonomy	9
4. Conclusions	10
Acknowledgements	11
References	11

2 The biological and molecular basis for cell tropism in HIV — 17
VICENTE PLANELLES, QI-XIANG LI, and IRVIN S. Y. CHEN

1. Introduction	17
2. Genetic variation of HIV	17
3. Species specificity of HIV	19
4. Tissue distribution of HIV	19
4.1 Infection of blood cells	20
4.2 Infection of the central nervous system	20
4.3 Infection of the gastrointestinal tract	20
4.4 Infection of the lung	21
4.5 Tropism and pathogenesis	21
5. The CD4 glycoprotein is the major receptor for HIV	22

6. Structure of the HIV envelope glycoprotein	22
7. Tropism of the AIDS virus: historical perspective	23
8. V3 loop as a determinant of tropism	24
8.1 Macrophage tropism	25
8.2 Tropism for immortalized T-cell lines	26
8.3 Neutralization by soluble CD4 correlates with viral tropism	26
9. Envelope regions other than V3 may be involved in tropism	27
9.1 The CD4-binding domain	27
9.2 A region N-terminal to V3 may modulate infection efficiency in macrophage-tropic isolates	28
9.3 Amino acid changes throughout envelope determine tropism in SIV	28
9.4 The cytoplasmic tail of SIV *env* may be associated with species specificity	28
10. Regions other than envelope may also control tropism	29
10.1 *tat*	29
10.2 *nef*	29
10.3 Reverse transcriptase	30
10.4 The LTR as a potential determinant for pathogenesis and tropism of SIV	30
11. Tropism for other cell types	31
11.1 Infection of fibroblasts	31
11.2 Infection through Fc and complement receptors	31
11.3 CD4-independent infection	32
12. Mechanisms	32
12.1 Important events controlling tropism act at the level of entry	32
12.2 The binding affinity between gp120 and CD4 may control tropism	33
12.3 The V3 loop may control tropism by regulating the interaction with ancillary cell surface molecules	33
13. Conclusion	35
Acknowledgements	36
References	36

3 The role of cellular transcription factors in the regulation of human immunodeficiency virus gene expression   49

GARY J. NABEL

1. Introduction	49

	2. RNA polymerases and associated transcription factors	51
	3. Transcriptional regulatory proteins and the HIV promoter	52
	4. Inducible gene expression and the HIV enhancer	53
	5. Negative regulation of the HIV enhancer	56
	6. The NF-κB family of transcriptional activators	57
	7. Regulation of NF-κB by IκB	59
	8. Anti-oxidants and the regulation of HIV transcription	60
	9. Macrophage regulation of HIV expression	60
	10. Transcriptional regulation of HIV-2	61
	11. Conclusions	63
	References	63

4 Tat *trans*-activator 75

B. MATIJA PETERLIN, MELANIE ADAMS, ALICIA ALONSO, ANDREAS BAUR, SUBIR GHOSH, XIAOBIN LU, and YING LUO

	1. Introduction	75
	2. Tat *trans*-activator	75
	2.1 Proteins that associate with Tat	79
	3. *Trans*-activation response element, TAR	80
	3.1 Interactions between Tat and TAR	80
	3.2 TAR decoys	83
	4. Mechanisms of *trans*-activation by Tat	83
	4.1 Transcriptional effects of Tat	83
	4.2 Post-transcriptional effects of Tat	87
	5. Tat and the HIV life cycle	88
	6. Other effects of Tat	88
	7. Conclusion	89
	References	90

5 Post-transcriptional regulation of human retroviral gene expression 101

TRISTRAM G. PARSLOW

	1. Introduction	101

2. Alternative splicing of retroviral RNAs   101
  2.1 Simple retroviruses   102
  2.2 Complex retroviruses   103
3. Post-transcriptional *trans*-activation of complex retroviruses   105
  3.1 The HIV-1 Rev protein   105
  3.2 Rev-like *trans*-activators in other complex retroviruses   108
  3.3 Rev-like *trans*-activators, latency, and the biphasic life cycle of complex retroviruses   110
  3.4 Experimental systems for studying Rev-like proteins   111
  3.5 Functional architecture of Rev-like proteins   112
  3.6 *Trans*-dominant inhibitory mutants   115
  3.7 Properties of the response elements   115
  3.8 The mechanism of post-transcriptional *trans*-activation   118
4. Translation of retroviral mRNAs   120
  4.1 Ribosomal frameshifting and the expression of Pol   120
  4.2 Translation of HIV-1 Env and other proteins from multicistronic mRNAs   123
5. Potential targets for antiretroviral therapy   125
Acknowledgements   126
References   126

6 Auxiliary proteins of the primate immunodeficiency viruses   137
JAMES S. GIBBS and RONALD C. DESROSIERS

1. Introduction   137
2. *vif*   141
3. *vpr*   142
4. *vpx*   144
5. *vpu*   145
6. *nef*   146
7. Concluding remarks   151
References   152

7 The molecular biology of human T-cell leukemia viruses 159

SCOTT D. GITLIN, JÜRGEN DITTMER, ROBERT L. REID, and JOHN N. BRADY

1. Introduction 159
2. HTLV genome structure 160
3. Transformation 161
4. Epidemiology 162
   4.1 Discriminating assays for HTLV-I and HTLV-II 162
   4.2 Transmission 162
5. HTLV-associated diseases 163
   5.1 ATL 163
   5.2 TSP/HAM 163
   5.3 Other diseases 164
   5.4 Animal models 164
6. Tax structure and functional domains 165
7. Rex 167
8. $p21^{x-III}$ 170
9. Transcriptional regulation of HTLV gene expression 170
   9.1 Tax-responsive element 1 (TRE-1) 172
   9.2 Tax-responsive element 2 (TRE-2) 173
   9.3 R region of the HTLV-I LTR 174
10. Tax *trans*-activation of cellular genes 174
11. Extracellular cytokine effects of Tax 177
12. Conclusion 179
References 180

8 The molecular biology of the human spumavirus 193

ROLF M. FLÜGEL

1. Introduction 193
2. Morphological and virological properties of spumaviruses 194
3. An overview of the replication cycle of HSRV in comparison with other retroviruses 194
4. The genetic structure of HSRV is complex 196

5. The transcriptional pattern of the HSRV genome is complex — 200
  5.1 HSRV *env* transcripts — 200
  5.2 *bel* and *bet* transcripts — 201
6. Bel 1 is the HSRV transcriptional *trans*-activator — 202
7. The HSRV LTR is unusually long and contains different DNA response elements for the *trans*-activator Bel 1 — 203
8. Is Bel 3 a spumaviral superantigen? — 205
9. HSRV-specific detection systems — 206
10. Sero-epidemiological studies on prevalence of HSRV antibodies — 207
11. Neuropathogenicity of HSRV genes in transgenic mice — 207
12. Concluding remarks — 209
Acknowledgements — 209
References — 209

Index — 215

Contributors

MELANIE ADAMS
Howard Hughes Medical Institute, Departments of Medicine and Microbiology and Immunology, University of California, SF, Third and Parnassus Avenues, San Francisco, CA 94143, USA.

ALICIA ALONSO
Howard Hughes Medical Institute, Departments of Medicine and Microbiology and Immunology, University of California, SF, Third and Parnassus Avenues, San Francisco, CA 94143, USA.

ANDREAS BAUR
Howard Hughes Medical Institute, Departments of Medicine and Microbiology and Immunology, University of California, SF, Third and Parnassus Avenues, San Francisco, CA 94143, USA.

JOHN N. BRADY
Laboratory of Molecular Virology, National Cancer Institute, National Institutes of Health, 9000 Rockville Pike, Bethesda, MD 20892, USA.

IRVIN S. Y. CHEN
Departments of Microbiology and Immunology and Medicine, UCLA School of Medicine and Jonsson Comprehensive Cancer Center, 10833 Leconte Avenue, Los Angeles, CA 90024, USA.

BRYAN R. CULLEN
Howard Hughes Medical Institute, Section of Genetics and Departments of Microbiology and Medicine, Duke University Medical Center, Box 3025, Durham, NC 27710, USA.

RONALD C. DESROSIERS
Harvard Medical School, New England Regional Primate Research Center, 1 Pinehill Drive, PO Box 9102, Southboro, MA 01772, USA.

JORGEN DITTMER
Laboratory of Molecular Virology, National Cancer Institute, National Institutes of Health, 9000 Rockville Pike, Bethesda, MD 20892, USA.

ROLF M. FLÜGEL
Deutsches Krebsforschungszentrum, Angewandte Tumorvirologie, Projektgruppe Humane Retroviren, Im Neuenheimer Feld 242, D-6900 Heidelberg, Germany.

SUBIR GHOSH
Howard Hughes Medical Institute, Departments of Medicine and Microbiology and

Immunology, University of California, SF, Third and Parnassus Avenues, San Francisco, CA 94143, USA.

JAMES S. GIBBS
Harvard Medical School, New England Regional Primate Research Center, 1 Pinehill Drive, PO Box 9102, Southboro, MA 01772, USA.

SCOTT D. GITLIN
Laboratory of Molecular Virology, National Cancer Institute, National Institutes of Health, 9000 Rockville Pike, Bethesda, MD 20892, USA.

QI-XIANG LI
Departments of Microbiology and Immunology and Medicine, UCLA School of Medicine and Jonsson Comprehensive Cancer Center, 10833 Leconte Avenue, Los Angeles, CA 90024, USA.

XIAOBIN LU
Howard Hughes Medical Institute, Departments of Medicine and Microbiology and Immunology, University of California, SF, Third and Parnassus Avenues, San Francisco, CA 94143, USA.

YING LUO
Howard Hughes Medical Institute, Departments of Medicine and Microbiology and Immunology, University of California, SF, Third and Parnassus Avenues, San Francisco, CA 94143, USA.

GARY J. NABEL
Howard Hughes Medical Institute, Section of Genetics and Departments of Microbiology and Medicine, Duke University Medical Center, Box 3025, Durham, NC 27710, USA.

TRISTRAM G. PARSLOW
Department of Pathology, University of California, SF, Third and Parnassus Avenues, San Francisco, CA 94143, USA.

B. MATIJA PETERLIN
Howard Hughes Medical Institute, Departments of Medicine and Microbiology and Immunology, University of California, SF, Third and Parnassus Avenues, San Francisco, CA 94143, USA.

VICENTE PLANELLES
Departments of Microbiology and Immunology and Medicine, UCLA School of Medicine and Jonsson Comprehensive Cancer Center, 10833 Leconte Avenue, Los Angeles, CA 90024, USA.

ROBERT L. REID
Laboratory of Molecular Virology, National Cancer Institute, National Institutes of Health, 9000 Rockville Pike, Bethesda, MD 20892, USA.

Abbreviations

AIDS	acquired immunodeficiency syndrome
ATL	adult T-cell leukemia
bel	between env and LTR
BLV	bovine leukemia virus
BRE	Bel-1 response element
CDR	complementarity-determining regions
CFS	chronic fatigue syndrome
CHO	Chinese hamster ovary
CNS	central nervous system
CREB	cyclic AMP response element binding protein
DRB	dichloro-1-β-D-ribofuranosylbenzimidazole
EIAV	equine infectious leukemia virus
env	envelope
gag	group-specific antigen
GALV	gibbon ape leukemia virus
GM-CSF	granulocyte-macrophage colony stimulating factor
HAM	HTLV-associated myelopathy
HFV	human foamy virus
HIP	HIV initiator protein
HIV	human immunodeficiency virus
HSRV	human spuma retrovirus
HTLV	human T-cell leukemia virus
IκB	inhibitor κB
IL	interleukin
IL-2R	IL-2 receptor
LBP	leader binding protein
LTR	long terminal repeat
MHC	major histocompatibility complex
MLV	murine leukemia virus
MMTV	mouse mammary tumour virus
Mo-MLV	Moloney murine leukemia virus
NAC	*N*-acetyl cysteine
nef	negative factor
NF-κB	nuclear factor κB
NRE	negative regulatory element
PBL	peripheral blood lymphocyte
PBMC	peripheral blood mononuclear cells
PCR	polymerase chain reaction

pol	polymerase
rev	regulator of expression of virion proteins (HIV)
rex	regulator of expression of virion proteins (HTLV)
RRE	Rev response element
RSV	Rous sarcoma virus
RxRE	Rex response element
sag	super antigen
SFV	simian foamy virus
SIV	simian immunodeficiency virus
SLII	Stem–loop II
TAR	*trans*-activation response (element)
tat	*trans*-activator (HIV)
tax	*trans*-activator (HTLV)
TCR	T-cell receptor
TM	transmembrane
TNF-α	tumour necrosis factor α
TRE	tax responsive element
TSP	tropical spastic paraparesis
UV	ultraviolet light
vif	virion infectivity factor

1 | An introduction to human retroviruses

BRYAN R. CULLEN

1. Introduction

The first retrovirus to be discovered, Rous sarcoma virus (RSV), was described in 1911 as a filterable agent that caused sarcomas in chickens (1). The association between retroviral infection and oncogenesis in animals is clearly the primary reason that this viral genus was to become such a focus of biological research over approximately the next 70 years (2). Indeed, retroviruses are still occasionally referred to as 'RNA tumour viruses' or 'oncornaviruses'. The various acutely transforming retroviruses, such as RSV, were subsequently shown to express novel gene products, termed oncogenes, that are critical for their transforming potential but that play no role in viral replication *per se* (3, 4). The demonstration that these viral oncogenes are actually mutant forms of cellular genes that had been acquired by retroviruses through a process of illegitimate recombination was the first, and most critical, step in the development of the proto-oncogene hypothesis of cellular transformation, i.e. the concept that oncogenesis initiates with the aberrant expression of cellular genes that regulate the response of eukaryotic cells to extracellular stimuli and/or that regulate gene expression directly (4, 5). Indeed, several classes of proteins now known to play critical roles in the signal transduction process, including tyrosine kinases (e.g. *src*), G-proteins (e.g. *ras*), and immediate-early response genes (e.g. *fos*), were first identified as retroviral oncogenes (4).

The demonstration in animals such as chickens and mice that certain retroviruses, in particular type C retroviruses, could act as etiologic agents of cancer led to an intense search for related viruses in man. It is, therefore, curious that the discovery of the first human retrovirus, the human foamy virus (HFV), by Achong *et al.* in 1971 (6) provoked only minimal scientific interest. One can only surmise that foamy viruses, a ubiquitous, apparently non-pathogenic subgroup of the retroviridae (Chapter 8), were not at that time considered important due to their evident lack of transforming potential. However, a very different response greeted the discovery in 1980 of human T-cell leukemia virus type I (HTLV-I), and the demonstration that HTLV-I was the etiologic agent of the human disease adult T-cell leukemia (7, 8) (Chapter 7). Research into the HTLV-I replication cycle led to the discovery of the first retrovirally encoded regulatory protein, the HTLV-I Tax gene product (9), and therefore served as the initial precedent for the concept of complex retroviruses (see Section 2.3). However, while the discovery of a transforming human retrovirus at first suggested that the direct subversion of cellular proto-oncogene expression characteristic of transformation by animal C-type retroviruses might also occur in humans, it has since become evident that this is not the case. Instead, transformation by HTLV-I appears to be mediated by the HTLV-I Tax gene product, a nuclear regulatory protein that is critical for HTLV-I replication (10, 11) (Chapter 7). Transformation by HTLV-I, therefore, appears more analogous to transformation by DNA tumour viruses, such as human papilloma virus, which also encode viral oncogenes that primarily function as nuclear *trans*-activators of viral gene expression but that can also induce transformation due to their effects on the expression of host cell genes (12).

The disease we now know as acquired immunodeficiency syndrome (AIDS) was first reported in 1981 as an outbreak of severe opportunistic infections among a group of previously healthy homosexual males (13, 14). The prediction, based on the pattern of disease transmission, that the etiologic agent of AIDS was likely to be a blood-borne viral pathogen was fulfilled in 1983 with the isolation of a third human retrovirus from the lymphocytes of AIDS patients (15, 16). This retrovirus, now known as human immunodeficiency virus type 1 (HIV-1) is the prototype of an extensive family of primate lentiviruses that also includes a second pathogenic human virus termed HIV-2. The severity of the disease caused by HIV-1 (Chapter 2), the continued rapid spread of HIV-1 in both the developed countries and the Third World, and the relative lack of effectiveness of current chemotherapeutic treatments for AIDS have made self-evident the need for better inhibitors of HIV-1 replication and, by extension, the need for a far more complete understanding of the replication cycle of HIV-1 and other human retroviruses.

2. The retroviral replication cycle
2.1 Receptor binding and entry

The first step in the retroviral replication cycle is the specific interaction of the outer membrane component of the virion envelope glycoprotein with a specific cell-surface receptor (Fig. 1). Frequently, this receptor is found only on cells of a particular species or tissue, and the target cell population or cell tropism of a particular retrovirus is therefore defined, in the first instance, by the prevalence of this receptor (17) (Chapter 2). Three distinct retroviral receptors have been defined thus far, i.e. the glycoprotein receptors for HIV-1 (18–21), for ecotropic murine leukemia virus (MLV) (22) and for gibbon ape leukemia virus (GALV) (23). The receptor for HIV-1 is CD4, a member of the immunoglobulin supergene family that has a single *trans*-membrane segment (18-21) (Chapter 2). CD4 is involved in mediating the interaction between the T-cell receptor complex and major histocompatibility complex class II molecules and in transducing the resultant signal to intracellular tyrosine kinases (21). The ecotropic MLV and GALV

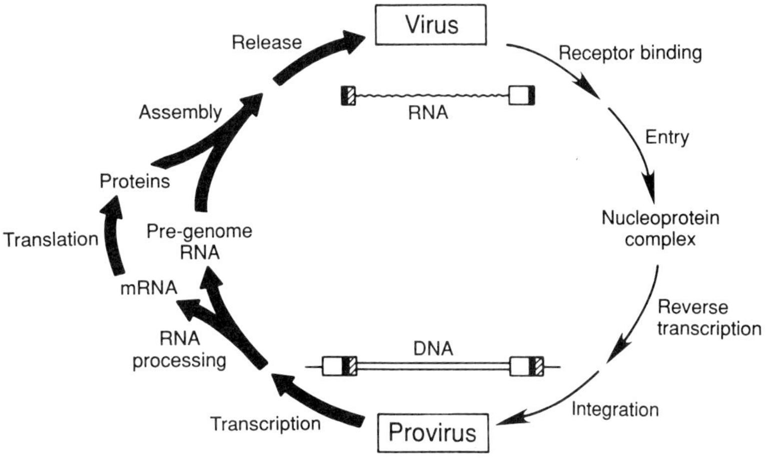

Fig. 1 An overview of the retroviral replication cycle. Steps prior to proviral integration are indicated by light arrows; postintegration events are denoted by thick arrows. A detailed discussion of each step is provided in Section 2

receptors both appear to contain multiple membrane-spanning regions and recent data suggest (22, 24) that the ecotropic MLV receptor is a cationic amino acid transporter protein. There is every reason to suspect that other retroviral receptors will be distinct both from the three that are currently known and from each other.

After receptor binding has occurred, the retroviral virion must enter the cell via a process of membrane fusion. Considerable evidence suggests that it is the hydrophobic amino terminus of the transmembrane component of the envelope protein that mediates this step (25, 26). Activation of the fusogenic potential of the retroviral envelope almost certainly requires complex conformational changes in the viral envelope proteins (27) (see Chapter 2). Retroviruses are known to enter cells via a pH-independent process, i.e. by direct fusion (28, 29). This conformational shift is, therefore, not induced by exposure to low pH subsequent to receptor-mediated endocytosis, as has been described for viruses such as influenza and vesicular stomatitis virus (30). The observation that CD4, while necessary, is not sufficient for infection of cells by HIV-1 (Chapter 2) has raised the possibility that the process of membrane fusion might well be mediated by the specific interaction of the retroviral envelope with a second cell-surface 'receptor'. Recent data suggesting that the functional interaction of the ecotropic MLV envelope with its cell-surface receptor might also require a cell-surface accessory factor have raised the possibility that this is a general phenomenon (24).

After fusion, the retroviral virion undergoes a process of uncoating which results in the partial or complete loss of the envelope proteins and the virion membrane and leads to the formation of a viral 'nucleoprotein complex' that is competent to initiate the process of reverse transcription (31). *In vitro*, this step can be partially reproduced by treating virions with non-ionic detergents or agents such as melittin (31, 32). While essentially nothing is currently known about retroviral uncoating *in vivo*, the recent description of compounds that appear specifically to inhibit the process of HIV-1 uncoating in culture (33) suggests that this process may be accessible to biochemical analysis.

2.2 Reverse transcription and integration

The next step in the retroviral replication cycle is the initiation of reverse transcription of the retroviral RNA genome within the viral nucleoprotein complex (Fig. 1). Normally, this process occurs in the cytoplasm of the infected cell (31). The primer for reverse transcription is a cellular tRNA molecule that has been specifically packaged into the retroviral virion. Interestingly, different retrovirus species are observed to use different tRNA primers, e.g. tRNA^{TRP} for RSV, tRNA^{PRO} for HTLV-I, and tRNA^{LYS} for HIV-1. However, in each case the 3' end of the tRNA primer is bound to the viral RNA genome by a perfectly complimentary stretch of approximately 18 nucleotides located immediately adjacent to the U5 region of the viral long terminal repeat (31, 34).

The complex and highly ordered process of reverse transcription is carried out by the virion enzymes reverse transcriptase and RNAse H (endonuclease) that are encoded by the retroviral *pol* gene. This key step of the retroviral replication cycle appears to be highly conserved between the various retroviral species and has been reviewed extensively elsewhere (31, 34). The final result of this process is the conversion of the single-stranded, positive-sense RNA genome into a double-stranded linear DNA proviral intermediate. Reverse transcription also generates the long terminal repeat (LTR) elements that are both characteristic of retroviral proviruses and crucial to its regulated expression (see below) (31).

As noted above, reverse transcription is normally performed in the cell cytoplasm within the viral nucleoprotein complex. The next step in the viral replication cycle is the migration of this complex into the cell nucleus (31). While it has been suggested that this may be facilitated by the breakdown of the nuclear membrane, recent data indicate that nuclear transport of the provirus is an active process that requires the continued association of the linear DNA proviral intermediate with the virion Gag proteins (35).

Once present in the nucleus, the next step in the retroviral replication cycle is the precise integration of the linear DNA proviral intermediate into the genome of the host cell (31) (Fig. 1). This process is mediated by the specific interaction of the viral integrase protein, the third gene product of the viral *pol* gene, with two short inverted DNA repeats located at the ends of the viral LTRs (36–39). In contrast, integration seems to be essentially random with respect to host DNA sequences. The process of proviral integration has been reproduced *in vitro* and the critical biochemical steps have been extensively charcterized (36–39). While integrase is clearly the only viral protein that is required *in vitro*, it remains possible that other retroviral proteins could play a role in mediating proviral integration in the infected cell. There also remains some debate as to whether proviral integration is an essential step in the retroviral replication cycle for all retroviral species (40–42). Clearly, the efficiency with which integration occurs appears to differ significantly among the various members of this viral genus. However, the conservation of the integrase function among all known retroviruses certainly implies that integration is likely to be critical for efficient viral replication *in vivo* (31).

An interesting aspect of retroviral reverse transcription and proviral integration is that these processes are extremely inefficient in resting cells such as serum-starved fibroblasts or quiescent T-lymphocytes (43–45) (Chapter 2). In contrast, macrophages, which are non-dividing but not quiescent, are fully capable of supporting reverse transcription and integration of the HIV-1 genome (46). These observations imply that resting cells either lack a cofactor required for provirus formation or express an inhibitor of this process.

A second aspect of the reverse transcription step that is worthy of note is that this process is the major source of the rapid generation of genomic variability that is characteristic of many retroviruses including, particularly, HIV-1 (47, 48). All retroviruses package two complete copies of their viral RNA genome into virion particles, i.e. retroviruses are diploid. During reverse transcription, a series of both random and non-random strand-switches occurs, leading to the generation of a single proviral DNA intermediate from these two RNA precursors (47, 48). This renders the retroviral virion highly resistant to agents that damage the RNA genome because such damage can be avoided by transfer of reverse transcriptase to the undamaged strand. More importantly, this process facilitates an extraordinarily high rate of recombination between the similar but distinct retroviral genomes that may exist in dually infected cells. In HIV-1, this capability, when combined with the high error rate of the reverse transcriptase (approximately 1 error per replication cycle), permits the rapid generation and dissemination of mutations in response to selective pressures imposed by the host immune system or by treatment of the infected individual with antiretroviral agents (49–51).

2.3 Regulation of retroviral gene expression

Integrated proviruses are flanked by the LTR elements that are generated during reverse transcription of the viral RNA genome (Fig. 2). The LTRs contain sequences required for efficient polyadenylation of viral transcripts within the 3' LTR and also contain essential enhancer and promoter elements that act as targets for cellular and viral regulatory proteins (31, 52). The interplay of these *cis*-acting sequences with the specific transcription factors available in infected cells determines the rate of transcription initiation from the 5' LTR promoter element and, therefore, strongly influences the level of viral replication.

Retroviruses can be broadly subdivided into 'simple' and 'complex' retroviruses (52) (Table 1). Simple retroviruses, such as RSV and MLV, rely entirely on host DNA sequence specific transcription factors to regulate the level of proviral transcription. Complex retroviruses, in contrast, encode potent transcriptional *trans*-activators of their LTR promoter elements (52). All human retroviruses thus far described fall into the complex retrovirus category. The specific transcriptional *trans*-activators encoded by HIV-1 (Chapter 4), HTLV-I (Chapter 7), and HFV

6 | AN INTRODUCTION TO HUMAN RETROVIRUSES

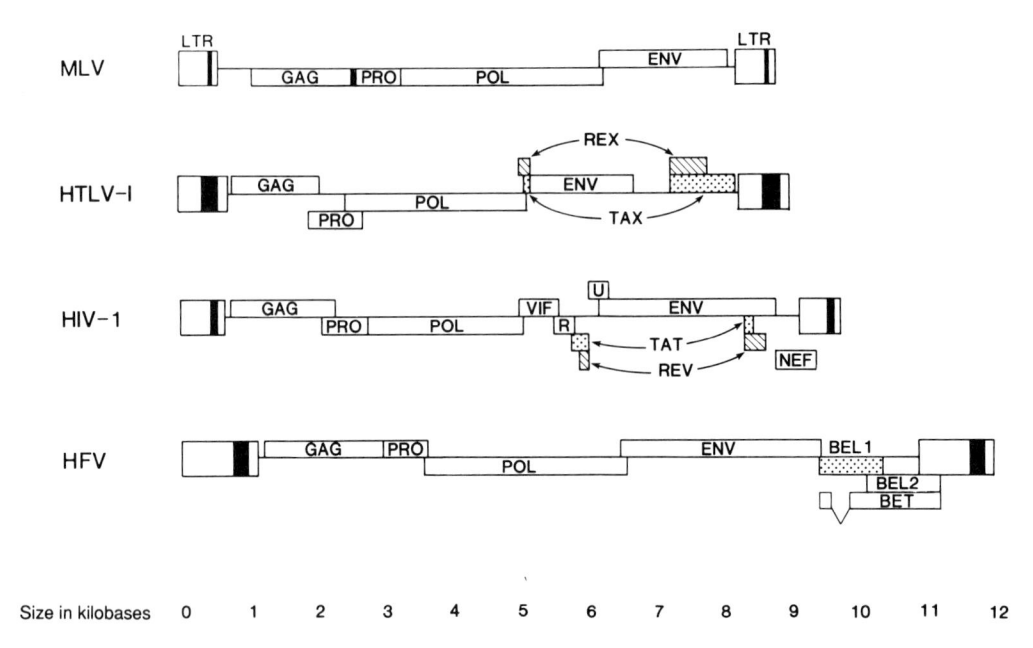

Fig. 2 Human retroviral genomic organization. Schematic representation of the proviral genomic organization seen in the various prototypic human retroviruses and in MLV, a representative simple retrovirus. While all known viral genes are named and drawn to scale, this should not be viewed as an exhaustive listing. Transcriptional activators are marked by stippling; known post-transcriptional regulators are indicated by hatching. The proviral LTRs are shown as large terminal boxes with the R region shown in black. PRO, protease; R, Vpr; U, Vpu

(Chapter 8) are discussed in detail elsewhere in this volume. In the absence of the homologous viral *trans*-activator, each of these LTR promoter elements induces only a low or basal level of proviral transcription that is insufficient to maintain a productive retroviral replication cycle. However, while these virally encoded transcription factors accord the provirus a degree of autonomy in the regulation of viral mRNA transcription, it must be emphasized that efficient retroviral gene expression remains critically dependent on the availability of specific cellular transcription factors and cofactors, regardless of whether the retrovirus is of the simple or complex class (Chapter 3).

All retroviruses share the same basic genomic organization (Fig. 2). In particular, the viral *gag* gene, which encodes the virion structural proteins, the *pol* gene, which encodes the various retroviral enzymes, and *env*, which encodes the envelope glycoprotein, are invariably present in the same 5' to 3' order within the viral genome (Fig. 2) (31, 52). Many simple retroviruses, including MLV, encode only these three characteristic gene products. However, complex retroviruses encode between two and six additional proteins that subserve critical regulatory or auxiliary functions within the viral life cycle (Fig. 2) (Chapter 6).

Transcription of the integrated provirus gives rise to a single genome length initial transcript that also serves as the mRNA for Gag and Pol (Chapter 5). All

Table 1 Major taxonomic divisions among retroviruses

	Subgroup	Prototype	Other examples
Simple retroviruses	B-type retroviruses	MMTV	
	C-type retroviruses		
	group A	RSV	ALV, AMV
	group B	MLV	FeLV, MSV, GALV, SNV, REV, SSV
	D-type retroviruses	MPMV	SRV-1
Complex retroviruses	Lentiviruses	HIV-1	HIV-2, SIV, Visna virus, FIV, EIAV
	T-cell leukemia viruses	HTLV-I	HTLV-II, STLV, BLV
	Foamy viruses	HFV	SFV, BFV

RSV, Rous sarcoma virus; ALV, avian leukemia virus; FeLV, feline leukemia virus; MSV, murine sarcoma virus; SNV, spleen necrosis virus; Rev, reticuloendotheliosis virus; SSV, simian sarcoma virus; MMTV, mouse mammary tumor virus; MPMV, Mason-Pfizer monkey virus; SRV-1, simian retrovirus type 1; STLV, simian T-cell leukemia virus; BLV, bovine leukemia virus; SFV, simian foamy virus; BFV, bovine foamy virus; HFV, human foamy virus; HIV-1, human immunodeficiency virus type 1; HTLV-I, human T-cell leukemia virus type I; FIV, feline immunodeficiency virus; EIAV, equine infectious anemia virus; MLV, murine leukemia virus; GALV, gibbon ape leukemia virus; AMV, avian myeloblastosis virus.

retroviruses also encode a singly spliced RNA species that acts as the mRNA for the viral envelope protein. These are the only two mRNA species encoded by the majority of simple retroviruses, although some members of this class do encode a second spliced RNA species that encodes an oncogene (e.g. *src* in RSV) or a gene involved in activating the replication of specific target cell subsets (e.g. *sag* in MMTV). In contrast, complex retroviruses express several additional multiply spliced mRNA species which encode the various regulatory and auxiliary proteins that are characteristic of this retroviral class (Fig. 2).

In the case of simple retroviruses, the ratio of unspliced to spliced viral mRNA is determined solely by the efficiency of the interaction of viral *cis*-acting RNA sequences with cellular splicing factors (53, 54). However, the intricate pattern of post-transcriptional RNA processing characteristic of complex retrovirus has led to the development of virally encoded post-transcriptional regulatory proteins that serve to orchestrate this process in infected cells. As discussed in detail in Chapter 5, these proteins, termed Rev in lentiviruses and Rex in T-cell leukemia viruses, are required for the cytoplasmic expression of the incompletely spliced mRNA species that encode the retroviral structural proteins (55). In contrast, the multiply spliced viral mRNA species that encode the retroviral regulatory proteins are expressed in a Rev/Rex-independent manner (Fig. 3). Currently, it remains unclear whether foamy viruses also encode a Rev-like activity or whether this retroviral subgroup has developed an alternative strategy to regulate the complex pattern of foamy virus mRNA expression (Chapter 8).

In simple retroviruses, the level of proviral transcription and the pattern of viral mRNA processing is entirely regulated by cellular proteins and, therefore, displays little or no temporal modulation. In contrast, efficient gene expression in complex retroviruses is dependent on the functional expression of regulatory proteins that

8 | AN INTRODUCTION TO HUMAN RETROVIRUSES

Fig. 3 Gene regulation in complex retroviruses. Complex retroviruses, including HIV-1 and HTLV-I, display a marked temporal regulation of their intracellular replication cycle marked by the sequential action of a viral 'Tat-like' transcriptional activator and of a 'rev-like' post-transcriptional regulatory protein. Progression to the late phase of the viral replication cycle, characterized by the extensive synthesis of viral structural proteins and by the release of progeny virions, is critically dependent on the functional expression of both of these virally encoded regulatory proteins

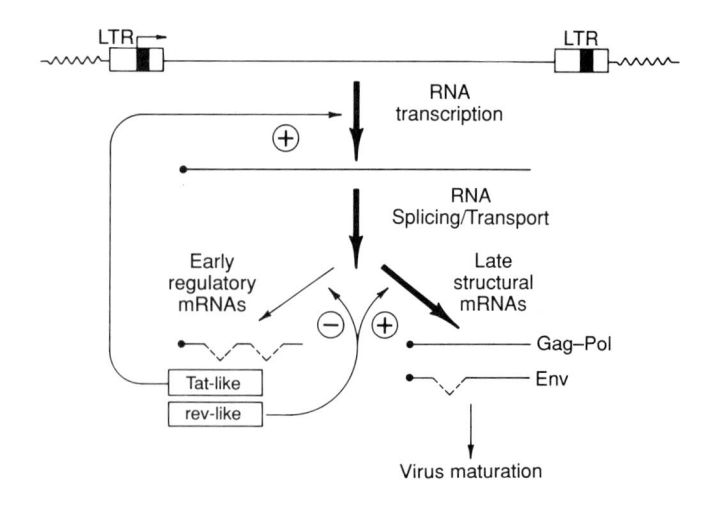

are encoded within the retroviral genome. As a result, complex retroviral gene expression is subject to a marked temporal regulation (55), as shown in Fig. 3. At early times after infection, the provirus gives rise to a low level of viral mRNA transcription that is mediated by available cellular transcription factors. Once mRNA specific for the retroviral 'Tat-like' *trans*-activator is expressed, a positive feed-back loop can be established, leading eventually to high-level transcription of the proviral genome (Fig. 3). At this time, a critical level of the 'Rev-like' post-transcriptional regulatory protein is achieved, thereby inducing viral structural protein expression and the release of progeny virions. This pattern of temporally regulated gene expression, featuring early, regulatory, and late structural phases, is also characteristic of DNA tumour viruses such as herpes simplex virus and human papilloma virus. It is perhaps no coincidence that complex retroviruses also share with DNA tumour viruses a predilection for establishing long-term, chronic infections marked by high levels of latently infected cells (56).

2.4 Virion assembly and release

Retroviral virion formation and release, the last step in the retroviral replication cycle, remain rather poorly understood (Fig. 1). In general, virion assembly occurs adjacent to the plasma membrane and involves the ordered assembly of the Gag and Gag–Pol polyproteins (31). The protein sequences involved in mediating this assembly process are known to be located in Gag but remain poorly defined. In the case of many retroviruses, including MLV and HIV-1, the Gag and Gag–Pol proteins are targeted to the inside of host cell membranes by myristoylation of their amino termini (57). This post-translational modification, which is performed by host-cell proteins, is essential for the formation of functional virions. During the process of virion assembly, two copies of the single-stranded retroviral RNA genome are incorporated into the virion by a process that generally involves the sequence-specific interaction of a zinc-finger nucleic-acid-binding motif located in the nucleocapsid component of the viral Gag protein, with a packaging signal located within the leader region of the genomic RNA (31, 58). An exception exists in the case of the foamy viruses, which lack this characteristic Gag zinc-finger motif and appear instead to contain an arginine-rich RNA-binding motif. Subsequently, the retroviral cores bud through regions of the plasma membrane that bear high concentrations of the viral Env proteins. While assembly of Env on to the virion core may be facilitated by an interaction between the transmembrane domain of Env and the viral Gag proteins (31, 52), this interaction is clearly not highly specific as retroviral virions can be readily pseudotyped using envelope proteins derived from other retroviruses or, indeed, from unrelated enveloped viruses (17, 59, 60). During or shortly after budding, the virion protease is activated, resulting in the ordered and highly specific cleavage of the Gag and Gag–Pol polyproteins into the mature virion structural proteins and enzymes (31, 61). This maturation event, which involves a detectable change in virion morphology, is essential for virion infectivity. The released virions are now capable of interacting with an appropriate cell-surface receptor and of initiating the next round of retroviral replication.

3. Retroviral taxonomy

Historically, retroviruses have been divided into three taxonomic subgroupings based primarily on the *in vivo* and *in vitro* consequences of infection (31, 62). The extensive oncovirus subgroup, which included all the type B, type C, and type D retroviruses as well as the T-cell leukemia viruses, originally comprised only viruses able to induce neoplastic disease *in vivo*, but has expanded to include several closely related but apparently non-pathogenic retroviruses (Table 1). The lentivirus subgroup was comprised of retroviruses that induced chronic diseases which generally lacked a neoplastic component. The spuma or foamy virus subgroup induces a characteristic 'foamy' cytopathic effect in culture but has generally been perceived as lacking disease potential.

Although the classification of retroviral subgroups based on their pathogenic potential remains in wide use, an alternative categorization based on the pattern of viral gene regulation in infected cells, has become increasingly popular (52). In this latter classification, retroviruses are defined as simple if they encode four or fewer gene products and lack retrovirally encoded regulatory proteins. Complex retroviruses, in contrast, are defined as encoding at least five gene products of which at least one is regulatory (Fig. 2). As noted above, complex retroviruses also display a characteristic temporal regulation of viral gene expression (Fig. 3), a feature lacking in simple retroviruses. Human retroviruses are all of the complex classification and, indeed, provide the prototype for each of the three subfamilies of complex retroviruses (Table 1).

Although the classification of retroviruses based on the pattern of viral gene regulation is extremely useful from the point of view of the molecular biologist,

Fig. 4 Evolutionary relationship of representative retroviruses. The phylogenetic tree shown here is based on a comparison of retroviral *pol* gene sequences and represents a compilation of several previously published reports (66–68). Subgroups that include human representatives, i.e. the foamy (FOAMY) and lentivirus (LENTI) subgroups as well as the T-cell leukemia (TLV) subgroup, are bracketed. However, prototypic members of retroviral subgroups that currently lack human representatives are also included. See Table 1 for classification by subgroup and for a listing of abbreviations

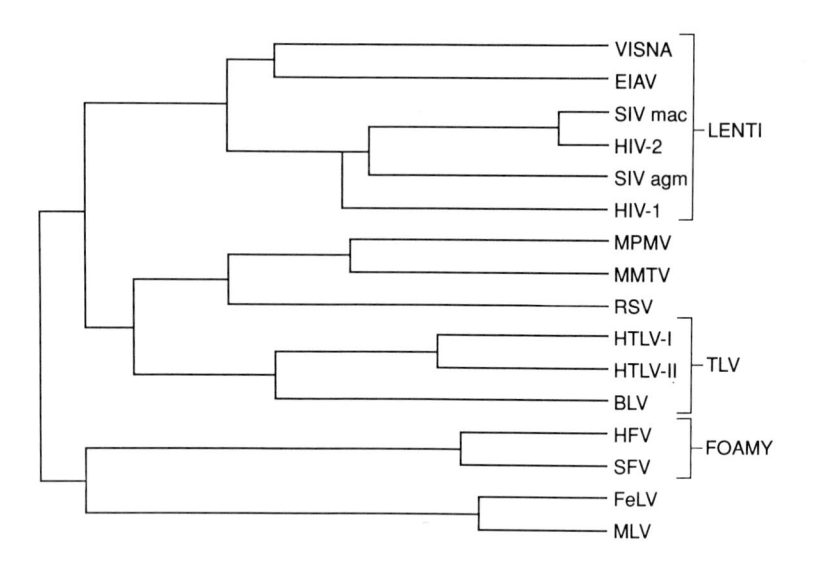

neither this classification, nor the more traditional retroviral classification based on pathogenic potential, can readily be justified based on phylogenetic analyses of retroviral genome sequences (Fig. 4). As retroviruses are generally thought to have evolved from simple intracellular transposable elements, termed retroposons, that are found widely dispersed in nature (31), this suggests that the intricate pattern of gene expression characteristic of each of the three complex retroviral subgroups (Table 1) may well have evolved independently in each case. It is interesting to speculate that the ability to regulate both the quantity and quality of retroviral gene expression in infected cells might well be key to the maintenance of an ongoing infection in the face of a potent host immune response (56) and might, therefore, offer a considerable selective advantage.

4. Conclusions

Retroviruses have long provided one of the most fruitful and rewarding experimental systems in eukaryotic molecular biology. For example, analysis of the retroviral replication cycle led directly to the discovery of reverse transcriptase (63, 64), a vital step in the development of recombinant DNA technology. Analysis of retroviral oncogenesis also led to the discovery of cellular proto-oncogenes, a key advance in the ongoing dissection of the molecular basis of cellular growth transformation and immortalization and a major step forward in the understanding of signal transduction in eukaryotes (3–5). However, the discovery of human retroviruses, and the realization that the human lentiviruses and, to a lesser extent, the human T-cell leukemia viruses pose a serious threat to public health has re-emphasized the importance of retroviruses as an experimental system. In fact, the rate of progress in understanding the replication cycle of HIV-1 since its initial discovery in 1983 has been remarkable. Among the most surprising discoveries to result from this research effort has been the extraordinary complexity of the replication cycle of HIV-1 and other human retroviruses when compared with prototypic simple retroviruses such as MLV and RSV. However, the complexity of these viruses may, in fact, provide novel targets for chemotherapeutic intervention that would not be available with simple retroviruses. This has already led to the development and clinical testing of agents able to interfere with the activity of the HIV-1 Tat *trans*-activator and hence, with HIV-1 replication (65). The more complete dissection of the critical steps in the replication cycle of HIV-1 and other human retroviruses, and the utilization of that information in the development of strategies to effectively interdict their pathogenic potential, remains the most important challenge in molecular virology today.

Acknowledgements

I thank Sharon Goodwin for her secretarial assistance. This work was supported by funds from the Howard Hughes Medical Institute.

References

1. Rous, P. (1911) A sarcoma of the fowl transmissible by an agent separable from the tumor cells. *J. Exp. Med.*, **13**, 397.
2. Weiss, R., Teich, N., Varmus, H., and Coffin, J. (1984) Origins of contemporary RNA tumor virus research. In *RNA Tumor Viruses, Molecular Biology of Tumor Viruses*, 2nd edn. Weiss, R., Teich, N., Varmus, H., and Coffin, J. (eds). Cold Spring Harbor Laboratory, Cold Spring Harbor, NY, p. 1.
3. Bishop, J. M. and Varmus, H. (1985) Supplement, Functions and origins of retroviral transforming genes. In *RNA Tumor Viruses, Molecular Biology of Tumor Viruses*, 2nd edn. Weiss, R., Teich, N., Varmus, H., and Coffin, J. (eds). Cold Spring Harbor Laboratory, Cold Spring Harbor, NY, p. 249.
4. Varmus, H. (1988) Retroviruses. *Science*, **240**, 1427.
5. Aaronson, S. A. (1991) Growth factors and cancer. *Science*, **254**, 1146.
6. Achong, B. G., Mansell, P. W. A., Epstein, M. A., and Clifford, P. (1971) An unusual virus in cultures from a human nasopharyngeal carcinoma. *J. Natl Cancer Inst.*, **46**, 299.
7. Poiesz, B. J., Ruscetti, F. W., Gazdar, A. F., Bunn, P. A., Minna, J. D., and Gallo, R. C. (1980) Detection and isolation of type C retrovirus particles from fresh cultured lymphocytes of a patient with cutaneous T-cell lymphoma. *Proc. Natl Acad. Sci. USA*, **77**, 7415.
8. Yoshida, M., Miyoshi, I., and Hinuma, Y. (1982) Isolation and characterization of retrovirus from cell lines of human adult T-cell leukemia and its implication in the disease. *Proc. Natl Acad. Sci. USA*, **79**, 2031.
9. Sodroski, J. G., Rosen, C. A., and Haseltine, W. A. (1984) Transacting transcriptional activation of the long terminal repeat of human T lymphotropic viruses in infected cells. *Science*, **225**, 381.
10. Tanaka, A., Takahashi, C., Yamaoka, S., Nosaka, T., Maki, M., and Hatanaka, M. (1990) Oncogenic transformation by the *tax* gene of human T-cell leukemia virus type I *in vitro*. *Proc. Natl Acad. Sci. USA*, **87**, 1071.

11. Grassman, R., Dengler, C., Muller-Fleckenstein, I., Fleckenstein, B., McGuire, K., Dokhelar, M.-C., Sodroski, J. G., and Haseltine, W. A. (1989) Transformation to continuous growth of primary human T lymphocytes by human T-cell leukemia virus type I X-region genes transduced by a Herpesvirus saimiri vector. *Proc. Natl Acad. Sci. USA,* **86,** 3351.
12. zur Hausen, H. (1991) Viruses in human cancers. *Science,* **254,** 1167.
13. Gottlieb, M. S., Schroff, R., Schanker, H. M., Weisman, J. D., Fan, P. T., Wolf, R. A., and Saxon, A. (1981) *Pneumocystis carinii* pneumonia and mucosal candidiasis in previously healthy homosexual men: evidence of a new acquired cellular immunodeficiency. *New Engl. J. Med.,* **305,** 1425.
14. Masur, H., Michelis, M. A., Greene, J. B., Onorato, I., Vande, Stouwe, R. A., Holzman, R. S., Wormser, G., Brettman, L., Lange, M., Murray, H. W., and Cunningham-Rundles, S. (1981) An outbreak of community-acquired *Pneumocystis carinii* pneumonia: initial manifestation of cellular immune dysfunction. *New Engl. J. Med.,* **305,** 1431.
15. Popovic, M., Sarngadharan, M. G., Read, E., and Gallo, R. C. (1984) Detection, isolation, and continuous production of cytopathic retroviruses (HTLV-III) from patients with AIDS and pre-AIDS. *Science,* **224,** 497.
16. Barre-Sinoussi, F., Chermann, J. C., Rey, F., Nugeyre, M. T., Chamaret, S., Gruest, T., Dauguet, C., Axler-Blin, C., Vezin-Brun, F., Rouzioux, C., Rosenbaum, W., and Montagnier, L. (1983) Isolation of a T-lymphotropic retrovirus from a patient at risk of acquired immune deficiency syndrome (AIDS). *Science,* **220,** 868.
17. Weiss, R. (1984) Experimental biology and assay of retroviruses. In *RNA Tumor Viruses, Molecular Biology of Tumor Viruses,* 2nd edn. Weiss, R., Teich, N., Varmus, H., and Coffin (eds). Cold Spring Harbor Laboratory, Cold Spring Harbor, NY, p. 209.
18. Dalgleish, A. G., Beverly, P. C. L., Clapham, P. R., Crawford, D. H., Greaves, M. F., and Weiss, R. A. (1984) The CD4 (T4) antigen is an essential component of the receptor for the AIDS retrovirus. *Nature (Lond.),* **312,** 763.
19. Klatzmann, D., Champagne, E., Chamaret, S., Gruest, J., Guetard, T., Hercend, T., Gluckman, J.-C., and Montagnier, L. (1984) T-lymphocyte T4 molecule behaves as the receptor for human retrovirus LAV. *Nature (Lond.),* **312,** 767.
20. McDouglas, J. S., Kennedy, M., Sligh, J., Cort, S., Mawle, A., and Nicholson, J. K. A. (1986) Binding of the HTLV-III/LAV to T4$^+$ T cells by a complex of the 100 K viral protein and the T4 molecule. *Science,* **231,** 382.
21. Maddon, P. J., Dalgleish, A. G., McDougal, J. S., Clapham, P. R., Weiss, R. A., and Axel, R. (1986) The T4 gene encodes the AIDS virus receptor and is expressed in the immune system and the brain. *Cell,* **47,** 333.
22. Albritton, L. M., Tseng, L., Scadden, D., and Cunningham, J. M. (1989) A putative murine ecotropic retrovirus receptor gene encodes a multiple membrane-spanning protein and confers susceptibility to virus infection. *Cell,* **57,** 659.
23. O'Hara, B., Johann, S. V., Klinger, H. P., Blair, D. G., Rubinson, H., Dunn, K. J., Sass, P., Vitek, S. M., and Robbins, T. (1990) Characterization of a human gene conferring sensitivity to infection by Gibbon ape leukemia virus. *Cell Growth Differ.,* **1,** 119.
24. Wang, H., Paul, R., Burgeson, R. E., Keene, D. R., and Kabat, D. (1991) Plasma membrane receptors for ecotropic murine retroviruses require a limiting accessory factor. *J. Virol.,* **65,** 6468.
25. Freed, E. O., Myers, D. J., and Risser, R. (1990) Characterization of the fusion domain of the human immunodeficiency virus type 1 envelope glycoprotein gp41. *Proc. Natl Acad. Sci. USA,* **87,** 4650.

26. Kowalski, M., Potz, J., Basiripour, L., Dorfman, T., Goh, W. C., Terwilliger, E., Dayton, A., Rosen, C., Haseltine, W., and Sodroski, J. (1987) Functional regions of the envelope glycoprotein of the human immunodeficiency virus type 1. *Science,* **237,** 1351.
27. Moore, J. P. and Nara, P. L. (1992) The role of the V3 loop of gp120 in HIV infection. *AIDS,* **6,** S21.
28. McClure, M. O., Sommerfelt, M. A., Marsh, M., and Weiss, R. A. (1990) The pH independence of mammalian retrovirus infection. *J. Gen. Virol.,* **71,** 767.
29. Maddon, P. J., McDougal, J. S., Clapham, P. R., Dalgleish, A. G., Jamal, S., Weiss, R. A., and Axel, R. (1988) HIV infection does not require endocytosis of its receptor, CD4. *Cell,* **54,** 865.
30. Marsh, M. and Helenius, A. (1989) Virus entry into animal cells. *Advances Virus Res.,* **36,** 107.
31. Varmus, H. and Brown, P. (1989) Retroviruses. In *Mobile DNA.* Berg, D. E. and Howe, M. M. (eds). American Society for Microbiology, Washington, DC, p. 53.
32. Boone, L. R. and Skalka, A. M. (1980) Two species of full-length cDNA are synthesized in high yield by mellitin-treated avian retrovirus particles. *Proc. Natl Acad. Sci. USA,* **77,** 847.
33. De Clercq, E., Yamamoto, N., Pauwels, R., Baba, M., Schols, D., Nakashima, H., Balzarini, J., Debyser, Z., Murrer, B. A., Schwartz, D., Thornton, D., Bridger, G., Fricker, S., Henson, G., Abrams, M., and Picker, D. (1992) Potent and selective inhibition of human immunodeficiency virus (HIV)-1 and HIV-2 replication by a class of bicyclams interacting with a viral uncoating event. *Proc. Natl Acad. Sci. USA,* **89,** 5286.
34. Varmus, H. and Swanstrom, R. (1984) Replication of retroviruses. In *RNA Tumor Viruses, Molecular Biology of Tumor Viruses,* 2nd edn. Weiss, R., Teich, N., Varmus, H., and Coffin, J. (eds). Cold Spring Harbor Laboratory, Cold Spring Harbor, NY, p. 369.
35. Bukrinsky, M. I., Sharovat, N., Dempsey, M. P., Stanwick, T. L., Bukrinskaya, A. G., Haggerty, S., and Stevenson, M. (1992) Active nuclear import of HIV-1 preintegration complexes. *Proc. Natl Acad. Sci. USA,* **89,** 6580.
36. Katz, R. A., Merkel, G., Kulkosky, J., Leis, J., and Skalka, A. M. (1990) The avian retroviral IN protein is both necessary and sufficient for integrative recombination *in vitro. Cell,* **63,** 87.
37. Craigie, R., Fujiwara, T., and Bushman, F. (1990) The IN protein of moloney murine leukemia virus processes the viral DNA ends and accomplishes their integration *in vitro. Cell,* **62,** 829.
38. Brown, P. O., Bowerman, B., Varmus, H. E., and Bishop, J. M. (1987) Correct integration of retroviral DNA *in vitro. Cell,* **49,** 347.
39. Brown, P. O., Bowerman, B., Varmus, H. E., and Bishop, J. M. (1989) Retroviral integration: structure of the initial covalent product and its precursor, and a role for the viral IN protein. *Proc. Natl Acad. Sci. USA,* **86,** 2525.
40. Harris, J. D., Blum, H., Scott, J., Traynor, B., Ventura, P., and Haase, A. (1984) Slow virus visna: reproduction *in vitro* of virus from extrachromosomal DNA. *Proc. Natl Acad. Sci. USA,* **81,** 7212.
41. Prakash, K., Ranganathan, P. N., Mettus, R., Reddy, P., Srinivasan, A., and Plotkin, S. (1992) Generation of deletion mutants of simian immunodeficiency virus incapable of proviral integration. *J. Virol.,* **66,** 167.
42. Stevenson, M., Haggerty, S., Lamonica, C. A., Meier, C. M., Welch, S.-K., and Wasiak, A. J. (1990) Integration is not necessary for expression of human immunodeficiency virus type 1 protein products. *J. Virol.,* **64,** 2421.
43. Miller, D. G., Adam, M. A., and Miller, A. D. (1990) Gene transfer by retrovirus vectors

occurs only in cells that are actively replicating at the time of infection. *Mol. Cell. Biol.*, **10**, 4239.
44. Stevenson, M., Stanwick, T. L., Dempsey, M. P., and Lamonica, C. A. (1990) HIV-1 replication is controlled at the level of T cell activation and proviral integration. *EMBO J.*, **9**, 1551.
45. Zack, J. A., Arrigo, S. J., Weitsman, S. R., Go, A. S., Haislip, A., and Chen, I. S. Y. (1990) HIV-1 entry into quiescent primary lymphocytes: molecular analysis reveals a labile, latent viral structure. *Cell*, **61**, 213.
46. Weinberg, J. B., Matthews, T. J., Cullen, B. R., and Malim, M. H. (1991) Productive human immunodeficiency virus type 1 (HIV-1) infection of nonproliferating human monocytes. *J. Exp. Med.*, **174**, 1477.
47. Katz, R. A. and Skalka, A. M. (1990) Generation of diversity in retroviruses. *Annu. Rev. Genet.*, **24**, 409.
48. Hu, W.-S. and Temin, H. M. (1990) Retroviral recombination and reverse transcription. *Science*, **250**, 1227.
49. Meyerhans, A., Cheynier, R., Albert, J., Seth, M., Kwok, S., Sninsky, J., Morfeldt-Månson, L., Asjö, B., and Wain-Hobson, S. (1989) Temporal fluctuations in HIV quasispecies *in vivo* are not reflected by sequential HIV isolations. *Cell*, **58**, 901.
50. Burns, D. P. W. and Desrosiers, R. C. (1991) Selection of genetic variants of simian immunodeficiency virus in persistently infected rhesus monkeys. *J. Virol.*, **65**, 1843.
51. Holmes, E. C., Zhang, L. Q., Simmonds, P., Ludlam, C. A., and Brown, A. J. L. (1992) Convergent and divergent sequence evolution in the surface envelope glycoprotein of human immunodeficiency virus type 1 within a single infected patient. *Proc. Natl Acad. Sci. USA*, **89**, 4835.
52. Cullen, B. R. (1991) Regulation of human immunodeficiency virus replication. *Annu. Rev. Microbiol.*, **45**, 219.
53. Arrigo, S. and Beemon, K. (1988) Regulation of Rous sarcoma virus RNA splicing and stability. *Mol. Cell. Biol.*, **8**, 4858.
54. Katz, R. A. and Skalka, A. M. (1990) Control of retroviral RNA splicing through maintenance of suboptimal processing signals. *Mol. Cell. Biol.*, **10**, 696.
55. Greene, W. C. and Cullen, B. R. (1990) The Rev–Rex connection: convergent strategies for the post-transcriptional regulation of HIV-1 and HTLV-I gene expression. *Semin. Virol.*, **1**, 195.
56. Garcia-Blanco, M. A. and Cullen, B. R. (1991) Molecular basis of latency in pathogenic human viruses. *Science*, **254**, 815.
57. Gottlinger, H. G., Sodroski, J. G., and Haseltine, W. A. (1989) Role of capsid precursor processing and myristoylation in morphogenesis and infectivity of human immunodeficiency virus type 1. *Proc. Natl Acad. Sci. USA*, **86**, 5781.
58. Aldovini, A. and Young, R. A. (1990) Mutations of RNA and protein sequences involved in human immunodeficiency virus type 1 packaging result in production of noninfectious virus. *J. Virol.*, **64**, 1920.
59. Chesebro, B., Wehrly, K., and Maury, W. (1990) Differential expression in human and mouse cells of human immunodeficiency virus pseudotyped by murine retroviruses. *J. Virol.*, **64**, 4553.
60. Spector, D. H., Wade, E., Wright, D. A., Koval, V., Clark, C., Jaquish, D., and Spector, S. A. (1990) Human immunodeficiency virus pseudotypes with expanded cellular and species tropism. *J. Virol.*, **64**, 2298.

Skalka, A. M. (1989) Retroviral proteases: first glimpses at the anatomy of a processing machine. *Cell*, **56**, 911.

Teich, N. (1984) Taxonomy of retroviruses. In *RNA Tumor Viruses, Molecular Biology of Tumor Viruses*, 2nd edn. Weiss, R., Teich, N., Varmus, H., and Coffin, J. (eds), Cold Spring Harbor Laboratory, Cold Spring Harbor, NY, p. 25.

Baltimore, D. (1970). RNA-dependent DNA polymerase in virions of RNA tumour viruses. *Nature*, **226**, 1209.

Temin, H. M. and Mizutani, S. (1970) RNA-directed DNA polymerase in virions of Rous sarcoma virus. *Nature*, **226**, 1211.

Hsu, M.-C., Schutt, A. D., Holly, M., Slice, L. W., Sherman, M. I., Richman, D. D., Potash, M. J., and Volsky, D. J. (1991) Inhibition of HIV replication in acute and chronic infections *in vitro* by a Tat antagonist. *Science*, **254**, 1799.

Renne, R., Friedl, E., Schweizer, M., Fleps, U., Turek, R., and Neumann-Haefelin, D. (1992) Genomic organization and expression of simian foamy virus type 3 (SFV-3). *Virology*, **186**, 597.

Garvey, K. J., Oberste, M. S., Elser, J. E., Braun, M. J., and Gonda, M. A. (1990) Nucleotide sequence and genome organization of biologically active proviruses of the bovine immunodeficiency-like virus. *Virology*, **175**, 391.

Gojobori, T., Moriyama, E. N., Ina, Y., Ikeo, K., Miura, T., Tsujimoto, H., Hayami, M., and Yokoyama, S. (1990) Evolutionary origin of human and simian immunodeficiency viruses. *Proc. Natl Acad. Sci. USA*, **87**, 4108.

2 | The biological and molecular basis for cell tropism in HIV

VICENTE PLANELLES, QI-XIANG LI, and IRVIN S. Y. CHEN

1. Introduction

Human immunodeficiency virus (HIV) is the primary etiologic agent for the acquired immunodeficiency syndrome (AIDS). HIV exhibits high genetic variation, which results in a wide variety of biological phenotypes displayed by different strains of this virus. Phenotypic heterogeneity of HIV is found in replication kinetics, susceptiblity to serum neutralization, antiviral drug resistance, induction of cytopathicity, and host range specificity. The host range of HIV, also termed 'tropism', constitutes a complicated scenario. Different strains of this virus show varying abilities to infect a spectrum of cell types *in vivo* as well as *in vitro*. The ability of HIV isolates to infect certain tissues and cell types has been associated with different pathologic manifestations. Tropism can arise from viral sequence variation in combination with environmental selection mechanisms. This chapter will discuss the current knowledge of tropism of HIV and its molecular determinants.

2. Genetic variation of HIV

There are two major HIV subtypes, HIV type 1 (HIV-1), the most prevalent throughout the world, and HIV type 2 (HIV-2), an AIDS virus most common in West Africa (1). HIV-1 and HIV-2 are members of a group of closely related human and non-human primate lentiviruses, which also includes the simian immunodeficiency viruses (SIV) (2, Chapter 6). SIV constitutes a group of lentiviruses that naturally infect several species of African monkeys. Experimental infection of Asian macaques with certain SIV strains causes an AIDS-like disease that closely resembles human AIDS (3). Primate lentiviruses have similar genetic organization, significant nucleotide sequence homology, and may have evolved from common ancestors (2, Chapter 6).

HIV shows extensive genetic and phenotypic variation at various levels, even within a particular subtype (1). Thus, viral sequences vary among isolates from different AIDS patients, among isolates from an individual patient at different stages of disease, and among different tissues of the same individual (4–12). Usually, there is less variation between isolates from one individual than among those from different individuals (4, 6, 10). Although not completely proven *in vivo*, the ability of HIV to change rapidly should enable it to adapt to different microenvironments (i.e. the genetic background of the host, different tissues and cell types within the host, changes in the specificity of immune responses, etc.). The reported low silent mutation rate observed *in vivo* constitutes an indication of the presence of phenotypic selection (12).

The degree of variability is not uniformly distributed throughout the HIV genome. The envelope gene (*env*) is the most variable (1, 13), while the *gag* and *pol* genes are relatively conserved. Variation in *env* occurs with higher frequency in certain regions. On the basis of this clustered variation, gp120 is divided into alternating conserved (C) and variable (V) domains (14) (Fig. 1). Within the third variable domain of gp120, a cluster of amino acid residues flanked by cysteines (the V3 loop) shows high variability and constitutes a major immunodominant epitope (15, 16). Sequence variation in this region has been associated with generation of viral mutants *in vitro* and *in vivo*, and results in alteration of several important phenotypes of HIV strains, such as cell tropism (see Section 8), susceptibility to neutralization, and syncytium induction (see Sections 6 and 8).

Thus, viruses isolated from different tissues of an HIV-infected individual display biological phenotypes that may enable them to best propagate themselves in such locations. Additionally, *in vitro* culture conditions will select for variants with certain phenotypes and with the ability to infect certain cell types. *In vitro* passage of viruses can also result in expansion of minor species that subsequently

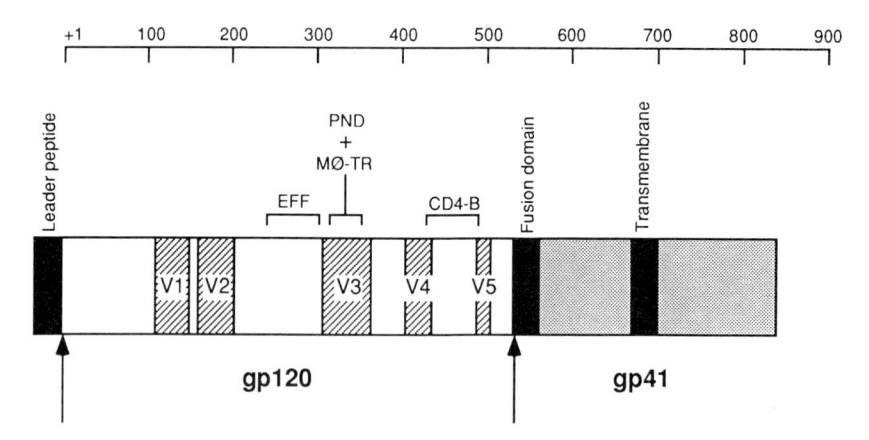

Fig. 1 Structure of the envelope glycoprotein of HIV-1. Proteolytic processing sites are indicated as vertical arrows. Numbers represent predicted amino acid residues starting at the leader peptide processing site. Conserved and variable (V1 to V5) areas of gp120 are depicted as white and hatched boxes, respectively. The principal neutralizing domain (PND, see Section 6), macrophage-tropism domain (MØ-TR, see Sections 6 and 8), efficiency domain (EFF, see Section 9.2), and principal CD4-binding region (CD4-B, see Sections 6 and 9.1) are indicated

become predominant (5, 17, 18). Therefore, care must always be taken in the interpretation of results using virus isolated from bulk cultures, particularly those involving passage through cell lines.

3. Species specificity of HIV

Humans are the natural hosts for HIV-1 and HIV-2. However, these viruses can infect certain non-human primates after experimental inoculation. Thus, some strains of HIV-1 infect chimpanzees (19–21) and gibbon apes (22). In addition, HIV-2 has been shown to transiently infect macaques (23–25). Infection of non-human primates with HIV isolates is accompanied by seroconversion, viremia, and, in some cases, the infection becomes persistent (3). However, infection of non-human primates with HIV has not been shown to cause AIDS-like pathology (3).

Rabbits can also be experimentally infected with HIV. Infection of rabbits with HIV-1 leads to seroconversion and low-level persistent infection (26). In these animals, HIV-1 could be recovered from circulating mononuclear cells several months postinfection (26). As for infection of non-human primates, HIV-1 infection of rabbits does not lead to disease (26).

Early experiments with mouse cells genetically engineered to express the receptor for HIV, the human CD4 glycoprotein, showed that these cells can bind virus but are not capable of supporting HIV infection (27). Introduction of HIV proviral DNA into mouse cells by transfection, however, led to the production of infectious virus by these cells (28). In addition, pseudotype virions containing the HIV genome and the murine leukemia virus envelope glycoprotein can infect murine 3T3 cells (29). These findings, taken together, indicate that expression of human CD4 is not sufficient to confer susceptibility to HIV infection into non-human cells, and that the restriction in mouse cells must be at the level of entry. In spite of the previous evidence that mice are not susceptible to HIV, a recent report describes the establishment of a persistent infection of mice with HIV-1 (30).

4. Tissue distribution of HIV

In HIV-infected individuals, virus can be detected in a wide variety of tissues and cell types. HIV has been shown to infect $CD4^+$ lymphocytes and monocyte-derived macrophages in peripheral blood and lymph nodes, as well as a series of tissues in the perivascular compartment including brain, lung, gastrointestinal epithelium, skin, heart, and kidney (31–38). Common hallmarks include immunodeficiency, wasting, neurological and gastrointestinal dysfunction, and opportunistic infections. Some of the pathologic manifestations associated with AIDS are believed to result from primary HIV infection in the corresponding tissues, as evidenced by the correlation between pathogenesis and tissue distribution of the virus (7, 8, 39–42). Immunodysfunction is likely related to a progressive decline in $CD4^+$ lymphocytes, the key T-cell subset for competent immune function and one of the major target cell types for HIV. AIDS dementia complex may be associated with HIV-1 infection of mononuclear phagocytes in brain tissue (34, 43). Within the bowel, direct infection of infiltrating macrophages and intestinal epithelial cells may be related to gastrointestinal dysfunction (36, 44).

4.1 Infection of blood cells

HIV in the blood has been associated with infection of $CD4^+$ T-lymphocytes and monocyte/macrophages (45–47). Recently, dendritic cells from the blood have also been shown to support HIV replication (48). Lymphocytes and macrophages can leave the vascular compartment and infiltrate peripheral tissues. In addition, monocyte/macrophages infiltrating tissues can further differentiate to become resident phagocytes in those tissues. Therefore, infection of blood-borne cells may be a means of dissemination of the virus throughout the organism. A main route of transmission of HIV-1 is through blood.

4.2 Infection of the central nervous system

A substantial proportion of individuals infected with HIV develop disorders of the central nervous system (CNS) (49, 50). These disorders are associated with the presence of HIV in the CNS (51). In brain autopsies from individuals with AIDS-related neurological disorders, HIV can be detected by virus isolation, immunohistochemistry of viral antigens, and hybridization techniques (34, 52). The cells predominantly infected in the CNS are the microglial cells (51, 53). However, some observations suggest that other cell types in the brain, including endothelial cells, astrocytes, and oligodendrocytes, can also be infected by HIV (54). Blood-borne monocytes are thought to have the ability to leave the vascular compartment through the blood–brain barrier and infiltrate tissues of the CNS (51, 55). Microglial cells are derived from bone marrow precursors that migrate into the CNS during development, where they differentiate and become resident cells (56). Primary cultures derived from adult human brain tissue can be infected with HIV *in vitro* by macrophage-tropic isolates but not by isolates adapted to grow in T-cell lines (53, 57). Moreover, most brain-derived strains productively infect macrophages *in vitro* (7, 47, 58–60). These findings suggest that HIV infection of the CNS and macrophage tropism may be highly related.

4.3 Infection of the gastrointestinal tract

Gastrointestinal dysfunctions such as malabsorption and chronic diarrhoea are frequent in AIDS patients (44). Thus, disease could be caused by direct infection of intestinal epithelial cells by HIV or, alternatively, by indirect mechanisms. It is not yet clear whether the infected cells in the gastrointestinal tract are the resident gut epithelial cells or infiltrating mononuclear cells (including lymphocytes and mononuclear phagocytes). Further experiments must be performed to determine the major cell type infected in the gut. One report describes HIV-1 infection of purified normal epithelial and submesenchymal cells (61). Among various strains tested, the most efficient infection was achieved with macrophage-derived virus strains (61). Persistent, long-term infection of these cells *in vitro* was achieved. These findings suggest that the gastrointestinal tract is a site of primary infection as well as a reservoir of HIV.

4.4 Infection of the lung

The lung tissue is a frequent target of HIV infection and AIDS-associated pulmonary complications. Alveolar macrophages of AIDS patients with pulmonary complications were found to be infected by HIV (62–64). *In vitro* experiments indicate that alveolar macrophages are more susceptible to HIV than their less differentiated parental blood monocytes or bone marrow macrophages (65). These observations suggest a role of HIV infection in pulmonary complications of AIDS patients.

4.5 Tropism and pathogenesis

The major target cell types of HIV infection *in vivo* are CD4$^+$ primary T-lymphocytes and cells of the mononuclear phagocyte system. Genetic and phenotypic variation also occurs within a single individual as the disease progresses (37, 66). Isolates from patients can be divided into two major categories based on their phenotype. Early isolates, obtained from asymptomatic individuals, usually display slow replication kinetics, are non-syncytium-inducing (NSI), and are tropic for primary cells but not for transformed T-cell lines (37, 66). Isolates from later stages of disease show more diverse phenotypes. In addition to the phenotypes present in the early group, some HIV strains from the late group are able to infect immortalized T-cell lines *in vitro*, have a high rate of replication, and are syncytium-inducing (SI). These observations suggest that the appearance of more 'virulent' strains of HIV-1 over time may correlate with progression from the asymptomatic state to AIDS (9, 11, 40, 41). It has been hypothesized that SI strains may be present during acute infection but are suppressed by the competent anti-HIV immune response mounted by the host, while macrophage-tropic NSI strains escape the host immune surveillance and establish a persistent infection throughout the asymptomatic period of the disease. At the late stages of disease, SI variants appear and rapid CD4$^+$ T-cell depletion ensues (67).

HIV has mechanisms for establishing viral persistence and eluding host immune surveillance. Thus, HIV induction of AIDS is preceded by a long period of clinical latency, with an average of 10 to 15 years for adults (for review, see ref. 68). Non-productive infection of certain tissues may help viruses avoid host immune surveillance during latent or low-level persistent infection periods. Quiescent T-cells and macrophages have been suggested to be reservoirs of HIV-1 (66, 67, 69–71). Potential mechanisms regulating tropism of HIV for macrophages and quiescent T-cells will be discussed below (Sections 8–9 and 10.3).

A correlation between increased virulence and wider tropism has also been observed for SIV (72). SIV$_{SMM-9}$ is an isolate of SIV which causes AIDS in several species of macaques. *In vivo* passage of this virus resulted in the production of a variant, designated SIV$_{SMM-PBj14}$, which caused acute AIDS-like disease and death of animals about 15 days postinoculation. Comparison of the acutely pathogenic virus, SIV$_{SMM-PBj14}$, with the parent isolate, SIV$_{SMM-9}$, showed that SIV$_{SMM-PBj14}$ replicated more efficiently than SIV$_{SMM-9}$ in human peripheral blood mononuclear cells (PMBC) and also replicated in chimpanzee PBMC and T-cell lines, whereas the parental virus did not (72). Molecular determinants of this phenotype will be discussed in Section 10.4.

5. The CD4 glycoprotein is the major receptor for HIV

The CD4 antigen is the major receptor for HIV (73, 74). Infection of cells with HIV requires viral attachment to CD4, which is mediated by the interaction between the envelope glycoprotein and CD4 antigen. A physical interaction between the HIV-1 surface glycoprotein, gp120, and CD4 was demonstrated by immunoprecipitation of gp120–CD4 complexes from lysates from infected cells (75). Additional evidence for the role of CD4 as the HIV receptor came from transfection experiments (27). In these studies, human CD4$^-$ cells that were refractory to infection became permissive when a cloned CD4 cDNA was introduced by transfection.

CD4 is a 55 kDa glycoprotein found on the surface of mature, thymus-derived T-lymphocytes (76) and cells of the monocyte/macrophage lineage (77, 78). CD4 is a type I membrane glycoprotein of the immunoglobulin gene family (79). This glycoprotein consists of four immunoglobulin-like extracellular domains, a membrane-spanning region, and a short cytoplasmic tail. The amino-terminal immunoglobulin-like domain of CD4 contains three loops which are homologous to the complementarity-determining regions (CDR) of immunoglobulin molecules, termed CDR1, 2, and 3. CDR2 and 3 have been identified as the points of interaction with the HIV glycoprotein gp120 (80–84).

The CD4-bearing T-lymphocytes primarily represent the T-helper subset. On the surface of these cells, CD4 is an accessory part of the T-cell antigen receptor complex. The extracellular portion of CD4 interacts with a non-polymorphic region of major histocompatibility complex (MHC) class II molecules on antigen-presenting cells, thereby stabilizing the MHC class II molecule–T-cell receptor complex interaction (for review, see ref. 85). In addition, the cytoplasmic domain of CD4 is associated with the lymphocyte-specific tyrosine kinase p56^{lck} (86, 87), and consequently, may be involved in signal transduction.

6. Structure of the HIV envelope glycoprotein

The HIV envelope glycoprotein is synthesized as a precursor glycoprotein (gp160), which is cleaved to give rise to an external surface domain (gp120) and a transmembrane domain (gp41). The mature cleavage products remain associated on the surface of the virion, where they appear to form dimers or tetramers (88). The main function of gp120 on the virion surface is to provide attachment to the host cell receptor, the CD4 molecule. The dissociation constant for binding between the viral surface glycoprotein and CD4 is 10^{-9} M (89). The receptor-binding domain of the HIV envelope has been partially mapped by mutational analysis (90, 91) and monoclonal antibodies (92) to conserved regions near the carboxy terminus of gp120 (see Fig. 1). Amino acid changes outside this region may also result in loss of binding capacity (90, 93–95). This observation suggests that the binding site may involve additional areas and, perhaps, be conformational. Binding of gp120 to CD4 appears to be dependent on the overall conformation of the molecule; reduction, denaturation, or synthesis of gp120 in bacteria or yeast result in proteins devoid of capacity to bind to the CD4 antigen (91, 96, 97).

Within the third variable region (V3) of HIV gp120, a cluster of amino acid residues flanked by cysteine residues constitute a major immunodominant epitope (14) (see Fig. 1). Mutations in this region occur with high frequency and, therefore, it has been termed a hypervariable loop (14, 16). Sequence variation in this region has been associated with generation of viral mutants that escape immune surveillance (98). In addition, antibodies that bind to V3 and mutations in this region block syncytium formation and neutralize viral infectivity, but do not affect CD4 binding (99–104).

Upon binding of gp120 to CD4, the transmembrane glycoprotein, gp41, induces fusion between the viral envelope and the plasma membrane (105). Fusion is mediated by a hydrophobic N-terminal domain within gp41, termed the fusion domain. Although the mechanism is not fully understood, it is thought that the binding between gp120 and CD4 induces a conformational change which triggers the fusion domain to be exposed and display its fusogenic capacity (88). The fusion domain of HIV has homology with the corresponding region of other fusogenic viruses (88). Mutations in the amino terminal region of gp41 abolish the ability of the envelope glycoprotein to induce syncytia and reduce or abolish virus infectivity *in vitro* (106).

7. Tropism of the AIDS virus: historical perspective

The enormous genetic variability of HIV both *in vivo* and *in vitro* is reflected in a wide spectrum of biological phenotypes displayed by different strains. Phenotypes that vary among HIV strains include kinetics of replication, preferential infectivity for certain cell types (tropism), induction of cytopathic effect, and susceptibility to virus neutralization by antibodies or sCD4 (7–9, 39–42, 107–110). These features are usually assessed in *in vitro* systems. In particular, the tropism of a given HIV strain is defined on the basis of its ability to infect a variety of cell types in tissue culture. The types of cells utilized to study HIV tropism fall into two major categories: primary cells and immortalized cell lines. Primary cells are commonly derived from fresh blood and include lectin-activated PBMC, purified T-lymphocyte populations, and monocyte-derived macrophages. Other sources of primary cells are lung lavages (to obtain alveolar macrophages), microglial cells of the brain, and other biopsies. A variety of immortalized cell lines of human origin have also been utilized in tropism studies; these include T-cell lines and monocytic cell lines. Examples of particular cell types and their ability to support HIV replication will be discussed below.

Initial *in vitro* infectivity studies showed that HIV replicated efficiently in immortalized T-cell lines and lectin-stimulated peripheral blood T-lymphocytes (111–113), and led to the conclusion that T-helper lymphocytes were the major reservoir for HIV. The idea that HIV infects non-T-cells was first put forward by reports that Epstein–Barr virus-transformed B-cell lines and the monocytic line U937 were capable of supporting HIV replication (35, 112, 114). Subsequently, it was shown that monocyte-derived macrophages from fresh blood could also support replication of HIV both *in vivo* and *in vitro* (46, 115). Macrophages were identified as the major cell type expressing viral antigens and RNA in tissues such as the CNS (51, 54, 55, 58), spinal cord (54), lung (62–64), and skin (38).

Given the high degree of genetic heterogeneity among HIV strains coexisting at any given time within an individual as well as following culture, it was crucial to obtain infectious molecular clones to study the molecular basis of HIV tropism. Two key techniques have been used to identify viral determinants of tropism: generation of recombinant viruses and site-directed mutagenesis.

8. V3 loop as a determinant of tropism

Although CD4-independent infection by HIV has been reported (for references, see Section 11.3), it is widely accepted that the primary pathway of entry into target cells involves interaction of the virus with CD4 (116, 117). Therefore, infection with most isolates of HIV, regardless of their tropism phenotypes, can be blocked by certain monoclonal antibodies against CD4 such as Leu3A (116, 117). Unless otherwise stated, the experiments described below were performed with prototypic HIV isolates which infect via CD4.

Tropism studies with recombinant viruses initially dealt with exchanges of subgenomic regions encompassing the envelope genes but also containing short open reading frames with putative regulatory function such as *tat, rev, vpu,* and *vif* (6, 118, 119). These studies concluded that qualitative and quantitative differences in the ability of HIV recombinants to infect a spectrum of cell types (including cord blood T-cells, monocyte-derived macrophages, and the T-cell lines CEM, Jurkat, MOLT-3, and H9) correlated with those of the parental virus from which the 3′ region including *env* had been obtained. Thus, recombinant viruses carrying 3′ subgenomic portions from their parental clones exhibited similar kinetics of replication, ability to induce cytopathology, and host range as the parental viruses. Similar experiments with SIV strains yielded parallel results (120). These experiments could not exclude the possibility that alterations in regulatory genes encoded within the 3' subgenomic fragments might account for some of the observed phenotypic differences.

8.1 Macrophage tropism

Further experiments were aimed at defining the minimal regions of the genome capable of conferring macrophage tropism (121–125). HIV-1$_{JR-FL}$ (8) is an HIV-1 brain isolate which replicates efficiently in PBL and monocyte-derived macrophages but is restricted in its ability to infect immortalized T-cell lines (123). HIV-1$_{NL4-3}$ is a molecularly constructed virus which does not infect macrophages but replicates in PBMC (126). The 3' region of the HIV-1$_{JR-FL}$ genome was cloned by the polymerase chain reaction, and the entire gene and progressively smaller fragments thereof were substituted into the molecular clone HIV-1$_{NL4-3}$ by use of shared restriction endonuclease sites. These experiments revealed that a *Stu*I to *Mst*II restriction fragment encoding amino acids 202 to 358 of the HIV-1$_{JR-FL}$ envelope region was sufficient to confer the macrophage-tropic phenotype when incorporated into the background of the non-macrophage-tropic clone. Similar results were subsequently obtained by other groups using other cloned macrophage-tropic strains (122, 124, 125). In one study, the minimal region capable of controlling macrophage tropism was determined to be a 20 amino acid sequence contained within the gp120 V3 loop (122) (see Table 1). Taken together, these results indicate that the role of the V3 loop in macrophage tropism is not an exclusive property of the viral strain HIV$_{JR-FL}$ and constitutes a general phenomenon.

It is noteworthy that all recombinant viruses used in these studies are capable of

Table 1 Amino acid sequence comparison of V3 loop regions from various macrophage-tropic and T-cell line-tropic HIV-1 isolates. Shaded areas show minimal regions determined to be sufficient to confer macrophage tropism by genetic exchange between HIV isolates HTLV-IIIB and BaL (see Section 8.2). Note that requirements may differ for other HIV-1 strains (see Section 9)

Cell tropism	HIV-1 clone	Predicted amino acid sequence
MØ	JR-FL (consensus)	**CTRPNNNTRKSIHI··GPGRAFYTTGEIIGDIRQAHC**
MØ	Ada	---------------··----------------------
MØ	BaL	---------------··--------L-------------
MØ	SF162	-------------T-··--------A--D----------
T-cell line	NL4-3	-------------R-QR------V-I-K·--NM------
T-cell line	SF-2	-------------Y-··-------H---R------K---
T-cell line	HTLV-IIIB	----------K-R-QR------V-I-K·--NM------

productively infecting activated PBMC (6, 118, 119, 121–125). This indicates that the lack of infectivity of certain clones in certain cell types is not a result of loss of biological activity during the construction of recombinants.

8.2 Tropism for immortalized T-cell lines

Although all HIV strains replicate in PBMC, only certain isolates display specific tropisms for CD4-bearing T-cell lines as a consequence of *in vitro* selection (9, 11, 40, 41). A number of studies showed that the inability to productively infect immortalized T-cell lines is a common property of most viral isolates obtained by short-term culture in PBMC after isolation from patients. In contrast, viral strains that have been passaged in immortalized T-cell lines are selected to replicate efficiently in this cell type. This is true for the prototypical T-cell line-adapted strains of HIV-1 such as HIV-1_{LAV} (111), HIV-$1_{HTLVIIIB}$ (113), and HIV-1_{ARV-2} (127).

To investigate the molecular determinants for the transformed T-cell line tropism of laboratory isolates, molecular genetic analyses were performed in a fashion similar to that previously utilized to map determinants of macrophage tropism (121, 124, 128). HIV-1_{SF162} (7), an isolate from the brain of an HIV-infected individual, was used as the recipient clone for exchanges of subgenomic portions from a T-cell line-tropic isolate, HIV-1_{SF2} (127). HIV-1_{SF162} is a prototypic primary isolate which replicates efficiently in PBMC but does not infect T-cell lines such as HUT-78. In contrast, HIV-1_{SF2} is a laboratory clone that is adapted for growth in many immortalized T-cell lines. Substitution of a DNA fragment encoding 321 amino acid residues from a region of gp120 transformed the tropism phenotype of HIV-1_{SF162} to resemble that of HIV-1_{SF2} (124). Conversely, transfer of a segment included within the previous region representing a 159-amino acid sequence from HIV-1_{SF2} to HIV-1_{SF162} disrupted the ability of the recombinant to infect T-cell lines. Both genetic exchanges involved the region of gp120 containing the V3 loop. Experiments on T-cell line tropism performed with independent molecular clones yielded very similar results (121, 128). It is noteworthy that, in some cases, macrophage tropism and T-cell line tropism are mutually exclusive; that is, recombinant viruses which infect macrophages could not infect T-cell lines, and vice versa (121, 123).

8.3 Neutralization by soluble CD4 correlates with viral tropism

Macrophage and T-cell tropic strains of HIV use the CD4 receptor in the infection process (116, 117). Binding of the virion via the viral envelope glycoprotein gp120 to CD4 is the first step leading to entry. Consequently, it was hypothesized that agents that interfere with the interaction of gp120 with CD4 might be useful in blocking infection. Several studies demonstrated that a non-membrane-attached recombinant version of CD4 (soluble CD4, sCD4) expressed in mammalian cells inhibits the replication of laboratory strains of HIV *in vitro* (129–133) and protects chimpanzees against experimental infection with HIV-1 (134). These findings pointed out the potential of sCD4 as a therapeutic agent in infected patients (135, 136).

Susceptibility to neutralization by sCD4 varies among different viral isolates. The concentrations of sCD4 that efficiently neutralize laboratory strains of HIV-1 (<1 μg/ml) fail to neutralize primary isolates. In fact, to achieve equivalent levels of neutralization, primary isolates require about 100- to 1000-fold higher concentrations of sCD4 than do laboratory strains (107, 109, 110, 137). For this reason, sCD4 has not been effective as a therapeutic agent.

It is possible that neutralization by sCD4 may be associated with tropism, since most laboratory strains of HIV are tropic for immortalized T-cell lines and most primary isolates are macrophage-tropic to some extent (67, 138). To test this hypothesis, recombinant viruses were constructed between laboratory strains that exhibited tropism for T-cell lines and were susceptible to neutralization by sCD4 and primary isolates that were tropic for macrophages and resistant to sCD4 neutralization (110, 137). In one study, the minimal region that defined resistance to neutralization by sCD4 was determined to encode amino acids 202 to 359 of the envelope of HIV-1$_{JR-FL}$ (110) and included the V3 loop, the same region found to be critical for macrophage tropism. Similarly, in another study, transfer of only part of the V3 sequences of HIV-1$_{Bal}$ was sufficient to confer resistance to sCD4 neutralization (137). Thus, macrophage tropism correlates with resistance to sCD4 and T-cell line tropism with sensitivity to sCD4 neutralization. The sequences thus identified as determinants of susceptibility to neutralization by sCD4 do not include the previously defined major CD4-binding domain (90, 91, 93, 139) or other amino acid residues important for CD4 binding (95).

In conclusion, both sensitivity to sCD4 neutralization and cell tropism are determined by the V3 loop of gp120. The linkage between these two phenomena is unknown at the moment, but suggests that both may be mediated by the same mechanisms (discussed further below).

9. Envelope regions other than V3 may be involved in tropism

9.1 The CD4-binding domain

Mutational analysis of HIV-1$_{Bru}$, an isolate which can infect T-cell lines and the monocytic cell line U937, revealed that a single amino acid change in the CD4-binding region was sufficient to abrogate the ability of the mutant virus to infect the monocytic cell line, but it maintained its ability to infect T-cell lines (CEM, SUP-T1) (93). When the CD4-binding properties of the mutant glycoproteins were tested and compared with the wild-type, no differences in affinity were detected. These observations suggest that in this particular experimental setting, tropism for a monocyte cell line (U937) is not influenced by the CD4 binding affinity.

9.2 A region N-terminal to V3 may modulate infection efficiency in macrophage-tropic isolates

In one of the studies mentioned earlier (124), it was demonstrated that the V3 loop was necessary to confer macrophage tropism; however, the levels of virus production of corresponding recombinant viruses were not as high as those of the parental clones. Therefore, it was concluded that the V3 region may not be the sole factor determining infectivity of macrophages. In fact, a second region of the envelope gene which maps immediately N-terminal to the V3-encoding sequences (amino acid residues 134 to 236 of HIV-1_{NL4-3}) has recently been identified (140) (see Fig. 1). This region influences the rate of replication in macrophages and has been termed an 'efficiency domain'. Thus, recombinant viruses carrying the same V3 loop from a macrophage-tropic isolate would infect macrophages, but the levels of replication were influenced by the sequences in the 'efficiency domain' (140).

9.3 Amino acid changes throughout envelope determine tropism in SIV

By the use of recombinant viruses, the envelope gene of SIV_{mac} was found to be responsible for macrophage tropism (120, 141). By construction of site-directed mutants, five positions within the predicted amino acid sequence, three in the external glycoprotein (residues 67, 176, and 382), and two within the transmembrane glycoprotein (residues 573 and 767) were found to contribute to macrophage tropism (141). However, none of the single-amino acid mutants displayed full capacity to replicate in macrophages compared with the macrophage-tropic parental or recombinant clones. Thus, macrophage tropism in SIV does not appear to be induced by specific regions of envelope but rather, it may be triggered by a combined effect of several point mutations throughout *env*.

9.4 The cytoplasmic tail of SIV *env* may be associated with species specificity

The transmembrane (TM) protein of SIV has been implicated in host-species tropism of these viruses (142, 143). Infectious clones of SIV passaged in macaques or macaque PBMC encode a full-length TM glycoprotein with a molecular weight of about 41 kDa. However, propagation of SIV in human cells (immortalized T-cell lines) gave rise to mutant viruses with a TM of about 28 kDa, as a result of truncation of the cytoplasmic tail. These results indicate that the natural form of TM in SIV_{mac} is the full-length 41-kDa glycoprotein, and viruses with truncated forms of TM appear to result from mutation and selection during propagation in human cells *in vitro* (142, 143).

10. Regions other than envelope may also control tropism

Several groups have attempted to define specific regions of HIV other than envelope that are responsible for differences in tropism. Regions thought to have a regulatory function have been investigated as potential regulators of viral kinetics, cytopathicity, and host-cell tropism. In particular, the *tat* (144, 145) and *nef* genes (146–148) and the long terminal repeat (LTR) (149) have been subjected to molecular analysis aimed at defining their role in various phenotypic traits.

10.1 *tat*

One group noted that HIV-1 isolates from asymptomatic individuals replicated poorly in the T-cell lines HUT-78 and MT4, but this deficiency could be complemented by passage in cell lines constitutively expressing a functional *tat* (144). In contrast, viral isolates from individuals in more advanced stages of disease (AIDS) replicated rapidly and displayed a wider host range because, unlike earlier isolates, they could infect the monocytic cell line U937 in addition to immortalized T-cell lines and PBMC. Moreover, the later isolates replicated to similar levels in cell lines expressing *tat* as in cell lines not expressing it. The molecular determinants of these phenotypes were investigated with recombinant viruses generated between two HIV-1 isolates recovered sequentially from the same individual. The results showed that infection of monocytic and T-cell lines is determined by two specific regions in the envelope gp120, one encompassing the V3 loop and a second one encompassing V4, V5, and the CD4-binding domain. In addition, the rate of HIV-1 replication in the HUT-78 and MT4 cells was controlled by the first coding exon of *tat*. Therefore, *tat* was shown to influence replication kinetics in specific cell types, and to modulate host-cell specificity.

The idea that variation in *tat* sequences might have an impact on the biological phenotype of distinct viral isolates was also tested by Meyerhans and collaborators (145), based on earlier studies comparing isolates with different replication kinetics (39). These studies took the approach of sequencing a number of *tat* clones generated by PCR and testing their ability to *trans*-activate the HIV-1 LTR in transient transfection assays. The results demonstrated that *tat* is subject to sequence variation; however, *tat* genes from HIV-1 isolates with different replication kinetics did not differ significantly in their ability to *trans*-activate the HIV-1 LTR. These results, however, cannot exclude the possibility that subtle changes in *tat* and/or sequences in the LTR which interact with the Tat protein (the TAR loop) may affect the virus life cycle without being detected in transient expression assays.

10.2 *nef*

The *nef* gene of HIV-1 and HIV-2 and the related SIV group share sequence homology and post-translational modifications (146). Although there are conflicting reports about the positive or negative influence of this gene on viral replication *in vitro* (150–157), the conservation of this gene among primate lentiviruses and the fact that antibodies against the Nef protein are found in HIV-1- and HIV-2-infected individuals and SIV-infected monkeys suggest that Nef plays an important role in the viral life cycle (see Chapter 6). Furthermore, a functional *nef* gene is required for high-level replication and induction of disease in macaques infected with SIV (147). These observations prompted studies aimed at understanding the role of *nef*. The isolate HIV$_{HXB2}$ is a laboratory-adapted virus that grows in immortalized T-cell lines and is restricted for infection of primary T-cells and macrophages. This clone has a defective *nef* gene. Insertion of a functional *nef* from the HIV$_{ELI}$ clone enabled the corresponding recombinant to efficiently infect primary T-cells and macrophages (148). In these experiments, the Nef protein itself seemed to be required for the phenotype switch since introduction of a frameshift mutation in the recombinant *nef* abrogated the switch. In conclusion, these experiments indicate that *nef* may influence host-cell specificity, although in a broader sense when compared with the action of V3 in T-cell-line versus macrophage tropism, as discussed above. The mechanisms that regulate such an effect are unknown at the moment. It would be very interesting to learn whether Nef and V3 can act in conjunction to provide a fine tuning of viral tropism.

10.3 Reverse transcriptase

In vivo, the majority of circulating T-lymphocytes are non-dividing quiescent cells. Infection of quiescent lymphocytes leads to non-productive but stable HIV-1 infection. In the absence of activation signals, quiescent lymphocytes can harbour the virus for some period of time without detectable virus production. Following activation of the resting lymphocytes, virus production is induced. Investigations of the molecular mechanisms regulating non-productive and productive infection in resting and stimulated lymphocytes, respectively, showed that cells in either state of activation did not differ in the ability to bind virus (158) or support virus entry and initiation of synthesis of DNA (159). However, whereas in stimulated lymphocytes reverse transcription goes to completion after initiating its activity, in quiescent cells, the majority of the DNA synthesized is an incomplete, labile DNA species (159). If subsequent stimulation of the quiescent cells occurs, virus production can follow. However, if the infected quiescent cells are not activated shortly after infection, the incompletely reverse-transcribed DNA species may be degraded. Therefore, the presence of this labile DNA form could be responsible for both a short-lived latency and a long-lived low-level persistent infection.

10.4 The LTR as a potential determinant for pathogenesis and tropism of SIV

The PBj14 isolate of SIV (SIV$_{SMM-PBj14}$) from sooty mangabey monkeys (*Cercocebus atys*) is the most acutely pathogenic primate lentivirus described thus far. It causes fatal disease in pig-tailed macaques (*Macaca nemestrina*) within several days of inoculation (24). Biological comparison of this acutely pathogenic virus with the parent isolate, SIV$_{SMM-9}$, showed that SIV$_{SMMP-PBj14}$ had a broader host range since, unlike the parental, it replicated in chimpanzee PBMC and immortalized T-cell lines. In addition, SIV$_{SMM-Pj14}$ replicated more efficiently than SIV$_{SMM-9}$ in human PBMC (72). Sequence analysis of both isolates revealed a high degree of conservation with other SIV isolates, except for a 22 bp duplication in the enhancer region of the LTR, which includes a second binding site for the transcription factor NF-κB (149). However, generation of specific recombinant viruses to prove the role of the enhancer duplication has not yet been performed.

11. Tropism for other cell types
11.1 Infection of fibroblasts

Fibroblasts and fibroblastoid cells are target cells for certain strains of HIV (160–162). Infection of these cells was shown to be mediated by CD4 (162), Fc receptors (161) (discussed below in Section 11.2), and through a potential alternative pathway (160) (discussed below in Section 11.3).

HIV$_{GUN-1}$ is an HIV isolate that is tropic for T-cells and produces a very low level of infection in BT cells (fibroblast-like cells derived from human brain and which express CD4) (162). Passage of this virus on BT cells generated a variant (HIV$_{GUN-1-V}$) that maintained its tropism for T-cells and, in addition, acquired the ability to productively infect BT cells. Infection of BT cells by HIV$_{GUN-1-V}$ was CD4-dependent because it was effectively blocked by the anti-CD4 antibody Leu3a. By genetic recombination and sequence analysis, it was deduced that a single point mutation (proline 311 to serine) within the V3 loop was responsible for the wider host range of the variant. These results are in accordance with the notion that the V3 loop is a major determinant of tropism. The findings that V3 regulates tropism of HIV for fibroblasts in addition to macrophages and T-cell lines, and that infection in all of these instances is mediated by CD4, suggest that HIV infects these cell types through similar pathways.

The fibroblast-like cell line HeLa, which is refractory to HIV infection, became permissive when a CD4 cDNA was stably introduced (121). This CD4-expressing HeLa cell line could efficiently propagate T-cell line-adapted strain HIV$_{NL4-3}$, but was not permissive for the primary isolates HIV$_{JRCSF}$, HIV$_{FR-FL}$, and HIV$_{BA-L}$ (121).

11.2 Infection through Fc and complement receptors

In several *in vitro* systems, HIV infection of cells through Fc and complement receptors has been observed (163–166). This phenomenon is termed 'antibody-mediated enhancement' because infection is more efficient in the presence of antiviral antibodies and/or complement factors.

Fc receptors are also involved in the infection of human embryonic lung fibroblasts, a cell type which is normally not susceptible to infection by HIV. Human embryonic lung cells can be rendered susceptible when infection with cytomegalovirus induces expression of Fc receptors for immunoglobulins (161). This type of infection was not blocked by anti-CD4 antibodies or sCD4, but it was abrogated by pretreatment of the cells with human immunoglobulin G. These observations suggest that the HIV tropism can be affected by infection with other viruses. Infections of HIV via Fc and complement receptors are, from a mechanistic point of view, very intriguing processes. Their biological relevance, however, has yet to be established in *in vivo* systems.

11.3 CD4-independent infection

Many human cell lines which do not detectably express CD4 antigen have been shown to be susceptible to infection, albeit less efficiently, with HIV. These include cell lines of various origins such as bone marrow precursors (167), colorectal carcinoma (168), glioma (169–173), hepatoma (174), neuroblastoma (175), rhabdomyosarcoma (171), and fibroblasts (160, 161). Infection of non-CD4-bearing cells is not blocked by anti-CD4 monoclonal antibodies (160, 173, 175) or sCD4 (175).

These results suggest that HIV can infect cells by means of an alternate pathway which is independent of the CD4 receptor. This idea prompted studies toward identifying potential surface molecules involved in HIV entry into non-CD4-bearing cells (176). This was approached by screening antibodies raised against a variety of CNS-related antigens for their ability to inhibit viral entry into certain neuronal cell lines. Antibodies against galactosyl ceramide, a glycolipid common to oligodendrocytes and Schwann cells, blocked infection of neuronal cell lines. In addition, gp120 was shown to bind to galactosyl ceramide but not to other glycolipids (176). Thus, galactosyl ceramide or a highly related molecule may be involved in infection of neural cells. The biological relevance of these findings is as yet not clear.

12. Mechanisms

In view of the results presented in the previous sections of this chapter, we formulate several hypotheses to explain the molecular mechanisms controlling host cell specificity of HIV-1. In this section, we will focus the discussion on instances of tropism where infection of the cell is mediated by the envelope–CD4 interaction. The hypotheses outlined below are not mutually exclusive, nor are they intended to explain tropism in a general way. In fact, it is likely that a variety of mechanisms modulate tropism at different steps in the infection cycle of HIV.

12.1 Important events controlling tropism act at the level of entry

The finding that the region encoding the external glycoprotein is sufficient to determine the ability of viral strains to infect T-cell lines or macrophages (see

Sections 6 and 7) suggests that the major events controlling cell tropism may be acting at the level of viral entry. Several lines of evidence further support this hypothesis. First, transfection of proviral clones into non-permissive cells leads to the production of infectious virus in many cases (28, 126, 177). This finding rules out that events subsequent to proviral formation (e.g. transcription, protein processing, assembly, and release) are responsible for the block in productive infection of non-susceptible cells. Secondly, studies based on detection of viral nucleic acids early after infection revealed that in T-cell lines (177) as well as in macrophages (123), the block occurs prior to the synthesis of viral DNA by the reverse transcriptase.

12.2 The binding affinity between gp120 and CD4 may control tropism

The first event in virus infection is attachment of virions to the target cell surface via CD4. The finding that the coding region for envelope gp120 or portions of it are sufficient to confer various tropism phenotypes, together with the finding that differential sensitivity to neutralization by sCD4 map to the same regions (Sections 7 and 8), led to the hypothesis that perhaps the affinity of different envelope glycoproteins for CD4 was a determining factor of viral tropism. This was tested by measuring the affinity of binding to CD4 by HIV gp120 glycoproteins produced as recombinant soluble polypeptides. The binding experiments with soluble viral glycoproteins showed that gp120 polypeptides from various HIV strains with diverse tropisms display no significant differences in their affinity for CD4 (109, 178, 179). However, binding experiments performed with intact virions suggest that virions from primary isolates have a lower affinity for sCD4 than those from laboratory strains (109). Recombinant gp120 is usually expressed as a monomer (180, 181), whereas the envelope glycoprotein on the surface of virions is thought to form multimers (180, 182). Multimerization of the envelope polypeptides may thus be a possible explanation for the observed differences in binding affinity, which may play a role in determining cell tropism. According to this hypothesis, laboratory strains would be more efficient in infecting T-cell lines because of the higher affinity of the virions for CD4. This, however, would not explain why these strains do not efficiently infect primary macrophages. Further experiments should be performed using virions and cells. This would allow evaluation of the contributions of both the virus and the cell to the binding affinity and its relation to host range.

12.3 The V3 loop may control tropism by regulating the interaction with ancillary cell surface molecules

Following virion attachment to CD4 and prior to viral DNA formation, the steps of fusion, entry, and uncoating take place. Little is known about molecular events involved in fusion, entry, and uncoating of retroviruses (183). For HIV, these

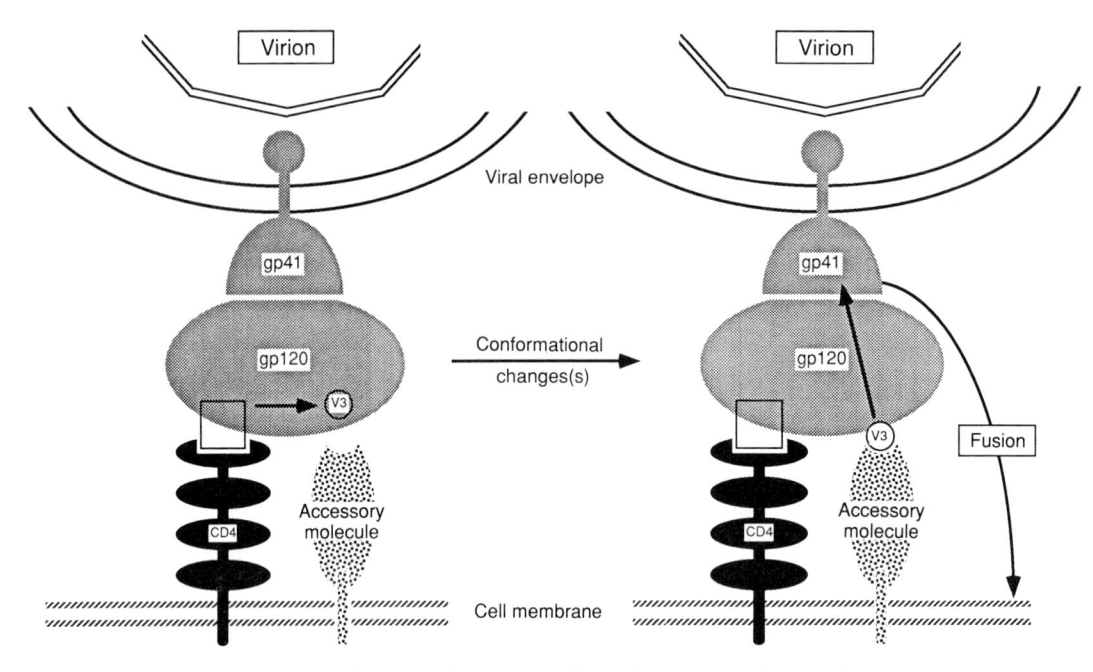

Fig. 2 Schematic representation of a model for the role of the V3 loop in HIV-1 entry. See Sections 6 and 12 for further details

processes, at the molecular level, may include conformational changes in the envelope glycoprotein and/or interactions with additional surface antigens other than CD4 (184).

We propose the following model of entry to explain how tropism may be controlled by V3 (see Fig. 2). Entry of virus into susceptible cells is a multi-step process which includes:

(1) attachment of virions to the cell surface, mediated by the interaction between gp120 and CD4;
(2) as a consequence of binding, V3 undergoes a conformational change;
(3) the conformational change allows the interaction of V3 with a cell-surface ancillary factor, possibly a secondary viral receptor or a protease that acts upon the envelope glycoprotein;
(4) the previous sequence of events leads to membrane fusion;
(5) translocation of the nucleocapsid.

The V3 region spans amino acid residues 293 to 327 (according to the sequence of HIV$_{JR-FL}$ isolate), is flanked by cysteine residues, is believed to form a loop structure (185), and is the principal domain for type-specific antibody neutralization (185) (also see Section 5). A role of the V3 loop in viral entry can be inferred from a number of experiments. Deletion or mutation of the V3 loop reduces or abolishes infection without affecting CD4 binding (99–101). In addition, antibodies that bind to V3 do not prevent binding, yet they block syncytium formation and neutralize virus infectivity (102, 103) and, therefore, presumably interrupt a post-CD4-binding step in the infection cycle (102–104).

Although little is kown about the mechanisms by which V3 controls these functions, recent data suggest that conformational alterations of gp120 take place after CD4 binding (184). Evidence supporting this idea has come from several fronts:

(1) complexing of sCD4 and HIV envelope glycoproteins results in increased exposure of antibody epitopes on the V3 loop of gp120 as well as on gp41 (184);
(2) the association of gp120 and sCD4 is followed by increased cleavage of the V3 loop by exogenous proteinases (184, 186, 187);
(3) binding of gp120 to sCD4 causes dissociation of gp120 from the surface of the virions and envelope glycoprotein-expressing cells, although this phenomenon has not been directly attributed to V3 (109, 184, 188, 189).

We propose that binding of gp120 to CD4 triggers the interaction of the viral envelope glycoprotein with cell surface molecules other than CD4, which is a necessary event for virion entry into the cell. Differences in host range of HIV strains can then be accounted for by the heterogeneous distribution and/or density of ancillary surface molecules among different cell types and, perhaps, species. Accessory surface molecules involved in viral entry could be proteins, lipids, or lipopolysaccharides. In addition, these molecules might function as coreceptors with CD4, or enzymatically, in the case of a surface protease. To date, however, there is no direct genetic or biochemical evidence for the existence of these putative accessory molecules.

13. Conclusion

HIV tropism and, in a broad sense, viral tropism are known to be related to pathogenesis, induction of immune responses by the host, escape from immune surveillance, transmission, and other processes. Variation in the ability to infect certain cell types *in vitro* has been shown to be associated with properties of particular strains. Molecular analysis of HIV strains has allowed the identification of specific areas of the viral genome which are involved in influencing tropism. Furthermore, there is substantial evidence that the mechanisms regulating tropism are complex and act at various steps of the retroviral life cycle.

A great deal of work has pointed to the V3 region of gp120 as the principal domain responsible for tropism, yet, the mechanism(s) by which V3 accomplishes that function is not clear. To gain an understanding of the molecular mechanism through which the V3 loop regulates entry and/or host range specificity, future experiments should be aimed at answering the following questions:

1. What are the cellular determinants for viral tropism?
2. What is the nature of the interaction between V3 and surface molecules on the target cell?
3. Do other regions of gp120 or gp41 interact with V3 to control tropism?

Finally, how do changes in the tropism phenotype of HIV relate to pathogenesis? Understanding these issues will provide insight into the dynamics of disease progression and, perhaps, a rational basis for development of therapeutic strategies.

Acknowledgements

We are grateful to David Camerini, William O'Brien, Jerome Zack, Paul Krogstad, and Wendy Aft for valuable discussions and helpful comments on the manuscript.

References

1. Coffin, J. M. (1986) Genetic variation in AIDS virus. *Cell*, **4,** 1.
2. Gardner, M. B. and Luciw, P. A. (1988) Simian immunodeficiency viruses and their relationship to the human immunodeficiency viruses. *AIDS,* **2 (Suppl. 1),** S3.
3. Gardner, M. B. and Luciw, P. A. (1989) Animal models for AIDS. *FASEB J.,* **3,** 2593.
4. Hahn, B. H., Shaw, G. M., Taylor, M. E., Redfield, R. R., Markham, P. D., Salahudin, S. Z., Wong-Staal, F., Gallo, R. C., Parks, E. S., and Parks, W. P. (1986) Genetic variation in HTLV-III/LAV over time in patients with AIDS or at risk for AIDS. *Science,* **232,** 1548.
5. Goodenow, M., Huet, T., Saurin, W., Kwok, S., Sninsky, J., and Wain-Hobson, S. (1989) HIV-1 isolates are rapidly evolving quasispecies: evidence for viral mixtures and preferred nucleotide substitutions. *J. AIDS,* **2,** 344.
6. Fisher, A. G., Ensoli, B., Looney, D., Rose, A., Gallo, R. C, Saag, M. S., Shaw, G. M., Hahn, B. H., and Wong-Staal, F. (1988) Biologically diverse molecular variants within a single HIV-1 isolate. *Nature,* **334,** 444.
7. Cheng-Mayer, C., Weiss, C., Seto, D., and Levy, J. A. (1989) Isolates of human immunodeficiency virus type 1 from the brain may constitute a special group of the AIDS virus. *Proc. Natl Acad. Sci. USA,* **86,** 8575.
8. Koyanagi, Y., Miles, S., Mitsuyatsu, R. T., Merryl, J. E., Vinters, H. V., and Chen, I. S. Y. (1987) Dual infection of the central nervous system by AIDS viruses with distinct cellular tropisms. *Science,* **236,** 819.
9. Sakai, K., Dewhurst, S., Ma, X. Y., and Volsky, D. J. (1988) Differences in cytopathogenicity and host cell range among infectious molecular clones of human immunodeficiency virus type 1 simultaneously isolated from an individual. *J. Virol.,* **62,** 4078.
10. Saag, M. S., Hahn, B. H., Gibbons, J., Li, Y. X., Parks, E. S., Parks, W. P., and Shaw, G. M. (1988) Extensive variation of human immunodeficiency virus type-1 in vivo. *Nature,* **334,** 440.
11. Tersmette, M., Gruters, R. A., de Wolf, F., de Goede, R., Lange, J. M., Schellekens, P. T., Goudsmit, J., Huisman, H. G., and Miedema, F. (1989) Evidence for a role of virulent human immunodeficiency virus (HIV) variants in the pathogenesis of acquired immunodeficiency syndrome: studies on sequential HIV isolates. *J. Virol.,* **63,** 2118.
12. Pang, S., Vinters, H., Akashi, T., O'Brien, W. A., and Chen, I. S. Y. (1991) HIV-1 *env* sequence variation in brain tissue of patients with AIDS-related neurologic disease. *J. AIDS,* **4,** 1082.
13. Hahn, B. H., Gonda, M. A., Shaw, G. M., Popovic, M., Hoxie, J. A., Gallo, R. C., and

Wong-Staal, F. (1985) Genomic diversity of the acquired immune deficiency syndrome virus HTLV-III: Different viruses exhibit divergence in their envelope genes. *Proc. Natl Acad. Sci. USA,* **82,** 4813.

14. Modrow, S., Hahn, B. H., Shaw, G. M., Gallo, R. C., Wong-Staal, F., and Wolf, H. (1987) Computer-assisted analysis of envelope protein sequences of seven human immunodeficiency virus isolates: prediction of antigenic epitopes in conserved and variable regions. *J. Virol.,* **61,** 570.
15. Broliden, P. A., von Gegerfelt, A., Clapham, P., Rosen, J., Fenyo, E. M., Wahren, B., and Broliden, K. (1992) Identification of human neutralization-inducing regions of the human immunodeficiency virus type 1 envelope glycoproteins. *Proc. Natl Acad. Sci. USA,* **89,** 461.
16. La Rosa, G. J., Davide, J. P., Weinhold, K., Waterbury, J. A., Profy, A. T., Lewis, J. A., Langlois, A. J., Dressman, G. R., Boswell, R. N., Shadduck, P., Holey, L. H., Karplus, M., Bolognesi, D. P., Matthews, T. J., Emini, E. A., and Putney, S. D. (1990) Conserved sequence and structural elements in the HIV-1 principal neutralizing determinant. *Science,* **249,** 932.
17. Kusumi, K., Conway, B., Cunningham, S., Berson, A., Evans, C., Iversen, A. K. N., Colvin, D., Gallo, M. V., Coutre, S., Shpaer, E. G., Faulkner, D. V., deRonde, A., Volkman, S., Williams, C., Hirsch, M. S., and Mullins, J. I. (1992) Human immunodeficiency virus type 1 envelope gene structure and diversity *in vivo* and after cocultivation *in vitro. J. Virol.,* **66,** 875.
18. Meyerhans, A., Cheynier, R., Albert, J., Seth, M., Kwok, S., Sninsky, J., Morfeldt, M. L., Asjo, B., and Wain, H. S. (1989) Temporal fluctuations in HIV quasispecies in vivo are not reflected by sequential HIV isolations. *Cell,* **58,** 901.
19. Nara, P. L., Robey, W. G., Arthur, L. O., Asher, D. M., Wolff, A. V., Gibbs, C. J., Gajdusek, C., and Fischinger, P. J. (1987) Persistent infection of chimpanzees with human immunodeficiency virus. Serological responses and properties of the reisolated virus. *J. Virol.,* **61,** 3173.
20. Fultz, P. N., McClure, H. M., Swenson, R. B., McGrath, C. R., Brodie, A., Getchell, J. P., Jensen, F. C., Anderson, D. C., Broderson, J. R., and Francis, D. P. (1986) Persistent infection of chimpanzees with HTLV-III/LAV: a potential model for acquired immunodeficiency syndrome. *J. Virol.,* **58,** 116.
21. Alter, H. J., Eichberg, J. W., Masur, H., Saxinger, W. C., Gallo, R., Macher, A. M., Lane, H. C., and Fayci, A. S. (1984) Transmission of HTLV-III infection from human plasma to chimpanzees: an animal model for AIDS. *Science,* **226,** 549.
22. Lusso, P., Markham, P. D., Ranki, A., Earl, P., Moss, B., Dorner, F., Gallo, R. C., and Krohn, K. J. (1988) Cell-mediated immune response toward viral envelope and core antigens in gibbon apes (*Hylobates lar*) chronically infected with human immunodeficiency virus-1. *J. Immunol.,* **141,** 2467.
23. Dormont, D., Livartowski, J., Chamaret, S., Guetard, D., Henin, D., Levagueresse, R., van de Moortelle, P. F., Larke, B., Gourmelon, P., Vazeux, R., Metivier, H., Flageat, J., Court, L., Hauw, J. J., and Montagnier, L. (1989) HIV-2 in rhesus monkeys: serological, virological and clinical results. *Intervirology,* **1,** 59.
24. Fultz, P. N., Switzer, W., McClure, H. M., Anderson, D. C., and Montagnier, L. (1988) Simian models for AIDS: SIV/Smm and HIV-2 infection of macaques. In *Vaccines 88.* Ginsberg, H. S., Brown, F., Lerner, R. A., and Chanock, R. M. (ed.). Cold Spring Harbor Laboratory Press, Cold Spring Harbor, New York, p. 167.
25. Franchini, G., Fargnoli, K. A., Giombini, F., Jagodzinski, L., De Rossi, A., Bosch, M.,

Biberfeld, G., Fenyo, E. M., Albert, J., and Gallo, R. C. (1989) Molecular and biological characterization of a replication competent human immunodeficiency type 2 (HIV-2) proviral clone. *Proc. Natl Acad. Sci. USA*, **86,** 2433.
26. Filice, G., Cereda, P. M., and Varnier, O. E. (1988) Infection of rabbits with human immunodeficiency virus. *Nature*, **335,** 366.
27. Maddon, P. J., Dalgleish, A. G., McDougal, J. S., Clapham, P. R., Weiss, R. A., and Axel, R. (1986). The T4 gene encodes the AIDS virus receptor and is expressed in the immune system and the brain. *Cell*, **47,** 333.
28. Levy, J. A., Cheng-Mayer, C., Dina, D., and Luciw, P. A. (1986) AIDS retrovirus (ARV-2) clone replicates in transfected human and animal fibroblasts. *Science*, **232,** 998.
29. Landau, N. R., Page, K. A., and Littman, D. R. (1991) Pseudotyping with human T-cell leukemia virus type I broadens the human immunodeficiency virus host range. *J. Virol.*, **65,** 162.
30. Locardi, C., Puddu, P., Ferrantini, M., Parlanti, E., Sestili, P., Varano, F., and Belardelli, F. (1992) Persistent infection of normal mice with human immunodeficiency virus. *J. Virol.*, **66,** 1649.
31. Armstrong, J. A. and Horne, R. (1984) Follicular dendritic cells and virus-like particles in AIDS related lymphadenopathy. *Lancet*, **ii,** 370.
32. Cohen, A. H., Sun, N. C., Shapshak, P., and Imagawa, D. T. (1989) Demonstration of human immunodeficiency virus in renal epithelium in HIV-associated nephropathy. *Modern Pathol.*, **2,** 125.
33. Ho, D. D., Schooley, R. T., Rota, T. R., Kaplan, J. C., Flynn, T., Salahuddin, S. Z., Gonda, M. A., and Hirsch, M. S. (1984) HTLV-III in the semen and blood of a healthy homosexual man. *Science*, **226,** 451.
34. Ho, D. D., Rota, T. R., Schooley, R. T., Kaplan, J. C., Allan, J. D., Groopman, J. E., Resnick, L., Felsenstein, D., Andrews, C., and Hirsch, M. S. (1985) Isolation of HTLV-III from cerebrospinal fluid and neural tissues of patients with neurologic syndromes related to the acquired immunodeficiency syndrome. *N. Engl. J. Med.*, **313,** 1493.
35. Levy, J. A., Shimabukuro, J., Hollander, H., Mills, J., and Kaminsky, L. (1985) Isolation of AIDS-associated retroviruses from cerebrospinal fluid and brain of patients with neurological symptoms. *Lancet*, **ii,** 586.
36. Nelson, J. A., Wiley, C. A., Reynolds-Kohler, C., Reese, C. E., Margaretten, W., and Levy, J. A. (1988) Human immunodeficiency virus detected in bowel epithelium from patients with gastrointestinal symptoms. *Lancet*, **i,** 259.
37. Popovic, M. G. and Gartner, S. (1987) Isolation of HIV from monocytes but not from T lymphocytes. *Lancet*, **ii,** 916.
38. Tschachler, E., Groh, V., Popovic, M., Mann, D. L., Konrad, K., Safai, B., Eron, L., di Marzo Veronese, F., Wolff, K., and Stingl, G. (1987) Epidermal Langerhans cells—A target for HTLV-III/LAV infection. *J. Invest. Dermatol.*, **88,** 233.
39. Asjo, B., Morfeldt-Manson, L., Albert, J., Biberfeld, G., Karlsson, A., Lidman, K., and Fenyo, E. M. (1986) Replicative capacity of human immunodeficiency virus from patients with varying severity of HIV infection. *Lancet*, **ii,** 660.
40. Cheng-Mayer, C., Seta, D., Tateno, M., and Levy, J. (1988) Biologic features of HIV-1 that correlate with virulence in the host. *Science*, **240,** 80.
41. Fenyo, E. M., Morfeldt-Manson, L., Chiodi, F., Lind, B., von, G. A., Albert, J., Olausson, E., and Asjo, B. (1988) Distinct replicative and cytopathic characteristics of human immunodeficiency virus isolates. *J. Virol.*, **62,** 4414.
42. Tersmette, M., de Goede, R., Al, B. J. M., Winkel, I. N., Gruters, R. A., Cuypers,

H. T., Huisman, H. G., and Miedema, F. (1988). Differential syncytium inducing capacity of HIV isolates: Frequent detection of syncytium inducing isolates in patients with AIDS and ARC. *J. Virol.,* **62,** 2026.

43. Price, R. W., Brew, B., Sidtis, J., Rosenblum, M., Scheck, A. C., and Cleary, P. (1988) The brain in AIDS: central nervous system HIV-1 infection and AIDS dementia complex. *Science,* **239,** 586.
44. Rodgers, V. D. and Kagnoff, M. F. (1987). Gastrointestinal manifestations of the acquired immunodeficiency syndrome. *West. J. Med.,* **146,** 57.
45. Nicholson, J. K. A., Cross, G. D., Callaway, C. S., and McDougal, J. S. (1986) *In vitro* infection of human monocytes with human T lymphotropic virus type III/lymphadenopathy-associated virus (HTLV-III/LAV). *J. Immunol.,* **137,** 323.
46. Ho, D., Rota, D., and Hirsch, M. (1986) Infection of monocyte/macrophages by human T-lymphotropic virus type III. *J. Clin. Invest.,* **77,** 1712.
47. Gartner, S., Markovits, P., Markovits, D. M., Kaplan, M. H., Gallo, R. C., and Popovic, M. (1986) The role of mononuclear phagocytes in HTLV-III/LAV infection. *Science,* **233,** 215.
48. Langhoff, E., Terwilliger, E. F., Bos, H. J., Kalland, K. H., Poznansky, M. C., Bacon, O. M., and Haseltine, W. A. (1991) Replication of human immunodeficiency virus type 1 in primary dendritic cell cultures. *Proc. Natl Acad. Sci. USA,* **88,** 7998.
49. Snider, W. D., Simpson, D. M., Nielsen, S., Gold, J. W., Metroka, C. E., and Posner, J. B. (1983) Neurological complications of the acquired immunodeficiency syndrome: analysis of 50 patients. *Ann. Neurol.,* **14,** 403.
50. Nielson, S. L., Petito, C. K., Urmacher, C. D., and Posner, J. B. (1984) Subacute encephalitis in acquired immunodeficiency syndrome: a postmortem study. *Am. J. Clin. Pathol.,* **82,** 678.
51. Koenig, S., Gendelman, H. E., Orenstein, J. M., Dal Canto, M. C., Pezeshkpour, G. H., Yungbluth, M., Janotta, F., Aksamit, A., Martin, M., and Fauci, A. S. (1986) Detection of AIDS virus in macrophages in brain tissue from AIDS patients with encephalopathy. *Science,* **233,** 1089.
52. Shaw, G. M., Harper, M. E., Hahn, B. H., Epstein, L. G., Gadjusek, D. C., Price, R. W., Navia, B. A., Petito, C. K., O'Hara, C. J., Groopman, J. E., Cho, E.-S., Oleske, J. M., Wong-Staal, F., and Gallo, R. C. (1985) HTLV infection in brains of children and adults with AIDS encephalopathy. *Science,* **227,** 177.
53. Watkins, B. A., Dorn, H. H., Kelly, W. B., Armstrong, R. C., Potts, B. J., Michaels, F., Kufta, C. V., and Dubois-Dalcq, M. (1990) Specific tropism of HIV-1 for microglial cells in primary human brain cultures. *Science,* **249,** 549.
54. Wiley, C. A., Schrier, R. D., Nelson, J. A., Lampert, P. W., and Oldstone, M. B. A. (1986) Cellular localization of human immunodeficiency virus infection within the brains of acquired immune deficiency syndrome patients. *Proc. Natl Acad. Sci. USA,* **83,** 7089.
55. Gabuzda, D. H., Ho, D. D., de la Monte, S. M., Hirsh, M. S., Rota, T. R., and Sobel, R. A. (1986) Immunohistochemical identification of HTLV-III antigen in brains of patients with AIDS. *Ann. Neurol.,* **20,** 289.
56. Perry, V. H. and Gordon, S. (1988) Macrophages and microglia in the nervous system. *Trends Neurosci.,* **11,** 273.
57. Sharpless, N. E., O'Brien, W. A., Verdin, E., Kufta, C. V., Chen, I. S. Y., and Dubois-Dalcq, M. (1992) Human immunodeficiency virus type 1 tropism for brain microglial cells is determined by a region of the env glycoprotein that also controls macrophage tropism. *J. Virol.,* **66,** 2588.

58. Gartner, S., Markovits, P., Markovits, D. M., Betts, R. F., and Popovic, M. (1986) Virus isolation from and identification of HTLV-III/LAV-producing cells in brain tissue from a patient with AIDS. *J. Am. Med. Assoc.*, **256**, 2365.
59. Koyanagi, Y., O'Brien, W. A., Zhao, J. Q., Golde, D. W., Gasson, J. C., and Chen, I. S. (1988) Cytokines alter production of HIV-1 from primary mononuclear phagocytes. *Science*, **241**, 1673.
60. Popovic, M. G., Mellert, V., Erfle, V., and Gartner, S. (1988) Role of mononuclear phagocytes and accessory cells in human immunodeficiency virus type I infection of the brain. *Ann. Neurol.*, **23**, S74.
61. Moyer, M. P. and Gendelman, H. E. (1991) HIV replication and persistence in human gastrointestinal cells cultured in vitro. *J. Leukocyte Biol.*, **49**, 499.
62. Ziza, J.-M., Brun-Vezinet, F., Venet, A., Rouzioux, C. H., Traversat, J., Israel-Biet, B., Barré-Sinoussi, F., and Godeau, P. (1985) Lymphadenopathy-associated virus isolated from bronchoalveolar lavage fluid in AIDS-related complex with lymphoid interstitial pneumononitis. *N. Engl. J. Med.*, **313**, 183.
63. Chayt, K. J., Harper, M. E., Marselle, L. M., Lewin, E. B., Rose, J. M., Oleske, J. M., Epstein, L. G., Wong-Staal, F., and Gallo, R. C. (1986) Detection of HTLV-III RNA in lungs of patients with AIDS and pulmonary involvement. *J. Am. Med. Assoc.*, **256**, 2356.
64. Salahuddin, S. Z., Rose, R. M., Groopman, J. E., Markham, P. D., and Gallo, R. C. (1986) Human T lymphotropic virus type III infection of human alveolar macrophages. *Blood*, **62**, 281.
65. Rich, E. A., Chen, I. S. Y., Zack, J. A., Leonard, M. L., and O'Brien, W. A. (1992) Increased susceptibility of differentiated phagocytes to productive infection with human immunodeficiency virus-1 (HIV-1). *J. Clin. Invest.*, **89**, 176.
66. Schuitemaker, H., Kootstra, N. A., de Goede, R., de Wolf, F., Miedema, F., and Tersmette, M. (1991) Monocytotropic human immunodeficiency virus type 1 (HIV-1) variants detectable in all stages of HIV-1 infection lack T-cell line tropism and syncytium-inducing ability in primary T-cell culture. *J. Virol.*, **65**, 356.
67. Schuitemaker, H., Koot, M., Kootstra, N. A., Wouter Dercksen, M., de Goede, R. E. Y., van Steenwijk, P., Lange, J. M. A., Eeftink Schattenkerk, J. K. M., Miedema, F., and Tersmette, M. (1992) Biological phenotype of human immunodeficiency virus type 1 clones at different stages of infection: progression of disease is associated with a shift from monocytotropic to T-cell-tropic virus populations. *J. Virol.*, **66**, 1354.
68. Lifson, A. R., Rutherford, G. W., and Jaffe, H. W. (1988) The natural history of human immunodeficiency virus infection. *J. Infect. Dis.*, **158**, 1360.
69. Bukrinsky, M. I., Stanwick, T. L., Dempsey, M. P., and Stevenson, M. (1991) Quiescent T lymphocytes as an inducible virus reservoir in HIV-1 infection. *Science*, **254**, 427.
70. Garcia-Blanco, M. and Cullen, B. R. (1991) Molecular basis of latency in pathogenic human viruses. *Science*, **254**, 815.
71. Stevenson, M., Stanwick, T. L., Dempsey, M. P., and Lamonica, C. A. (1990) HIV-1 replication is controlled at the level of T cell activation and proviral integration. *Eur. Mol. Biol. J.*, **9**, 1551.
72. Fultz, P. N., McClure, H. M., Anderson, D. C., and Switzer, W. M. (1989) Identification and biologic characterization of an acutely lethal variant of simian immunodeficiency virus from sooty mangabeys (SIV/SMM). *AIDS Res. Hum. Retroviruses*, **5**, 397.

73. Maddon, P. J., McDougal, J. S., Clapham, P. R., Dalgleish, A. G., Jamal, S., Weiss, R. A., and Axel, R. (1988) HIV infection does not require endocytosis of its receptor, CD4. *Cell,* **54,** 865.
74. White, J. M. and Littman, D. R. (1989) Viral receptors of the immunoglobulin superfamily. *Cell,* **56,** 725.
75. McDougal, J. S., Kennedy, M. S., Sligh, J. M., Cort, S. P., Mawle, A., and Nicholson, J. K. A. (1986) Binding of HTLV-III/LAV to T4+ cells by a complex of the 110K viral protein and the T4 molecule. *Science,* **231,** 382.
76. Reinhertz, E. L., King, P. C., Golgstein, G., and Schlossman, S. F. (1979) Separation of functional subsets to human T cells by a monoclonal antibody. *Proc. Natl Acad. Sci. USA,* **76,** 4061.
77. Moscicki, R. A., Amento, E. P., Krane, S. M., Kurnick, J. T., and Colvin, R. B. (1983) Modulation of surface antigens of a human monocyte cell line, U937, during incubation with T lymphocyte-conditioned medium: detection of T4 antigen and its presence on normal blood monocytes. *J. Immunol.,* **131,** 743.
78. Wood, G. S., Warner, N. L., and Warnke, R. A. (1983) Anti-Leu-3/T4 antibodies react with cells of monocyte/macrophage and Langerhans lineage. *J. Immunol.,* **131,** 212.
79. Maddon, P. J., Molineaux, S. M., Maddon, D. E., Zimmerman, K. A., Godfrey, M., Alt, F. W., Chess, L., and Axel, R. (1987) Structure and expression of the human and mouse T4 genes. *Proc. Natl Acad. Sci. USA,* **84,** 9155.
80. Arthos, J., Deen, K. C., Chaikin, M. A., Fornwald, J. A., Sathe, G., Sattentau, Q. J., Clapham, P. R., Weiss, R. A., McDougal, J. S., and Pietropaolo, C. (1989) Identification of the residues in human CD4 critical for the binding of HIV. *Cell,* **57,** 469.
81. Camerini, D. and Seed, B. (1990) A CD4 domain important for HIV-mediated syncytium formation lies outside the virus binding site. *Cell,* **60,** 747.
82. Clayton, L. K., Hussey, R. E., Steinbrich, R., Ramachandran, H., Husain, Y., and Reinherz, E. L. (1988) Substitution of murine for human CD4 residues identifies amino acids critical for HIV-gp120 binding. *Nature,* **335,** 363.
83. Landau, N. R., Warton, M., and Littman, D. R. (1988) The envelope glycoprotein of the human immunodeficiency virus binds to the immunoglobulin-like domain of CD4. *Nature,* **334,** 159.
84. Peterson, A. and Seed, B. (1988) Genetic analysis of monoclonal antibody and HIV binding sites on the human lymphocyte antigen CD4. *Cell,* **54,** 65.
85. Sattentau, Q. J. and Weiss, R. A. (1988) The CD4 antigen: physiological ligand and HIV receptor. *Cell,* **52,** 631.
86. Rudd, C. E., Trevyllian, J. M., Dasgupta, J. D., Wong, L. L., and Slossman, S. F. (1988) The CD4 receptor is complexed in detergent lysates to a protein-tyrosine kinase (pp58) from human T lymphocytes. *Proc. Natl Acad. Sci. USA,* **85,** 5190.
87. Veillette, A., Bookman, M. A., Horak, E. M., and Bolen, J. B. (1988) The CD4 and CD8 T cell surface antigens are associated with the internal membrane tyrosine-protein kinase p56^{lck}. *Cell,* **55,** 301.
88. Putney, S. D. and Montelaro, R. C. (1990) Lentiviruses. In *Immunochemistry of viruses, II. The basis for serodiagnosis and vaccines.* Neurath, A. R. and Regenmortel, M. H. V. (ed.). Elsevier Science Publishers B.V., Amsterdam, p. 307.
89. Smith, D. H., Byrn, R. A., Masters, S. A., Gregory, T., Groopman, J. E., and Capon, D. J. (1987) Blocking of HIV infection by a soluble, secreted form of CD4 antigen. *Science,* **238,** 1704.
90. Kowalsky, M., Potz, J., Basiripour, L., Dorfman, T., Goh, W. C., Terwilliger, E.,

Dayton, A., Rosen, C., Haseltine, W., and Sodroski, J. (1987) Functional regions of the envelope glycoprotein of human immunodeficiency virus type 1. *Science,* **237,** 1351.

91. Lasky, L. A., Nakamura, G., Smith, D. H., Fennie, C., Shimasaki, C., Patzer, E., Berman, P., Gregory, T., and Capon, D. J. (1987) Delineation of a region of the human immunodeficiency virus type 1 gp120 glycoprotein critical for interaction with the CD4 receptor. *Cell,* **50,** 975.

92. Sun, N. C., Ho, D. D., Sun, C. R., Liou, R. S., Gordon, W., Fung, M. S., Li, X. L., Ting, R. C., Lee, T. H., and Chang, N. T. (1989) Generation and characterization of monoclonal antibodies to the putative CD4-binding domain of human immunodeficiency virus type 1 gp120. *J. Virol.,* **63,** 3579.

93. Cordonnier, A., Montagnier, L., and Emerman, M. (1989) Single amino acid changes in the HIV envelope affect viral tropism and receptor binding. *Nature,* **340,** 571.

94. Linsey, P. S., Ledbetter, J. A., Kinney-Thomas, E., and Hu, S. L. (1988) Effects of anti-gp120 monoclonal antibodies on CD4 receptor binding by the envelope protein of human immunodeficiency virus type 1. *J. Virol.,* **62,** 3695.

95. Olshevsky, U., Helseth, E., Furman, J., Li, J., Haseltine, W., and Sodroski, J. (1990) Identification of individual human immunodeficiency virus type 1 gp120 amino acids important for CD4 receptor binding. *J. Virol.,* **64,** 5701.

96. McDougal, J. S., Nicholson, J. K. A., Cross, G. D., Cort, S. P., Kennedy, M. S., and Mawle, A. C. (1986) Binding of the human retrovirus HTLV-III/LAV/ARV/HIV to the CD4 (T4) molecule: conformation dependence, epitope mapping, antibody inhibition, and potential for idiotypic mimicry. *J. Immunol.,* **137,** 2937.

97. Putney, S., Matthews, T., Robey, W., Lynn, D., Robert-Guroff, M., Mueller, W., Langlois, A., Ghrayeb, J., Petteway, S., Weinhold, K., Fischinger, P., Wong-Staal, F., Gallo, R., and Bolognesi, D. (1986) HTLV/III/LAV neutralizing antibodies to an *E. coli*-produced fragment of the virus envelope. *Science,* **234,** 1392.

98. Nara, P. L., Smit, L., Dunlop, N., Hatch, W., Merges, M., Waters, D., Kelliher, J., Gallo, R. C., Fischinger, P. J., and Goudsmit, J. (1990) Emergence of viruses resistant to neutralization by V3-specific antibodies in experimental human immunodeficiency type 1 IIIB infection of chimpanzees. *J. Virol.,* **64,** 3779.

99. Page, K. A., Stearns, S. M., and Littman, D. R. (1992) Analysis of mutations in the V3 domain of gp160 that affect fusion and infectivity. *J. Virol.,* **66,** 524.

100. Freed, E. O. and Risser, R. (1991) Identification of conserved residues in the human immunodeficiency virus type 1 principal neutralizing determinant that are involved in fusion. *AIDS Res. Hum. Retroviruses,* **7,** 807.

101. Ivanoff, L. A., Dubay, J. W., Morris, J. F., Roberts, S. J., Gutshall, L., Sternberg, E. J., Hunter, E., Matthews, T. J., and Petteway, Jr, S. R. (1992) V3 loop region of the HIV-1 gp120 envelope protein is essential for virus infectivity. *Virology,* **187,** 423.

102. Rusche, J. R., Javaherian, K., McDanal, C., Petro, J., Lynn, D. L., Grimaila, R., Langlois, A. J., Gallo, R. C., Arthur, L. O., Fischinger, P. J., Bolognesi, D. P., Putney, S. D., and Matthews, T. J. (1988) Antibodies that inhibit fusion of human immunodeficiency virus-infected cells bind a 24-amino acid sequence of the viral envelope, gp120. *Proc. Natl Acad. Sci. USA,* **85,** 3198.

103. Skinner, M. A., Langlois, A. J., McDanal, C. B., McDougal, J. S., Bolognesi, D. P., and Mathews, T. J. (1988) Neutralizing antibodies to an immunodominant envelope sequence do not prevent gp120 binding to CD4. *J. Virol.,* **62,** 4195.

104. McKeating, J. A. and Willey, R. A. (1989) Structure and function of the HIV envelope. *AIDS 89,* **3,** S35.

105. Daar, E. S. and Ho, D. D. (1990) The structure and function of retroviral envelope glycoproteins. *Semin. Virol.*, **1**, 205.
106. White, J. M. (1990) Viral and cellular membrane fusion proteins. *Annu. Rev. Physiol.*, **52**, 675.
107. Daar, E. S., Li, X. L., Moudgil, T., and Ho, D. D. (1990) High concentrations of recombinant soluble CD4 are required to neutralize primary HIV-1 isolates. *Proc. Natl Acad. Sci. USA*, **87**, 6574.
108. Hart, T. K., Kirsh, R., Ellens, H., Sweet, R. W., Lambert, D. M., Petteweay, Jr, S. R., Leary, J., and Bugdliski, P. J. (1991) Binding of soluble CD4 proteins to human immunodeficiency virus type-1 and infected cells induces release of envelope glycoprotein gp120. *Proc. Natl Acad. Sci. USA*, **88**, 2189.
109. Moore, J. P., McKeating, J. A., Huang, Y., Ashkenazi, A., and Ho, D. D. (1992) Virions of primary human immunodeficiency virus type 1 isolates resistant to soluble CD4 (sCD4) neutralization differ in sCD4 binding and glycoprotein gp120 retention from sCD4-sensitive isolates. *J. Virol.*, **66**, 235.
110. O'Brien, W. A., Chen, I. S. Y., Ho, D. D., and Daar, E. S. (1992) Mapping genetic determinants for human immunodeficiency virus type 1 resistance to soluble CD4. *J. Virol.*, **66**, 3125.
111. Barré-Sinoussi, F., Montagnier, L., Dauguet, C., Chermann, J. C., Rey, F., Nugeyre, S., Axler-Blin, C., Gruest, J., Chamaret, S., Vézinet-Brun, F., Rouzioux, C., and Rozenbaum, W. (1983) Isolation of a T-lymphotropic retrovirus from a patient at risk for acquired immune deficiency syndrome (AIDS). *Science*, **220**, 868.
112. Dalgleish, A. G., Beverley, P. C. L., Clapham, P. R., Crawford, D. H., Greaves, M. F., and Weiss, R. A. (1984) The CD4 (T4) antigen is an essential component of the receptor for the AIDS retrovirus. *Nature*, **312**, 763.
113. Popovic, M. G., Sarngadharan, E., Read, E., and Gallo, R. C. (1984) Detection, isolation, and continuous production of cytopathic retroviruses (HTLV-III) from patients with AIDS and pre-AIDS. *Science*, **224**, 497.
114. Montagnier, L., Gruest, J., Chamaret, S., Dauguet, C., Axler, C., Guétard, D., Nugeyre, M. T., Barré-Sinoussi, F., Chermann, J.-C., Brunet, J. B., Klatzmann, D., and Gluckman, J. C. (1984) Adaptation of lymphadenopathy associated virus (LAV) to replication in EBV transformed B lymphoblastoid cell lines. *Science*, **225**, 63.
115. Gyorkey, F., Melnik, J. L., Sinkovics, J. G., and Gyorkey, P. A. (1987) Human immunodeficiency virus in brain biopsies of patients with AIDS and progressive encephalopathy. *J. Infect. Dis.*, **155**, 870.
116. Collman, R., Godfrey, B., Cutilli, J., Rhodes, A., Hassan, N. F., Sweet, R., Douglas, S. D., Friedman, H., Nathanson, N., and Gonzalez-Scarano, F. (1990) Macrophage-tropic strains of human immunodeficiency virus type 1 utilize the CD4 receptor. *J. Virol.*, **64**, 4468.
117. Gomatos, P. J., Stamatos, N. M., Gendelman, H. E., Fowler, A., Hoover, D. L., Kalter, D. C., Burke, D. S., Tramont, E. C., and Meltzer, M. S. (1990) Relative inefficiency of soluble recombinant CD4 for inhibition of infection by monocyte-tropic HIV in monocytes and T cells. *J. Immunol.*, **144**, 4183.
118. Yourno, J., Fisher, A. G., Looney, D. J., Gallo, R. C., and Wong-Staal, F. (1989) A recombinant clone of HIV-1 preferentially transmitted in normal peripheral blood mononuclear cells. *AIDS Res. Hum. Retroviruses*, **5**, 565.
119. Cheng-Mayer, C., Quiroga, M., Tung, J. W., Dina, D., and Levy, J. A. (1990) Viral determinants of human immunodeficiency virus type 1 T-cell or macrophage tropism, cytopathogenicity, and CD4 antigen modulation. *J. Virol.*, **64**, 4390.

120. Banapour, B., Marthas, M. L., Ramos, R. A., Lohman, B. L., Unger, R. E., Gardner, M. B., Pedersen, N. C., and Luciw, P. A. (1991) Identification of viral determinants of macrophage tropism for simian immunodeficiency virus SIVmac. *J. Virol.*, **65**, 5798.
121. Chesebro, B., Nishio, J., Peryman, S., Cann, A., O'Brien, W., Chen, I. S. Y., and Wehrly, K. (1991) Identification of human immunodeficiency virus envelope gene sequences influencing viral entry into CD4-positive HeLa cells, T-leukemia cells, and macrophages. *J. Virol.*, **65**, 5782.
122. Hwang, S. S., Boyle, T. J., Lyerly, H. K., and Cullen, B. R. (1991) Identification of the V3 loop as the primary determinant of cell tropism in HIV-1. *Science*, **253**, 7174.
123. O'Brien, W. A., Koyanagi, Y., Namazie, A., Zhao, J.-Q., Diagne, A., Idler, K., Zack, J. A., and Chen, I. S. Y. (1991) HIV-1 tropism for mononuclear phagocytes can be determined by regions of gp120 ouside the CD4-binding domain. *Nature*, **349**, 69.
124. Shioda, T., Levy, J. A., and Cheng-Mayer, C. (1991) Macrophage and T-cell line tropisms of HIV-1 are determined by specific regions of the envelope gp120 gene. *Nature*, **349**, 167.
125. Westerbelt, P., Gendelman, H. E., and Ratner, L. (1991) Identification of a determinant within the human immunodeficiency virus 1 surface envelop glycoprotein critical for productive infection of primary monocytes. *Proc. Natl Acad. Sci. USA*, **88**, 3097.
126. Adachi, A., Gendelman, H. E., Koenig, S., et al. (1986) Production of acquired immunodeficiency syndrome-associated retrovirus in human and non-human cells transfected with an infectious molecular clone. *J. Virol.*, **59**, 284.
127. Levy, J. A., Hoffman, A. D., Kramer, S. M., Landis, J. A., Shimabukuro, J. M., and Oshiro, L. S. (1984) Isolation of lymphocytopathic retroviruses from San Francisco patients with AIDS. *Science*, **225**, 840.
128. Cann, A. J., Churcher, M. J., Boyd, M., O'Brien, W., Zhao, J., Zack, J. A., and Chen, I. S. Y. (1992) The region of the envelope gene of human immunodeficiency virus type 1 responsible for determination of cell tropism. *J. Virol.*, **66**, 305.
129. Watanabe, M., Reimann, K. A., DeLong, P. A., Liu, T., Fisher, R. A., and Letvin, N. L. (1989) Effect of recombinant soluble CD4 in rhesus monkeys infected with simian immunodeficiency virus of macaques. *Nature*, **337**, 267.
130. Hussey, R. E., Richardson, N. E., Kowalski, M., Brown, N. R., Chang, H. C., Siliciano, R. F., Dorfman, T., Walker, B., Sodroski, J., and Reinherz, E. L. (1988) A soluble CD4 protein selectively inhibits HIV replication and syncytium formation. *Nature*, **331**, 78.
131. Deen, K. C., McDougal, J. S., Inacker, R., Folena, W. G., Arthos, J., Rosenberg, J., Maddon, P. J., Axel, R., and Sweet, R. W. (1988) A soluble form of CD4 (T4) protein inhibits AIDS virus infection. *Nature*, **331**, 82.
132. Clapham, P. R., Weber, J. N., Whitby, D., McIntosh, K., Dalgleish, A. G., Maddon, P. J., Deen, K. C., Sweet, R. W., and Weiss, R. A. (1989) Soluble CD4 blocks the infectivity of diverse strains of HIV and SIV for T cells and monocytes but not for brain and muscle cells. *Nature*, **337**, 368.
133. Byrn, R. A., Sekigawa, I., Chamow, S. M., Johnson, J. S., Gregory, T. J., Capon, D. J., and Groopman, J. E. (1989) Characterization of in vitro inhibition of human immunodeficiency virus by purified recombinant CD4. *J. Virol.*, **63**, 4370.
134. Ward, R. H., Capon, D. J., Jett, C. M., Murthy, K. K., Mordenti, J., Lucas, C., Frie, S. W., Prince, A. M., Green, J. D., and Eichberg, J. W. (1991) Prevention of HIV-1 IIIB infection in chimpanzees by CD4 immunoadhesion. *Nature*, **352**, 434.

135. Kahn, J. O., Allan, J. D., Hodges, T. L., Kaplan, L. D., Arri, C. J., Fitch, H. F., Izu, A. E., Mordenti, J., Sherwin, S. A., Groopman, J. E., and Volberding, P. A. (1990) The safety and pharmacokinetics of recombinant soluble CD4 (rCD4) in subjects with the acquired immunodeficiency syndrome (AIDS) and AIDS-related complex. *Ann. Intern. Med.*, **112**, 254.

136. Schooley, R. T., Merigan, T. C., Gaut, P., Hirsch, M. S., Holodniy, M., Flynn, T., Liu, S., Byington, R. E., Henochowicz, S., and Gubish, E. (1990) Recombinant soluble CD4 therapy in patients with the acquired immunodeficiency syndrome (AIDS) and AIDS-related complex. A phase I-II escalating dosage trial. *Ann. Intern. Med.*, **112**, 247.

137. Hwang, S. S., Boyle, T. J., Lyerly, H. K., and Cullen, B. R. (1992) Identification of the envelope V3 loop as the major determinant of CD4 neutralization sensitivity of HIV-1. *Science*, **257**, 535.

138. Gartner, S. and Popovic, M. (1990) Macrophage tropism of HIV-1. *AIDS Res. Hum. Retroviruses*, **6**, 1017.

139. Cordonnier, A., Riviere, Y., Montagnier, L., and Emerman, M. (1989) Effects of mutations in hyperconserved regions of the extracellular glycoprotein of human immunodeficiency virus type 1 on receptor binding. *J. Virol.*, **63**, 4464.

140. Westervelt, P., Trowbridge, D. B., Epstein, L. G., Blumberg, B. M., Li, Y., Hahn, B. H., Shaw, G. M., Price, R. W., and Ratner, L. (1992) Macrophage tropism determinants of human immunodeficiency virus type 1 *in vivo*. *J. Virol.*, **66**, 2577.

141. Mori, K., Ringler, D. J., Kodama, T., and Desrosiers, R. C. (1992) Complex determinants of macrophage tropism in env of simian immunodeficiency virus. *J. Virol.*, **66**, 2067.

142. Hirsch, V. M., Edmondson, P., Murphey-Corb, M., Arbeille, B., Johnson, P. R., and Mullins, J. I. (1989) SIV adaptation to human cells. *Nature*, **341**, 573.

143. Kodama, T., Wooley, D. P., Naidu, Y. M., Kestler, H. W., Daniel, M. D., Li, Y., and Desrosiers, R. C. (1989) Significance of premature stop codons in env of simian immunodeficiency virus. *J. Virol.*, **63**, 4709.

144. Cheng-Mayer, C., Shioda, T., and Levy, J. A. (1991) Host range, replicative, and cytopathic properties of human immunodeficiency virus type 1 are determined by very few amino acid changes in tat and gp120. *J. Virol.*, **65**, 6931.

145. Meyerhans, R., Cheyner, R., Albert, J., Seth, M., Kwok, S., Sninsky, J., Morfeldt-Manson, L., Asjo, B., and Wain-Hobson, S. (1989) Temporal fluctuations in HIV quasispecies *in vivo* are not reflected by sequential HIV isolations. *Cell*, **58**, 901.

146. Delassus, S., Cheyner, R., and Wain-Hobson, S. (1991) Evolution of human immunodeficiency virus type 1 *nef* and long terminal repeat sequences over 4 years *in vivo* and *in vitro*. *J. Virol.*, **65**, 225.

147. Kestler, H. W., Ringler, D. J., Panicaly, D. L., Sehgal, P. K., Daniel, M. D., and Desrosiers, R. C. (1991) Importance of the *nef* gene for maintenance of high virus load and for development of AIDS. *Cell*, **65**, 651.

148. Terwilliger, E. F., Langhoff, E., Gabuzda, D., Zazopoulos, E., and Haseltine, W. A. (1991) Allelic variation in the effects of the viral *nef* gene on replication of human immunodeficiency virus type 1. *Proc. Natl Acad. Sci. USA*, **88**, 10971.

149. Dewhurst, S., Embretson, J. E., Anderson, D. C., Mullins, J. I., and Fultz, P. N. (1990) Sequence analysis and acute pathogenicity of molecularly cloned $SIV_{SMM-PBj14}$. *Nature*, **345**, 636.

150. Cheng-Mayer, C., Iannello, P., Shaw, K., Luciw, P. A., and Levy, J. A. (1989) Differential effects of nef on HIV replication: implications for viral pathogenesis in the host. *Science*, **246**, 1629.

151. Ahmad, N. and Venkatesan, S. (1988) Nef protein of HIV-1 is a transcriptional repressor of HIV-1 LTR. *Science*, **241**, 1481.
152. Niederman, T. M., Hu, W., and Ratner, L. (1991) Simian immunodeficiency virus negative factor suppresses the level of viral mRNA in COS cells. *J. Virol.*, **65**, 3538.
153. Niderman, T. M., Thielan, B. J., and Ratner, L. (1989) Human immunodeficiency virus type 1 negative factor is a transcriptional silencer. *Proc. Natl Acad. Sci. USA*, **86**, 1128.
154. Kim, S., Ikeuchi, K., Byrn, R., Groopman, J., and Baltimore, D. (1989) Lack of negative influence on viral growth by the nef gene of human immunodeficiency virus type 1. *Proc. Natl Acad. Sci. USA*, **86**, 9544.
155. Hammes, S. R., Dixon, E. P., Malim, M. H., Cullen, B. R., and Greene, W. C. (1989) Nef protein of human immunodeficiency virus type 1: evidence against its role as a transcriptional inhibitor. *Proc. Natl Acad. Sci. USA*, **86**, 9549.
156. Terwilliger, E., Sodroski, J. S., Rosen, C. A., and Haseltine, W. A. (1986) Effects of mutations within the 3' *orf* open reading frame region of human T-cell lymphotropic virus type III (HTLV-III/LAV) on replication and cytopathogenicity. *J. Virol.*, **60**, 754.
157. Luciw, P. A., Cheng-Mayer, C., and Levy, J. A. (1987) The *orf*-B region down-regulates viral replication. *Proc. Natl Acad. Sci. USA*, **84**, 1434.
158. McDougal, J. S., Mawle, A., Cort, S. P., Nicholson, J. K. A., Cross, G. D., Scheppler-Campbell, J. A., Hicks, D., and Sligh, J. (1985) Cellular tropism of the human retrovirus HTLV-III/LAV. *J. Immunol.*, **135**, 3151.
159. Zack, J. A., Arrigo, S. J., Weitsman, S. R., Go, A. S., Haislip, A., and Chen, I. S. Y. (1990) HIV-1 entry into quiescent primary lymphocytes: molecular analysis reveals a labile, latent viral structure. *Cell*, **61**, 213.
160. Tateno, M., Gonzalez-Scarano, F., and Levy, J. A. (1989) Human immunodeficiency virus can infect CD4-negative human fibroblastoid cells. *Proc. Natl Acad. Sci. USA*, **86**, 4287.
161. McKeating, J. A., Griffiths, P. D., and Weiss, R. A. (1990) HIV susceptibility conferred by cytomegalovirus-induced Fc receptor. *Nature*, **343**, 659.
162. Takeuchi, Y., Akutsu, M., Murayama, K., Shimizu, M., and Hoshino, H. (1991) Host range mutant of human immunodeficiency virus type 1: modification of cell tropism by a single point mutation at the neutralization epitope in the *env* gene. *J. Virol.*, **65**, 1710.
163. Homsy, J., Meyer, M., Tateno, M., Clarkson, S., and Levy, J. A. (1989) The Fc and not CD4 receptor mediates antibody enhancement of HIV infection in human cells. *Science*, **244**, 1357.
164. Haubrich, R. H., Takeda, A., Koff, W., Smith, G., and Ennis, F. A. (1992) Studies of antibody-dependent enhancement of human immunodeficiency virus (HIV) type 1 infection mediated by Fc receptors using sera from recipients of a recombinant gp160 experimental HIV-1 vaccine. *J. Infect. Dis.*, **165**, 545.
165. Robinson, Jr., W., Gorny, M. K., Xu, J. Y., Mitchell, W. M., and Zolla-Pazner, S. (1991) Two immunodominant domains of gp41 bind antibodies which enhance human immunodeficiency virus type 1 infection *in vitro*. *J. Virol.*, **65**, 4169.
166. Boyer, V., Desgranges, C., Trabaud, M. A., Fischer, E., and Kazatchkine, M. D. (1991) Complement mediates human immunodeficiency virus type 1 infection of a human T cell line in a CD4- and antibody-independent fashion. *J. Exp. Med.*, **173**, 1151.
167. Folks, T. M., Kessler, S. W., Orenstein, J. M., Justement, J. S., Jaffe, E. S., and Fauci, A. S. (1988) Infection and replication of HIV-1 in purified progenitor cells of human normal bone marrow. *Science*, **242**, 919.

168. Adachi, A., Koenig, S., Gendelman, H. E., Daugherty, D., Gattoni-Celli, S., Fauci, A. S., and Martin, M. A. (1987) Productive, persistent infection of human colorectal cell lines with human immunodeficiency virus. *J. Virol.*, **61**, 209.
169. Cheng-Mayer, C., Rutka, J. T., Rosenblum, M. L., McHugh, T., Sites, D. P., and Levy, J. A. (1987) Human immunodeficiency virus can productively infect cultured human glial cells. *Proc. Natl Acad. Sci. USA*, **84**, 3526.
170. Chiodi, F., Fuerstenberg, S., Gidlund, M., Asjo, B., and Fenyö, E. M. (1987) Infection of brain-derived cells with the human immunodeficiency virus. *J. Virol.*, **61**, 1244.
171. Clapham, P. R., Weber, J. N., Whitby, D., McKintosh, K., Dalgleish, A. G., Maddon, P. J., Deen, K. C., Sweet, R. W., and Weiss, R. A. (1989) Soluble CD4 blocks the infectivity of diverse strains of HIV and SIV for T cells and monocytes but not for brain and muscle cells. *Nature*, **337**, 369.
172. Dewhurst, S., Sakai, K., Bresser, J., Stevenson, M., Evinger-Hodges, M. J., and Volsky, D. J. (1987) Persistent productive infection of human glial cells by human immunodeficiency virus (HIV) and by infectious molecular clones of HIV. *J. Virol.*, **61**, 3774.
173. Harouse, J. M., Kunsch, C., Hartle, H. T., Laughlin, M. A., Hoxie, J. A., Wigdahl, B., and Gonzalez-Scarano, F. (1989) CD4-independent infection of human neural cells by human immunodeficiency virus type 1. *J. Virol.*, **63**, 2527.
174. Cao, Y., Friedman-Kien, A. E., Huang, Y., Li, X. L., Mirabile, M., Moudgil, T., Zucker-Franklin, D., and Ho, D. D. (1990) CD4-independent productive human immunodeficiency virus type 1 infection of hepatoma cell lines *in vitro*. *J. Virol.*, **64**, 2553.
175. Li, X. L., Moudgil, T., Vinters, H. V., and Ho, D. D. (1990) CD4-independent productive infection of a neuronal cell line by human immunodeficiency virus type 1. *J. Virol.*, **64**, 1383.
176. Harouse, J. M., Bhat, S., Spitalnik, S. L., Laughlin, M., Stefano, K., Silberberg, D. H., and Gonzalez, S. F. (1991) Inhibition of entry of HIV-1 in neural cell lines by antibodies against galactosyl ceramide. *Science*, **253**, 320.
177. Cann, A. J., Zack, J. A., Go, A. S., Arrigo, S. J., Koyanagi, Y., Green, P. L., Koyanagi, Y., Pang, S., and Chen, I. S. Y. (1990) Human immunodeficiency virus type 1 T-cell tropism is determined by events prior to provirus formation. *J. Virol.*, **64**, 4735.
178. Ashkenazi, A., Smith, D. H., Masters, S. A., Riddle, L., Gregory, T. J., Ho, D. D., and Capon, D. J. (1991) Resistance of primary isolates of human immunodeficiency virus type 1 to soluble CD4 is independent of CD4-gp120 binding affinity. *Proc. Natl Acad. Sci. USA*, **88**, 7056.
179. Brighty, D. W., Rosenberg, M., Chen, I. S. Y., and Ivey-Hoyle, M. (1991) Neutralization-resistant primary clinical isolates of HIV-1 possess gp120 glycoproteins with high affinity for recombinant sCD4. *Proc. Natl Acad. Sci. USA*, **88**, 7802.
180. Berman, P. W., Riddle, L., Nakamura, G., Haffar, O. K., Nunes, W. M., Skehel, P., Byrn, R., Groopman, J., Matthews, T., and Gregory, T. (1989) Expression and immunogenicity of the extracellular domain of the human immunodeficiency virus type 1 envelope glycoprotein, gp160. *J. Virol.*, **63**, 3489.
181. Planelles, V., Haigwood, N. L., Marthas, M. L., Mann, K. A., Scandella, C., Lidster, W. D., Shuster, J. R., Van, K. R., Marx, P. A., Gardner, M. B., and Luciw, P. A. (1991) Functional and immunological characterization of SIV envelope glycoprotein produced in genetically engineered mammalian cells. *AIDS Res. Hum. Retroviruses*, **7**, 889.
182. Doms, R. W., Earl, P. L., Chakrabarti, S., and Moss, B. (1990) Human immunodeficiency virus types 1 and 2 and simian immunodeficiency virus env proteins possess a functionally conserved assembly domain. *J. Virol.*, **64**, 3537.

183. Weiss, R. (1982) Experimental biology and assay and RNA tumor viruses. In *RNA Tumor Viruses*, 2nd edition. Weiss, R., Teich, N., Varmus, H., and Coffin, J. (ed.). Cold Spring Harbor Laboratories, Cold Spring Harbor, New York, p. 209.
184. Sattentau, Q. J. and Moore, J. P. (1991) Conformational changes induced in the human immunodeficiency virus envelope glycoprotein by soluble CD4 binding. *J. Exp. Med.*, **174**, 407.
185. Bolognesi, D. P. (1990) Immunobiology of the human immunodeficiency virus envelope and its relationship to vaccine strategies. *Mol. Biol. Med.*, **7**, 1.
186. Hattori, T., Koito, A., Takaatsuki, H., Kido, H., and Katunuma, N. (1989) Involvement of tryptase-related cellular protease(s) in human immunodeficiency virus type 1 infection. *FEBS Lett.*, **248**, 48.
187. Clements, G. J., Price-Jones, M., Stephens, P. E., Sutton, C., Schulz, T. F., Clapham, P. R., McKeating, J. A., McClure, M. O., Thomson, S., and Marsh, M. (1991) The V3 loops of the HIV-1 and HIV-2 surface glycoproteins contain proteolytic cleavage sites: a possible function in viral fusion? *AIDS Res. Hum. Retroviruses*, **7**, 3.
188. Habeshaw, J. A. and Dalgleish, A. G. (1989). The relevance of HIV env/CD4 interactions to the pathogenesis of acquired immune deficiency syndrome. *AIDS Res. Hum. Retroviruses*, **2**, 457.
189. Berger, E. A., Lifson, J. D., and Eiden, L. E. (1991) Stimulation of glycoprotein gp120 dissociation from the envelope glycoprotein complex of human immunodeficiency virus type 1 by soluble CD4 and CD4 peptide derivatives: implications for the role of the complementarity-determining region 3-like region in membrane fusion. *Proc. Natl Acad. Sci. USA*, **88**, 8082.

3 | The role of cellular transcription factors in the regulation of human immunodeficiency virus gene expression

GARY J. NABEL

1. Introduction

The human immunodeficiency virus (HIV) is dependent on its host cell for the synthesis and processing of viral RNA. Steady-state levels of viral RNA can be regulated by several mechanisms, including the rate of transcriptional initiation, elongation, processing, export, and packaging. Diverse regulatory mechanisms have, therefore, evolved to control viral RNA during HIV replication. These events play a critical role in the regulation of viral replication.

Relative to the human genome, the genomic complexity of HIV is relatively small, <0.001% in size. To regulate viral gene expression specifically, the virus has evolved its own regulatory gene products. These viral trans-activators are also dependent upon cellular factors that interact with viral proteins and cis-acting DNA and RNA regulatory elements to further modulate HIV gene expression. These mechanisms of regulation are diverse. For example, the Tat trans-activator affects the ability of a cellular transcription complex to elongate and inhibits premature termination of the HIV transcript (1, 2, Chapter 4). The Rev gene product, on the other hand, does not affect the synthesis of viral RNA but acts with cellular factors on an RNA structure after formation of the primary transcript (3, 4, Chapter 5). For these trans-activators, an intimate association has evolved between host cell factors and viral gene products. These essential trans-activators are synthesized only after viral transcription has been initiated. The regulation of the initiation of HIV transcription also plays an important role in the viral life cycle.

Studies of virus replication in cell culture first provided insight into the events that regulate HIV transcriptional initiation. Early studies suggested a link between cellular activation and HIV replication: HIV was difficult to propagate *in vitro*, and viral replication significantly increased when HIV-infected T-cells were stimulated by mitogens (5, 6). Such mitogens also induce the synthesis of a variety of coordinately activated T-cell gene products (Fig. 1), including interleukin-2 (IL-2), IL-3, IL-4, γ-interferon, and granulocyte-macrophage colony stimulating factor (GM-CSF) (7–19). These studies suggested that HIV replication was dependent on cellular transcription factors which normally function in T-cell activation (Fig. 1). In the case of HIV-infected cells, however, cellular activation and viral gene expression lead to cell death (Fig. 1).

Another hallmark of HIV infection is the variable length between the time of initial infection and the onset of symptoms in the patient. During this time, seropositive patients contain detectable virus in the blood (20, 21) but also contain a significant amount of latent provirus, approximately 1 per cent of circulating T-cells (22), of which 10 per cent appear to synthesize virus (0.1 per cent of total CD4$^+$ cells). Although several factors are likely to influence this process, inducible cellular transcription factors, such as NF-κB, may contribute to viral gene activation. These cellular factors play a role in inducible activation in both the T-cell and monocyte lineage (23, 24). Once viral transcription has begun, viral *trans*-activators, including Tat and Rev, have been further implicated as important regulators of viral replica-

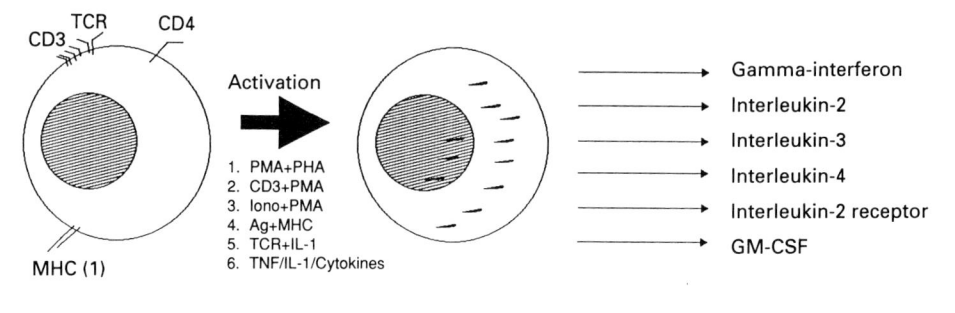

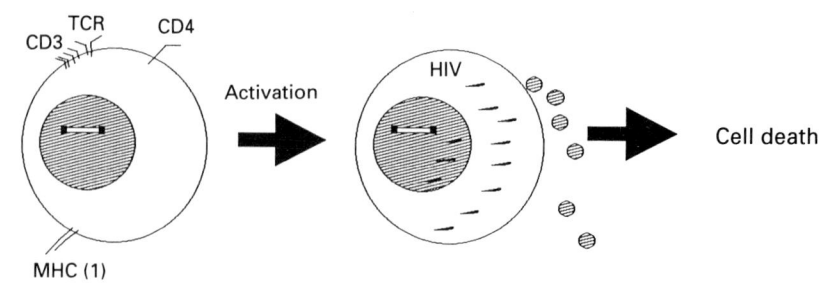

Fig. 1 Cellular activation and HIV infection in T-lymphocytes. Schematic representation of changes in gene expression and cell viability after T-cell activation. Through mechanisms which are not yet defined, T-cell activation and virus synthesis in latently infected cells result in cell death

tion through their effects on transcriptional elongation (1, 2, Chapter 4) or their ability to activate the cytoplasmic appearance of unspliced viral RNA (3, 4, 25, 26, Chapter 5). It is also possible that other viruses or pathogens act as cofactors to activate HIV replication. Several viral *trans*-activators have been implicated in stimulation of the HIV enhancer in transfected cells (27), although the role of these viruses in patients has yet to be rigorously defined.

Cytokines, including tumour necrosis factor-α (TNF-α) and possibly IL-1, likely serve as the physiologic activators that stimulate cellular factors to increase HIV gene expression (28–30). Each of these cellular activation pathways induced by extracellular stimuli induce distinct sets of genes in response to cellular stimulation (31). The relevant signal transduction pathways stimulate discrete sets of transcription factors which activate viral gene expression. In the case of HIV-1, activation of the viral enhancer, for example, is much more responsive to TNF-α, but it is much less responsive to signalling through the T-cell antigen receptor in Jurkat T-leukemia cells (32) (M. Hannibal, D. Markovitz, and G. Nabel, manuscript in preparation). In contrast, the HIV-2 enhancer is more responsive to stimulation of the T-cell antigen receptor and less responsive to these cytokines. Further examination of these enhancers reveals fundamental differences in their organizational structure and their interaction with cellular transcription factors. In part, these structures may account for the differences in latency between these viruses. The essential regulatory elements and sequence-specific DNA binding proteins that contribute to the regulation of transcriptional initiation of these retroviral genomes are reviewed in this chapter.

2. RNA polymerases and associated transcription factors

Gene expression in eukaryotic cells is regulated by different cellular RNA polymerases. RNA polymerase acts in combination with other basal transcription factors to initiate, elongate, and terminate the synthesis of RNA. The transcription complex is biochemically complex and is estimated to consist of more than 30 subunits (33). Multiple factors associate with different RNA polymerases which associate with coactivators required for transcriptional activation *in vitro* (34–36). Recently, several of the essential factors constituting the transcription complex have been cloned, and the steps leading to formation of the preinitiation complex defined (reviewed in ref. 33).

Eukaryotic cells contain three types of RNA polymerases that were defined based on their differential sensitivity to α-amanitin. RNA polymerase I resides in the nucleolus, is largely responsible for the transcription of ribosomal RNA genes, and is relatively insensitive to this inhibitor. In contrast, RNA polymerase II is localized in the nucleoplasm, is responsible for the synthesis of a large number of diverse messenger RNAs, and is inhibited by low concentrations of α-amanitin (~1 μg/ml). RNA polymerase III is also localized in the nucleoplasm, controls the synthesis of tRNA and small nuclear RNAs, and is insensitive to α-amanitin. Because HIV transcription is sensitive to small amounts of α-amanitin (2), viral replication is thought to be characteristic of a type II promoter. Several features of this transcription unit are unusual for a pol II promoter, however. For example, viral *trans*-activators such as Rev localize to the nucleolus (37–39), which is normally associated with pol I. This finding has raised the possibility that RNA processing pathways other than those normally attributed to RNA polymerase II may be utilized by HIV. The recent finding that the TFIID transcription factor can associate with RNA polymerase I (40), when it was previously thought to associate only with RNA polymerase II, has raised the possibility that transcription complex associated factors (TAFs) may affect both the class of RNA polymerase that is utilized, as well as the regulatory properties of that subunit. Another unusual feature of the HIV transcription unit is its propensity to terminate transcription prematurely. It is likely that a unique set of factors are associated with the basal transcription complex and are responsible for this effect.

3. Transcriptional regulatory proteins and the HIV promoter

The *cis*-acting regulatory elements that control retroviral gene expression are thought normally to reside in the long terminal repeat (LTR) of the virus, particularly within the U3 region (Fig. 2). Potential regulatory sites have been defined in several ways. In some cases, sequence similarity to known sites has been used. In others, protein binding, by electrophoretic mobility shift gels or DNA footprinting, has been used to define important elements. A variety of different cellular transcription factors appear to contribute to the constitutive expression of HIV-1 (Fig. 2). Among these cellular factors, Sp1 appears to play an important role (41), and three Sp1 sites have been described that are proximal to the TATA box. Of these three sites, the most upstream site represents the best consensus sequence and is active by *in vitro* transcription assays (41). Deletion of the three Sp1 sites reduces enhancer activity by approximately 5- to 10-fold *in vitro* or *in vivo* (27, 41) and mutation of the Sp1 site in intact virus results in markedly diminished replication in susceptible cell lines (42). More recently, the possibility has been raised that Sp1 is required not only for basal transcription but also for inducible gene expression through an interaction with NF-κB (43).

Several other *cis*-acting regulatory elements are found near the TATA box and transcriptional initiation site. Binding to the promoter by the TATA factor, TFIID, is not absolutely required for NF-κB-mediated inducibility of the HIV enhancer, but mutation of the TATA box results in a substantial loss (10- to 50-fold) of basal activity (44) and a marked reduction in Tat *trans*-activation (44, 45). Mutations in the TATA element also severely impair the ability of the virus to replicate in cell lines.

Another important function of the TATA element is its role in facilitating the assembly of other members of the transcription initiation complex. An essential component in this process is the site of transcriptional initiation. Several studies have suggested that a specific DNA-binding protein is required for efficient transcriptional initiation at this position (44, 46) (R. Roeder, personal communication), through a site similar to the initiator sequence described in the TdT promoter (46). Recently, this initiator-binding protein has been highly purified (36) (R. Roeder, personal communication) and is distinguishable from a fourth DNA-binding protein which can bind to this region. The fourth protein, leader-binding protein-1 (LBP-1), (47) binds to two to three discrete binding sites repeated between −17 to +27 of the promoter (Fig. 2). This protein is likely to be identical to another activity, termed upstream-binding protein-1 (UBP-1) (48, 49), which is not as readily separated from *cis*-acting sites which fall within the region encoding the *trans*-activation response element (TAR) RNA structure.

The functional role of LBP-1 in constitutive and inducible HIV-1 gene expression is not yet clear. One study has shown that LPB-1 can prevent binding of TFIID to the HIV promoter *in vitro* and inhibit transcription initiation (50). However, this repressive effect has not been observed *in vivo*. Several other potential regulatory sites have been described within the U3/R region of the HIV LTR. These include sites for the T cell-specific transcription factor (TCF-1) DNA-binding protein (51), the HIV initiator protein (HIP) (47), and CAAT transcription factor/nuclear factor 1 (CTF/NF1) located (+32 to +52) in this region (Fig. 2). Studies with *in vitro* transcription suggest that CTF has no effect on initiation from its cognate binding site (47). The role of LBP-1 is more uncertain since conflicting results have been described regarding its role *in vitro* (47, 50). HIP has been cloned by its ability to bind to a *cis*-acting regulatory element proximal to the transcriptional initiation site (−37 → +2) and is functional in cotransfection assays with an HIV reporter plasmid (K. A. Jones, personal communication). A similar site is also found in the E2 promoter which is a target for transcriptional activation by E1A (47). Whether this site plays a role in activation of the HIV by E1A (27) is unknown and will require further studies.

4. Inducible gene expression and the HIV enhancer

A T-cell responds to activation signals by inducing a specific set of transcription factors within the nucleus. Among a variety of different stimuli examined, the HIV enhancer has been found to be responsive to stimulation by specific cytokines. Among a panel of more than 10 cytokines which have been studied, two cytokines, TNF-α and IL-1, have been suggested as potential activators of HIV transcription (29). These cytokines act to induce the NF-κB transcription factor within activated T-cells. Although the complexes induced by these cytokines are indistinguishable from those induced by phorbol 12-myristate 13-acetate (PMA), they have otherwise different effects on T-cell function. For example, in EL-4 cells, TNF-α stimulated the HIV enhancer but had no effect on Ap1 or IL-2 promoters (31). In contrast, IL-1 was able to activate IL-2 secretion and the HIV enhancer, but did not affect Ap1 reporters. PMA, in contrast, stimulated all three of these targets (31). These findings suggest that multiple distinct activators can stimulate NF-κB binding and HIV gene expression.

54 | ROLE OF CELLULAR TRANSCRIPTION FACTORS IN THE REGULATION OF HIV GENE EXPRESSION

(A) Site A, ?Ap1, myb, USF(NRF), kB, Sp1, TATA box, Cap site, CTF (top labels); Steroid (TRE), NFAT IL-2 hom., Ap2, LBP-1 (UBP-1), HIP-1, TCF-1, Initiator (bottom labels); positions −453, −90, 1.

(B) kB, Sp1, TATA box, Cap site (top); Ap2, (HIP-1) initiator (bottom); positions −453, −104, −90, 1.

Nabel
Figure 2C

(C)

```
      -450       -440       -430       -420       -410       -400
       ↑          ↑          ↑          ↑          ↑          ↑
     TGGA AGGGCTAATT CACTCCCAAC GAAGACAAGA TATCCTTGAT CTGTGGATCT ACCACACACA


      -390       -380       -370       -360       -350       -340
       ↑          ↑          ↑          ↑          ↑          ↑
     AGGCTACTTC CCTGATTAGC AGAACTACAC ACCAGGGCCA GGGATCAGAT ATCCACTGAC
                -384  Site A protein  -367        -350 AP-1 -343  -336 AP-1
                         (63,145)                    (52)            (52)

                                               -357  Steroid Receptor Superfamily
                                                            (63,145)


      -330       -320       -310       -300       -290       -280
       ↑          ↑          ↑          ↑          ↑          ↑
     CTTTGGATGG TGCTACAAGC TAGTACCAGT TGAGCCAGAG AAGTTAGAAG AAGCCAACAA
       -330                           -303  IL-2 homology  -281   -274
                                                 (146)

              -325    -314    Myb    -293
                             (147)

                        -301  AP-1  -293
                               (52)


      -270       -260       -250       -240       -230       -220
       ↑          ↑          ↑          ↑          ↑          ↑
     AGGAGAGAAC ACCAGCTTGT TACACCCTGT GAGCCTGCAT GGAATGGATG ACCCGGAGAG
     IL-2 homology  -256         NFAT DNase footprint        -218
        (53,146)                       (53,54)


      -210       -200       -190       -180       -170       -160
       ↑          ↑          ↑          ↑          ↑          ↑
     AGAAGTGTTA GAGTGGAGGT TTGACAGCCG CCTAGCATTT CATCACATGG CCCCGAGAGCT
                     -193  mouse factor  -178  -175  Neg.Reg.Factor  -161
                                (148)              HIV-TF1
                                                   USF/MLTF
                                                   SP-50
                                                   (48,149-151)
```

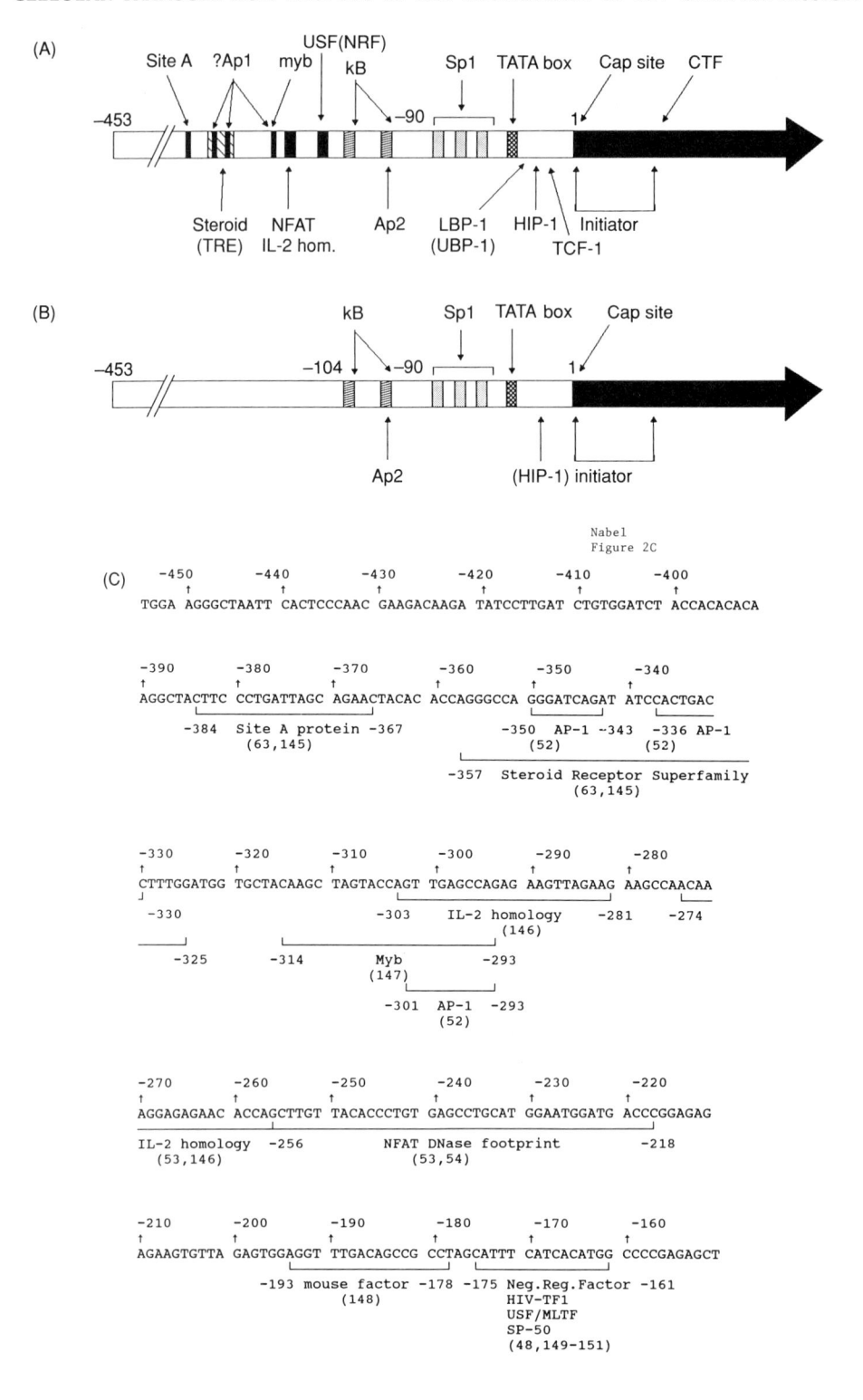

INDUCIBLE GENE EXPRESSION AND THE HIV ENHANCER | 55

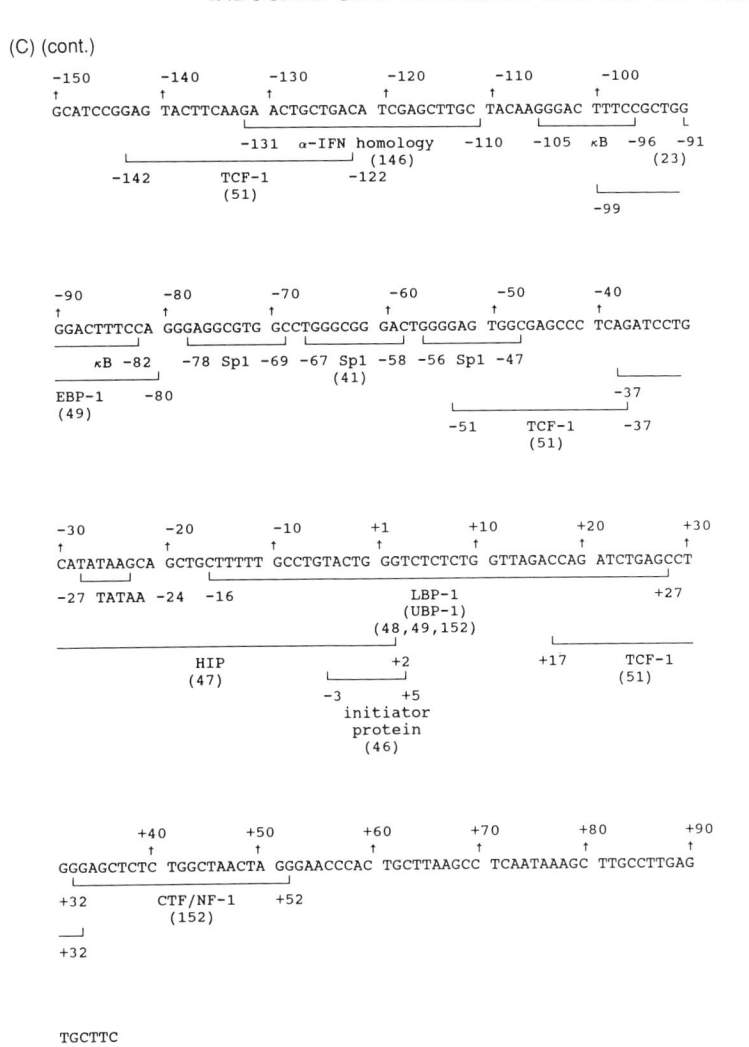

(C) (cont.)

```
        -150        -140        -130        -120        -110        -100
         ↑           ↑           ↑           ↑           ↑           ↑
        GCATCCGGAG TACTTCAAGA ACTGCTGACA TCGAGCTTGC TACAAGGGAC TTTCCGCTGG
                   └──────────────────────────────┘ └────────┘          └
                    -131  α-IFN homology   -110   -105   κB  -96     -91
                    └──────────────────────┘                (23)
               -142          TCF-1         -122
                             (51)
                                                                 └─────────
                                                                  -99
```

```
         -90         -80         -70         -60         -50         -40
          ↑           ↑           ↑           ↑           ↑           ↑
         GGACTTTCCA GGGAGGCGTG GCCTGGGCGG GACTGGGGAG TGGCGAGCCC TCAGATCCTG
         └────────┘ └────────┘ └────────┘ └────────┘
           κB -82   -78 Sp1 -69 -67 Sp1 -58 -56 Sp1 -47
                               (41)
         └────┘     └──┘                                        └─────────
         EBP-1      -80                                                 -37
         (49)
                                           └─────────────────────────┘
                                          -51          TCF-1          -37
                                                       (51)
```

```
         -30         -20         -10         +1         +10         +20         +30
          ↑           ↑           ↑          ↑          ↑           ↑           ↑
         CATATAAGCA GCTGCTTTTT GCCTGTACTG GGTCTCTCTG GTTAGACCAG ATCTGAGCCT
         └───────┘  └─────┘                └─────────────────┘              └
         -27 TATAA  -24   -16                   LBP-1                       +27
                                               (UBP-1)
                                             (48,49,152)
         └─────────────────────┘
                    HIP                    +2        +17      TCF-1
                   (47)                  └────┘                (51)
                                          -3  +5
                                        initiator
                                         protein
                                          (46)
```

```
         +40         +50         +60         +70         +80         +90
          ↑           ↑           ↑           ↑           ↑           ↑
         GGGAGCTCTC TGGCTAACTA GGGAACCCAC TGCTTAAGCC TCAATAAAGC TTGCCTTGAG
         └────┘     └────────┘
          +32       CTF/NF-1  +52
                    (152)
         └──┘
         +32
```

```
         TGCTTC
```

Fig. 2 Cis-acting regulatory elements in the HIV-1 LTR. (A) Potential cis-acting regulatory sites and (B) cis-acting regulatory sites known to contribute to HIV-1 enhancer function are shown. (C) Location and sequence of relevant sites are indicated

These findings also suggest that NF-κB is the major inducible activator of the HIV-1 enhancer. Several reports have described other cis-acting regulatory elements that are potentially responsive to cellular activation, including a non-standard binding site for the Fos/Jun complex (52). However, the functional effect of this latter site, which is not highly conserved among isolates, is questionable. In addition, a potential consensus binding site for the NFAT-1 complex has been described (Fig. 2). Although a DNAse footprint had been shown in the vicinity of this site (53, 54), further DNA sequence analysis has shown that the footprint in this region resides downstream of NFAT-1 related cis-acting regulatory element

(32) (G. R. Crabtree, personal communication). An exhaustive study using site-specific mutations in each of these regions in the HIV enhancer has revealed that NFAT-1 is not likely to be responsible for basal or induced activity of the HIV-1 enhancer following T-cell activation (32). In addition, this site does not appear to mediate the negative regulatory effects previously generally ascribed to this region (54). Although gross deletions in this region can result in an increase in enhancer activity or viral replication *in vitro*, such deletion studies grossly distort the overall LTR structure of the virus and are not necessarily reflective of transcriptional regulatory effects. There is thus no indication that potential sites such as NFAT-1 or Ap1 are active in regulating inducible gene expression of the HIV-1 enhancer, and it is likely that the NF-κB family of transcription factors are the primary inducible activators of HIV gene expression, both in activated T-cells and in the monocyte macrophage lineage (23, 24).

5. Negative regulation of the HIV enhancer

Several claims have been made that negative regulatory elements in the HIV enhancer might suppress viral transcription. Although such regulatory elements would be important in the maintenance of viral latency, inhibitory *cis*-acting sites have been difficult to document in the HIV enhancer. Deletion of a region upstream of the κB regulatory sites resulted in a two- to threefold increase in HIV enhancer activity in the presence of the Tat *trans*-activator (55). Although this effect has been consistently observed, its magnitude is small, and a specific *cis*-acting regulatory element that mediates this effect is yet to be identified. Because another deletion within a similar region which contains a putative NFAT-1 site has resulted in increased replication of a mutant virus, it has been claimed that a site related to NFAT-1 might be responsible for this phenomenon (56, 57). However, point mutations within the NFAT-1 site or other protein-binding sites in this region have revealed no such effect (54). In addition, linker scanning mutations in this region have shown minimal effects, either in the presence or absence of the *tat* gene (54).

It has also been suggested that the Nef gene product exerts a negative effect on viral transcription (see Chapter 6). One study reported a marked negative regulatory effect on HIV gene expression (58). Several subsequent studies, however, have not confirmed this finding (59–61). Using the simian immunodeficiency virus, Desrosiers and colleagues have demonstrated that point mutations within the *nef* open reading frame revert to wild type after inoculation of cloned viral isolates into recipients (62). These data suggest that Nef performs a function essential to viral replication and is, therefore, unlikely to repress viral gene expression *in vivo*.

Finally, at least one other potential negative regulatory site has been described in the HIV enhancer (63). Mutation of a site related to the thyroid hormone receptor element increased viral gene expression two- to threefold in the presence of the Tat gene product. Point mutations in this site (Fig. 2) eliminated binding of a specific DNA-binding protein at the same time that expression increased. Although this site has been defined more definitively, it is not yet certain whether it significantly contributes to the regulation of viral replication *in vivo* because of the low magnitude of this effect.

6. The NF-κB family of transcriptional activators

The 11 bp κB regulatory element is twice repeated in the HIV enhancer immediately upstream of three Sp1 sites (Fig. 2). These *cis*-acting regulatory elements are essential for inducible gene expression by the HIV-1 enhancer. Mutation of the κB sites in the enhancer abolishes both the inducibility of the HIV-1 LTR in T-leukemia cells and other cells and the ability of NF-κB to bind to this site (27, 29). An interaction between NF-κB and the Tat gene product has also been suggested since these activators act in concert to stimulate HIV transcription (23, 64–66).

Since its original description, it has become increasingly apparent that NF-κB regulation of transcription displays several levels of complexity. At the DNA level, several distinct κB-like sites have been found in association with various cellular and viral genes. Although the complexes which bind to these sites are competed by κB, they exhibit considerable sequence variation (67–74). At the same time, the proteins in the NF-κB complex consist of several distinct protein species. These include a 50 kDa DNA-binding subunit and an associated *trans*-activation protein of approximately 65 kDa (Fig. 3). Both subunits are related to the *rel* oncogene and the *Drosophila* maternal effect gene Dorsal (75–78). Additional DNA-binding proteins have been defined by cross-linking experiments (31, 79, 80).

Among the 50 kDa proteins, at least two distinct gene products have been identified. Each of these products is encoded by a larger precursor molecule, p105 (75, 76, 78, 81) or p100 (82–84). These precursors show sequence similarity but map to distinct chromosomal loci (83, 85, 86). Proteolytic processing of these larger proteins presumably generates the 50 kDa DNA-binding form. A 49 kDa form of p100 can be generated by an independent mechanism, by formation of an alternatively spliced mRNA (82). p49 or processed p100, similar to processed p105, specifically bind to κB and associate with either p65 or the proto-oncogene c-*rel* (82, 87).

Fig. 3 Potential members of the NF-κB complex. The NF-κB DNA-binding complex can be composed of alternative NF-κB/Rel/Dorsal family members which can be broadly defined as primarily DNA-binding subunits (p50) or transcription activation subunits (p65)

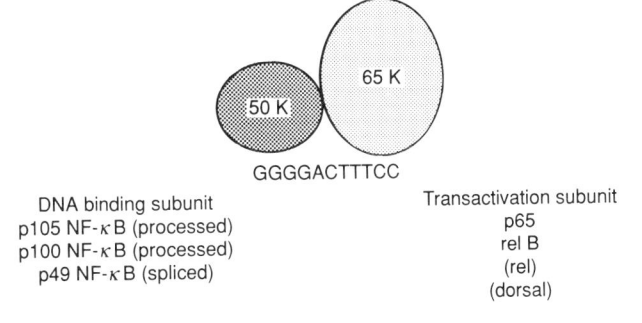

DNA binding subunit
p105 NF-κB (processed)
p100 NF-κB (processed)
p49 NF-κB (spliced)

GGGGACTTTCC

Transactivation subunit
p65
rel B
(rel)
(dorsal)

In addition to these DNA-binding subunits of NF-κB, heterogeneity has also been detected among the associated p65-like subunits. p65 is similar to Rel and Dorsal in two respects. First, the amino terminal regions of the gene products display significant amino acid sequence similarity (75, 76, 88–90). In addition, each of these proteins contain *trans*-activation domains which can act to stimulate transcriptional initiation (91, 92). The Rel B gene product is highly related to Rel and associates with other NF-κB family members (89).

These diverse subunits play an important role in transcriptional activation of HIV, and it is likely that NF-κB regulates viral replication by several mechanisms (Fig. 4; see also ref. 93). Both p49 (100) and p50 (105) can act in concert with p65 to stimulate HIV transcription; however, the combination of p49 (100) and p65 is more efficient in stimulating transcriptional activation in Jurkat cells than are p50. (105) and p65 (82). Large amounts of transfected p65 alone can also stimulate HIV transcription in these cells, although it is likely that the transfected p65 interacts with endogenous 50 kDa DNA-binding subunits to achieve this effect. Interestingly, an alternatively spliced p65 protein has also been described which does not interact with I-κB and shows transforming activity (92). The Rel B gene product is also capable of stimulating κB-dependent transcription (89), although an alternatively spliced form of this gene product, I rel, inhibits such transcriptional activation (S. Ruben, C. Rosen, personal communication). In addition to the NF-κB subunits related to *rel* and Dorsal, other cDNAs have been defined for gene products to bind to κB. These include MBP-1 (94, 95) and HIVen 86 (96). The latter is likely to be identical to the c-Rel protein (97). No functional role in activation of the HIV

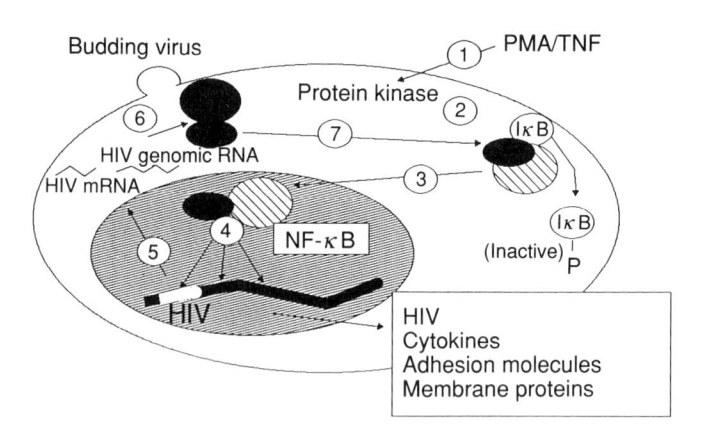

Fig. 4 Mechanism of NF-κB activation and its relationship to HIV infection. A hypothetical pathway of HIV activation by NF-κB in an infected T-cell. Cytokines or mitogens activate cells (1). The phosphorylated form of IκB does not complex to p50/p65 (2), and dissociation of p50/p65 from IκB is associated with translocation to the nucleus (3). In the nucleus, NF-κB stimulates transcription of HIV and other cellular genes, including cytokine, adhesion molecule, and cell-surface glycoprotein mRNA (4). At the same time, it stimulates viral RNA synthesis, including messenger RNAs for structural proteins and genomic HIV RNA (5). Proteases presumably stimulate cleavage of the p100 precursor of the DNA-binding subunit of NF-κB (6). Infection or cellular activation could also induce cellular proteases. The DNA-binding subunit is incorporated into the p50/p65 complex (7).

enhancer has yet been demonstrated for these two gene products. Similarly, EBP1 has been previously shown to bind to the HIV enhancer (98), but its functional role and its relation to these other gene products is not known.

7. Regulation of NF-κB by IκB

In addition to members of the NF-κB family of transcription factors that bind to κB sites, other cellular proteins can also regulate NF-κB-mediated transcriptional activation. In unstimulated cells, NF-κB can be detected in the cytoplasm associated with an inhibitor protein (IκB) that complexes to NF-κB and inhibits DNA-binding activity (99–101). This complex can be disassociated by treatment with different protein kinases *in vitro* (102, 103). This phosphorylation likely results in modification of the inhibitor (104), although it may also affect phosphorylation of NF-κB, and it has been suggested that release form IκB may facilitate translocation into the nucleus. It is thus possible that different activators of NF-κB could act on the different proteins in this signal transduction pathway. Molecular cloning of IκB has now revealed significant heterogeneity among different IκB family members. The MAD-3 gene, for example, is highly related to the c-terminal region of p105 and p100 (105, 106). The MAD-3 gene product binds to the p65 subunit of NF-κB and inhibits functional *trans*-activation (105) while overexpression of the MAD-3 cDNA in transfected cells specifically inhibits HIV transcriptional activation (Duckett *et al.*, in preparation). The C-terminal region of p105 binds preferentially to the 50 kDa form of p105 (107). In addition to these cDNAs, at least one other IκB has been identified. The *Bcl-3* proto-oncogene also displays amino acid sequence similar to MAD-3 and complexes with NF-κB to affect its DNA-binding activity (88). The significance of these different IκBs and their interactions with various NF-κB subunits has not yet been fully established, although it appears that specific IκBs may selectively target either the p65-like subunits or the 50 kDa subunits derived from p100 or p105 (Table 1).

Table 1 Specificity of different IκB gene products for NF-κB family members. Inhibition of DNA-binding activity of different NF-κB family members by IκBs determined by adsorption to immobilized protein. (+), weakly inhibitory; (+++), strongly inhibitory; nd, not determined; (−), stimulatory

	Inhibition					
	p50 (105)	p49 (100)	p65	Rel	p49/p65	p50/p65
IκB (MAD-3)	−	−	+++	+++	+++	+++
IκB (p105)	+	+	−	+	nd	+
IκB (Bcl-3)	+	+	−	nd	+	+

8. Anti-oxidants and the regulation of HIV transcription

Because transcriptional initiation is required for viral gene expression, considerable attention has been paid to methods to inhibit activation of the HIV enhancer. Although a variety of agents have been used to inhibit viral replication *in vitro*, most antivirals have been screened for their effect on specific viral regulatory proteins, including reverse transcriptase, Tat, or viral protease. Recently, attention has been focused on anti-oxidants as a means to regulate inducible HIV gene expression. Initially, it was noted that treatment of T-leukemia cells with *N*-acetyl cysteine (NAC) suppressed activation of NF-κB DNA-binding activity and suppressed the stimulation of an HIV reported plasmid (108, 109). It was therefore suggested that this anti-oxidant affected NF-κB. Further studies showed that addition of hydrogen peroxide could directly induce NF-κB and stimulate the HIV enhancer. This effect could be inhibited by several different anti-oxidants (110). Taken together, these data raised the possibility that the redox potential of a host cell affects induction of NF-κB through a specific signal transduction mechanism (109, 110). Despite these provocative results, the specificity and mechanism of action of these anti-oxidants are not yet completely understood. The concentration required for these anti-oxidants to suppress NF-κB are sufficiently high to affect transcription globally, and it is not certain that these agents act solely on NF-κB. Another possibility is that anti-oxidants affect more general factors involved in transcriptional regulation. In any case, the addition of *N*-acetyl cysteine *in vitro* to infected cells does inhibit viral replication (111). Based on these studies, phase I clinical trials have begun to ascertain the potential benefits of this treatment to HIV infection. While the toxicity and clinical efficacy of this treatment are being assessed, further studies will be needed to determine whether these agents are acting directly on NF-κB, IκB, the signal transduction pathway, or through a more general mechanism to achieve this effect.

9. Macrophage regulation of HIV expression

Because HIV infection is not restricted to T-lymphocytes, cellular factors in the monocyte/macrophage lineage also play an important role in the pathogenesis of AIDS (1, 112–118). The contribution of different cellular transcription factors in the monocyte lineage has been analysed. These studies have shown that HIV enhancer activity is stimulated by PMA or TNF-α in immature monocyte leukemia lines, but not in mature monocyte macrophage lines. This enhancer stimulation is dependent on κB. Moreover, as cells were induced to differentiate *in vitro* using phorbol esters or TNF-α, NF-κB binding activity became constitutive and correlated with induction of differentiation. Analysis of nuclear extracts from human peripheral blood monocytes or adherent cells, as well as mouse peritoneal macrophages, confirmed the expression of NF-κB within normal cells (24). Induction of virus production in the monocytic line also correlated with induction of NF-κB when the latently-infected U1 monocytic cell line was used as a model system (24). Because several studies have suggested that immature bone marrow progenitor cells can become infected with HIV (119, 120), these findings raised the possibility that promonocytic stem cells may provide a source of latent provirus which could be subsequently activated during the course of monocyte differentiation. In addition to the role of NF-κB in this cell type, other cytokines, including GM-CSF and IL-6, also have induced viral gene expression in this lineage (28, 121). In the case of IL-6, this stimulation is independent of NF-κB or transcriptional initiation, suggesting that additional cellular factors may cooperate to activate viral replication (Figs 1, 4) in this cell type (121).

10. Transcriptional regulation of HIV-2

HIV-2 represents a retrovirus distinct from HIV-1 but which shares considerable nucleic acid and protein sequence similarity with HIV-1 (122–125). Endemic to Western Africa, HIV-2 has also been implicated as a causative agent in acquired immunodeficiency syndrome (AIDS) and has begun to appear throughout the world (126–133). Despite the ability of both viruses to cause disease, the length of the asymptomatic phase following infection differs between the two viruses (126, 134). As with HIV-1, activation may be influenced by the rate of transcriptional initiation and associated cofactors which stimulate cellular transcription factors to stimulate viral gene expression.

Despite some differences in primary sequence, the organization of HIV-2 is similar to that of HIV-1. Both viruses contain *tat* and *rev* open reading frames and relevant target structures in the RNA. There are also several similarities between HIV-1 and HIV-2 within the LTR. Like HIV-1, HIV-2 contains three Sp1-binding sites upstream of the TATA box. Although one κB site is conserved among HIV-2 isolates, the site present in HIV-1 immediately upstream of the Sp1 site has mutated from the consensus κB sequence. This modified κB site does not function as a κB regulatory element, although it may be recognized by an Ap3-like binding activity (135).

Cis-acting regulatory sequences in HIV-2 have now been analysed in considerable detail. It is now evident that the HIV-2 enhancer differs from HIV-1 in several respects (Fig. 5). Within T-cells, activation of the HIV-2 enhancer is achieved effectively through stimulation of the T-cell receptor complex, either through antibodies to CD3 or by stimulation with major histocompatibility complex (MHC) presented antigens (32) (Hannibal *et al.*, in preparation). In contrast, the HIV-1 enhancer is much more responsive to stimulation of T-cells by TNF-α, mediated through the two κB regulatory sites. Although the κB site in the HIV-2 enhancer is important for responsiveness to stimulation by several agents, its requirement for other *cis*-acting regulatory sites is strikingly different from HIV-1. In particular, the single κB site within the HIV-2 enhancer is not responsive to mitogen stimulation in the absence of other upstream *cis*-acting regulatory sites. Rather, it functions in

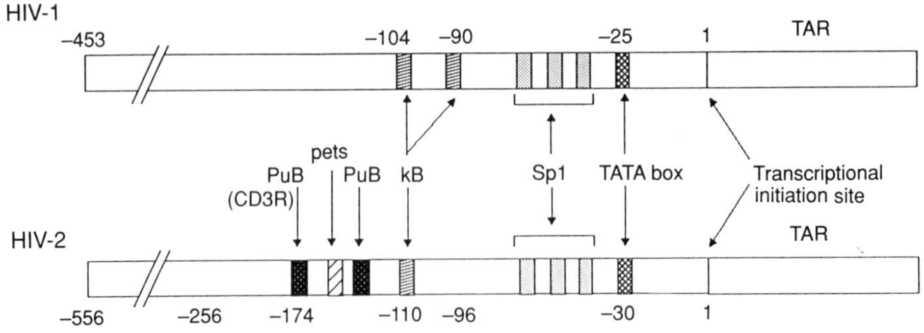

Fig. 5 Structural organization and major transcriptional regulatory elements in the HIV-1 and HIV-2 enhancers. The major cis-acting regulatory sequences which contribute to enhance function are indicated. Comparison of these regulatory elements in HIV-1 and HIV-2 reveals fundamental differences in their structural organization and response to T-cell activation

association with three other cis-acting regulatory elements located within 50 bp upstream of the site (136, 137). These sites include two purine boxes recognized by Ets-like binding proteins. Of significance, these regulatory elements are recognized by a recently described Ets-like protein, termed Elf-1 (138), which binds these sites and is found in activated T-cells after stimulation with anti-CD3 (136). It is likely that this transcriptional regulatory protein acts in concert with NF-κB and other factors to induce HIV-2 transcription following activation of the T-cell antigen receptor. In addition to these three regulatory elements, a fourth site provides additional stimulatory effects on the enhancer and is located adjacent to the downstream purine box recognized by Elf-1 (137). An additional potential cis-acting regulatory element which could affect the HIV-2 enhancer includes an element related to Ap1, which can stimulate inducible gene expression of a heterologous promoter (139); however, a site-specific mutation in this region does not alter the inducibility of the intact HIV-2 enhancer (D. Markovitz and G. Nabel, unpublished observations). Despite the similarity in organization of the HIV-1 and HIV-2 enhancers, these subtle changes in LTR structure have resulted in a markedly different response to cellular activation and interactions between transcription factors which induce transcriptional activation in T-cells.

Finally, it is also likely that the response of the HIV-2 LTR to different viral trans-activators differs from HIV-1. Although both the LTRs contain a consensus TATA box, the HIV-2 enhancer is minimally responsive to trans-activation by E1A (D. Markovitz and G. Nabel, unpublished observations). In contrast, the HIV-1 enhancer is quite responsive to this transcriptional regulatory element (27). The role of constitutive transcription factors relevant to HIV-1 transcription in the HIV-2 enhancer has not been investigated in detail. Although the Sp1 sites are likely to be important for inducible HIV-2 activation, whether LBP-1, TCF, HIP, or initiator binding proteins function in a similar fashion in the assembly of the transcription complex is unknown. It is also possible that the elements downstream of the transcriptional initiation site other than TAR can affect the stability of the transcription initiation complex, since the length of the sequence in this region can contribute to differences in the efficiency of transcription following transfection of reporter plasmids into host cells (140).

11. Conclusions

In summary, despite their relative simplicity, the structure of the HIV enhancers shows considerable complexity and has achieved a remarkable specificity of induction in response to cellular activation. The arrangement of *cis*-acting regulatory elements in two types of HIV differs significantly and is, in turn, largely unrelated to the elements which control human T-cell leukemia virus replication (see Chapter 7). A notable feature of the HIV enhancers is their ability to stimulate transcriptional initiation following cellular activation. Although this stimulation may play a role in activating viral transcription and induction from latency, the possibility also remains that the cellular transcription factors also act upon other cellular genes that, in turn, mediate the activity of viral regulatory or structural gene products. Such molecules might include cell-surface molecules required for efficient virion budding or essential cellular proteins which interact with the critical viral *trans*-activators Tat or Rev (Figs 1 and 4). It is noteworthy that a variety of different primate viruses including HIV-1, HIV-2, cytomegalovirus, Epstein–Barr virus, and herpesvirus, all stimulate multiple cellular transcription factors, and the process of viral infection itself can in some cases induce the production of such binding proteins (141, 142). Which of these inductions is direct or indirect is not yet clear; however, the marked association of NF-κB and other cellular transcription factors with infection suggests that similar strategies may be utilized by different primate viruses to mobilize cellular gene products required for successful gene replication (93). An understanding of these cellular factors and their specific mechanisms of interaction with viral *trans*-activators may lead to effective strategies in inhibiting the replication of these primate viruses in eukaryotic cells (143, 144). If such strategies can be achieved without affecting normal cellular function, this information may well prove useful in the development of novel therapeutic approaches to the treatment of HIV infection.

References

1. Kao, S. Y., Calman, A. F., Luciw, P. A., and Peterlin, B. M. (1987) Anti-termination of transcription within the long terminal repeat of HIV-1 by tat gene product. *Nature,* **330,** 489.
2. Laspia, M. F., Rice, A. P., and Mathews, M. B. (1989) HIV-1 Tat protein increases transcriptional initiation and stabilized elongation. *Cell,* **59,** 283.
3. Cullen, B. R. and Malim, M. H. (1991) The HIV-1 Rev protein: prototype of a novel class of eukaryotic post-transcriptional regulators. *Trends Biochem. Sci.,* **16,** 346.
4. Garcia-Blanco, M. A. and Cullen, B. R. (1991) Molecular basis of latency in pathogenic human viruses. *Science,* **254,** 815.

5. Zagury, D., Bernard, J., Leonard, R., Cheynier, R., Fledman, M., Sarin, P. S., and Gallo, R. C. (1986) Long-term cultures of HTLV-III-infected T cells: a model of cytopathology of T-cell depletion in AIDS. *Science*, **231**, 850.
6. Harada, S., Koyanagi, Y., Nakashima, H., Kobayashi, N., and Yamamoto, N. (1986) Tumor promoter, TPA, enhances replication of HTLV-III/LAV. *Virology*, **154**, 249.
7. Smith, K. A. (1980) T-cell growth factor. *Immunol. Rev.*, **51**, 337.
8. Taniguchi, T., Matsui, H., Fujita, T., Takaoka, C., Kashima, N., Yoshimotor, R., and Hamuro, J. (1983) Structure and expression of a cloned cDNA for human interleukin-2. *Nature*, **302**, 305.
9. Fujita, T., Takaoka, C., Matsui, H., and Taniguchi, T. (1983) Structure of the human interleukin 2 gene. *Proc. Natl Acad. Sci. USA*, **80**, 7437.
10. Degrave, W., Tavernia, J., Duerinck, F., Plaetinck, G., Devos, R., and Fiers, W. (1983) Cloning and structure of the human interleukin 2 chromosomal gene. *Eur. J. Immunol.*, **2**, 2349.
11. Yokota, T., Lee, F., Rennick, D., Hall, C., Arai, N., Mosmann, T., Nabel, G., Cantor, H., and Arai, K. (1984) Isolation and characterization of a mouse cDNA clone that expresses mast-cell growth-factor activity in monkey cells. *Proc. Natl Acad. Sci. USA*, **81**, 1070.
12. Fung, M. C., Hapel, A. J., Ymer, S., Cohen, D. R., Johnson, R. M., Campbell, H. D., and Young, I. G. (1984) Molecular cloning of cDNA for murine interleukin-3. *Nature*, **307**, 233.
13. Noma, Y., Sideras, P., Naito, T., Bergstedt-Lindquist, S., Azuma, C., Severinson, E., Tanabe, T., Kinashi, T., Matsuda, F., and Yaoita, Y. (1986) Cloning of cDNA encoding the murine IgG1 induction factor by a novel strategy using SP6 promoter. *Nature*, **319**, 640.
14. Lee, F., Yokota, T., Otsuka, T., Meyerson, P., Villaret, D., Coffman, R., Mosmann, T., Rennick, D., Roehm, N., and Smith, C. (1986) Isolation and characterization of a mouse interleukin cDNA clone that expresses B-cell stimulatory factor 1 activities and T-cell- and mast-cell-stimulating activities. *Proc. Natl Acad. Sci. USA*, **83**, 2061.
15. Gray, P. W., Leung, D. W., Pennica, D., Yelverton, E., Najarian, R., Simonsen, C. C., Derynck, R., Sherwood, P. J., Wallace, D. M., Berger, S. L., Levinson, A. D., and Goeddel, D. V. (1982) Expression of human immune interferon cDNA in *E. coli* and monkey cells. *Nature*, **295**, 503.
16. Devos, R., Cheroutre, H., Taya, Y., Degrave, W., Van Heuvererswyn, H., and Fiers, W. (1982) Molecular cloning of human immune interferon cDNA and its expression in eukaryotic cells. *Nucl. Acids Res.*, **10**, 2487.
17. Wong, G. G., Witek, J. S., Temple, P. A., Wilkens, K. M., Leary, A. C., Luxenberg, D. P., Jones, S. S., Brown, E. L., Kay, R. M., and Orr, E. C. (1985) Human GM-CSF: molecular DNA and purification of the natural and recombinant proteins. *Science*, **228**, 810.
18. Nagata, S., Tsuchiya, M., Asano, S., Kaziro, Y., Yamazaki, T., and Yamamoto, O. (1985) Molecular cloning and expression of cDNA for human granulocyte colony-stimulating factor. *Nature*, **319**, 415.
19. Gough, N. M., Metcalf, D., Gough, J., Grail, D., and Dunn, A. R. (1985) Structure and expression of the mRNA for murine granulocyte-macrophage colony stimulating factor. *Eur. J. Immunol.*, **4**, 645.
20. Ho, D. D., Moudgil, T., and Alan, M. (1989) Quantitation of human immunodeficiency virus type 1 in the blood of infected persons. *N. Engl. J. Med.*, **321**, 1621.

21. Coombs, R. W., Collier, A. C., Allain, J. P., Nikora, B., Leuther, M., Gjerset, G. F., and Corey, L. (1989) Plasma viremia in human immunodeficiency virus infection. *N. Engl. J. Med.*, **321**, 1626.
22. Schnittman, S. M., Greenhouse, J. J., Psallidopoulos, M. C., Baseler, M., Salzman, N. P., Fauci, A. S., and Lane, H. C. (1990) Increasing viral burden in CD4+ T cells from patients with human immunodeficiency virus (HIV) infection reflects rapidly progressive immunosuppression and clinical disease. *Ann. Intern. Med.*, **113**, 438.
23. Nabel, G. and Baltimore, D. (1987) An inducible transcription factor activates expression of human immunodeficiency virus in T cells [published erratum appears in *Nature* 1990 Mar 8; **344** (6262):178]. *Nature*, **326**, 711.
24. Griffin, G. E., Leung, K., Folks, T. M., Kunkel, S., and Nabel, G. J. (1989) Activation of HIV gene expression during monocyte differentiation by induction of NF-κB. *Nature*, **339**, 70.
25. Malim, M. H. and Cullen, B. R. (1991) HIV-1 structural gene expression requires the binding of multiple Rev monomers to the viral RRE-implications for HIV-1 latency. *Cell*, **65**, 241.
26. Pomerantz, R. J., Trono, D., Feinberg, M. B., and Baltimore, D. (1990) Cells non-productively infected with HIV-1 exhibit an aberrant pattern of viral RNA expression: a molecular model for latency. *Cell*, **61**, 1271.
27. Nabel, G. J., Rice, S. A., Knipe, D. M., and Baltimore, D. (1988) Alternative mechanisms for activation of human immunodeficiency virus enhancer in T cells. *Science*, **239**, 1299.
28. Folks, T. M., Justement, J., Kinter, A., Dinarello, C. A., and Fauci, A. S. (1988) Cytokine-induced expression of HIV-1 in a chronically infected promonocyte cell line. *Science*, **238**, 800.
29. Osborn, L., Kunkel, S., and Nabel, G. J. (1989) Tumor necrosis factor a and interleukin-1 stimulate the human immunodeficiency virus enhancer by activation of the nuclear factor kB. *Proc. Natl Acad. Sci. USA*, **86**, 2336.
30. Folks, T. M., Clouse, K. A., Justement, J., Rabson, A., Duh, E., Kehrl, J. H., and Fauci, A. S. (1989) Tumor necrosis factor alpha induces expression of human immunodeficiency virus in a chronically infected T-cell clone. *Proc. Natl Acad. Sci. USA*, **86**, 2365.
31. Krasnow, S. W., Zhang, L., Leung, K., Osborn, L., Kunkel, S., and Nabel, G. (1991) Tumor necrosis factor-a, interleukin 1, and phorbol myristate acetate are independent activators of NF-κB which differentially activate T cells. *Cytokine*, **3**, 372.
32. Markovitz, D. M., Hannibal, M., Perez, V. L., Gauntt, C., Folks, T. M., and Nabel, G. J. (1990) Differential regulation of human immunodeficiency viruses: a novel HIV-2 regulatory element responds to stimulation of the T-cell antigen receptor. *Proc. Natl Acad. Sci. USA*, **87**, 9098.
33. Roeder, R. G. (1991) The complexities of eukaryotic transcription initiation: regulation of preinitiation complex assembly. *Trends Biochem. Sci.*, **16**, 402.
34. Dynlacht, B. D., Hoey, T., and Tjian, R. (1991) Isolation of coactivators associated with the TATA-binding protein that mediate transcriptional activation. *Cell*, **66**, 563.
35. Tanese, N., Pugh, B. F., and Tjian, R. (1991) Coactivators for a proline-rich activator purified from the multisubunit human TFIID complex. *Genes Dev.*, **5**, 2212.
36. Roy, A. L., Meisterernst, M., Pognonec, P., and Roeder, R. G. (1991) Cooperative interaction of an initiator-binding transcription initiation factor and the helix-loop-helix activator USF. *Nature*, **354**, 245.

37. Kubota, S., Nosaka, T., Cullen, B. R., Maki, M., and Hatanaka, M. (1991) Effects of chimeric mutants of human immunodeficiency virus type 1 Rev and human T-cell leukemia virus type I Rex on nucleolar targeting signals. *J. Virol.*, **65**, 2452.
38. Dillon, P. J., Nelbock, P., Perkins, A., and Rosen, C. A. (1991) Structural and functional analysis of the human immunodeficiency virus type 2 Rev protein. *J. Virol.*, **65**, 445.
39. Cochrane, A. W., Perkins, A., and Rosen, C. A. (1990) Identification of sequences important in the nucleolar localization of human immunodeficiency virus Rev: relevance of nucleolar localization to function. *J. Virol.*, **64**, 881.
40. Comai, L., Tanese, N., and Tjian, R. (1992) The TATA-binding protein and associated factors are integral components of the RNA polymerase I transcription factor, SL1. *Cell*, **68**, 965.
41. Jones, K. A., Kadonaga, J. T., Luciw, P. A., and Tjian, R. (1986) Activation of the AIDS retrovirus promoter by the cellular transcription factor, Sp1. *Science*, **232**, 755.
42. Leonard, J., Parrott, C., Buckler-White, A. J., Turner, W., Ross, E. K., Martin, M. A., and Rabson, A. B. (1989) The NF-κB binding sites in the human immunodeficiency virus type 1 long terminal repeat are not required for virus infectivity. *J. Virol.*, **63**, 4919.
43. Perkins, N. D., Edwards, N. L., Duckett, C. S., Schmid, R. M., Agranoff, A., and Nabel, G. J. (1992) A cooperative interaction between NF-κB and Sp1 is required for HIV enhancer activation. (Unpublished.)
44. Bielinska, A., Krasnow, S., and Nabel, G. J. (1989) NF-κB-mediated activation of the human immunodeficiency virus enhancer: site of transcriptional initiation is independent of the TATA box. *J. Virol.*, **63**, 4097.
45. Berkhout, B. and Jeang, K.-T. (1992) Functional roles for the TATA promoter and enhancers in basal and tat-induced expression of the human immunodeficiency virus type 1 long terminal repeat. *J. Virol.*, **66**, 139.
46. Smale, S. T. and Baltimore, D. (1989) The 'initiator' as a transcription control element. *Cell*, **57**, 103.
47. Jones, K. A. (1989) HIV trans-activation and transcription control mechanisms. *New Biol.*, **1**, 127.
48. Garcia, J. A., Wu, F. K., Mitsuyasu, R., and Gaynor, R. B. (1987) Interactions of cellular proteins involved in the transcriptional regulation of the human immunodeficiency virus. *EMBO J.*, **6**, 3761.
49. Wu, F. K., Garcia, J. A., Harrick, D., and Gaynor, R B. (1988) Purification of the human immunodeficiency virus type 1 enhancer and TAR binding proteins EBP-1 and UBP-1. *EMBO J.*, **7**, 2117.
50. Kato, H., Horikoshi, M., and Roeder, R. G. (1991) Repression of HIV-1 transcription by a cellular protein. *Science*, **251**, 1476.
51. Waterman, M. L. and Jones, K. A. (1990) Purification of TCF-1 alpha, a T-cell-specific transcription factor that activates the T-cell receptor C alpha gene enhancer in a context-dependent manner. *New Biol.*, **2**, 621.
52. Franza, B. R., Jr., Rauscher, F. J., III, Josephs, S. F., and Curran, T. (1988) The fos complex and fos-related antigens recognize sequence elements that contain AP-1 binding sites. *Science*, **239**, 1150.
53. Shaw, J. P., Utz, P. J., Durand, D. B., Toole, J. J., Emmel, E. A., and Crabtree, G. R. (1988) Identification of a putative regulator of early T cell activation genes. *Science*, **241**, 202.
54. Markovitz, D. M., Hannibal, M. C., Smith, M. J., Cossman, R., and Nabel, G. J. (1992)

Activation of the human immunodeficiency virus type 1 enhancer is not dependent on NFAT-1. *J. Virol.*, **66**, 3961.

55. Rosen, C. A., Sodroski, J. G., and Haseltine, W. A. (1985) The location of *cis*-acting regulatory sequences in the human T cell lymphotropic virus type III (HTLV-III/LAV) long terminal repeat. *Cell,* **41**, 813.
56. Lu, Y., Touzjian, N., Stenzel, M., Dorfman, T., Sodroski, J. G., and Haseltine, W. A. (1990) Identification of *cis*-acting repressive sequences within the negative regulatory element of human immunodeficiency virus type 1. *J. Virol.*, **64**, 5226.
57. Lu, Y., Touzjian, N., Stenzel, M., Dorfman, T., Sodroski, J. G., and Haseltine, W. A. (1991) NF-κB independent cis-acting sequences in HIV-1 LTR responsive to T-cell activation. *J. Acquired Immune Deficiency Syndr.*, **4**, 173.
58. Ahmad, N. and Venkatesan, S. (1988) Nef protein of HIV-1 is a transcriptional repressor of HIV-1 LTR. *Science*, **241**, 1481.
59. Hammes, S. R., Dixon, E. P., Malim, M. H., Cullen, B. R., and Greene, W. C. (1989) Nef protein of human immunodeficiency virus type 1: evidence against its role as a transcriptional inhibitor. *Proc. Natl Acad. Sci. USA*, **86**, 9549.
60. Mori, S., Takada, R., Shimotohno, K., and Okamoto, T. (1990) Repressive effect of the *nef* cDNA of human immunodeficiency virus type 1 on the promoter activity of the viral long terminal repeat. *Jpn. J. Cancer Res.*, **81**, 1124.
61. Cullen, B. R. (1991) AIDS: the positive effect of the negative factor. *Nature,* **351**, 698.
62. Kestler, H. W., Ringler, D. J., Mori, K., Panicali, D. L., Sehgal, P. K., Daniel, M. D., and Desrosiers, R. C. (1991) Importance of the *nef* gene for maintenance of high virus loads and for development of AIDS. *Cell,* **65**, 651.
63. Orchard, K., Perkins, N., Chapman, C., Harris, J., Emery, V., Goodwin, G., Latchman, D., and Collins, M. (1990) A novel T-cell protein which recognizes a palindromic sequence in the negative regulatory element of the human immunodeficiency virus long terminal repeat. *J. Virol.*, **64**, 3234.
64. Tong-Starksen, S. E., Luciw, P. A., and Peterlin, B. M. (1987) Human immunodeficiency virus long terminal repeat responds to T-cell activation signals. *Proc. Natl Acad. Sci. USA*, **84**, 6845.
65. Berkhout, B., Gatignol, A., Rabson, A. B., and Jeang, K. T. (1990) TAR-independent activation of the HIV-1 LTR: evidence that tat requires specific regions of the promoter. *Cell,* **62**, 757.
66. Liu, J., Perkins, N. D., Schmid, R. M., and Nabel, G. J. (1992) Specific NF-κB subunits act in concert with Tat to stimulate Human Immunodeficiency Virus-1 transcription. *J. Virol.*, **66**, 3883.
67. Lenardo, M. J. and Baltimore, D. (1989) NF-κB: a pleiotropic mediator of inducible and tissue specific gene control. *Cell,* **58**, 227.
68. Israel, A., Kimura, A., Kieran, M., Yano, O., Kanellopoulos, J., LeBail, O., and Kourilsky, P. (1987) A common positive transacting factor binds to enhancer sequences in the promoters of mouse H-2 and b-2 microglobulin genes. *Proc. Natl Acad. Sci. USA*, **84**, 2653.
69. Israel, A., LeBail, O., Hatat, D., Piette, J., Kieran, M., Logeat, F., Wallach, D., Fellous, M., and Kourilsky, P. (1989) TNF stimulates expression of mouse MHC class I genes by inducing an NF kappa B-like enhancer binding activity which displaces constitutive factors. *Eur. J. Immunol.*, **8**, 3793.
70. Baldwin, A. S. and Sharp, P. A. (1987) Binding of a nuclear factor to a regulatory

sequence in the promoter of the mouse H-2kB class I major histocompatibility gene. *Mol. Cell. Biol.*, **7**, 305.
71. Baldwin, A. S. and Sharp, P. A. (1988) Two transcription factors, NF-κB and H2TF1, interact with a single regulatory sequence in the class I major histocompatibility complex promoter. *Proc. Natl Acad. Sci. USA*, **85**, 723.
72. Leung, K. and Nabel, G. J. (1988) HTLV-I transactivator induces interleukin-2 receptor expression through an NF-κB-like factor. *Nature*, **333**, 776.
73. Cross, S. L., Halden, N. F., Lenardo, M. J., and Leonard, W. J. (1989) Functionally distinct NF-κB binding sites in the immunoglobulin k and IL-2 receptor a chain genes. *Science*, **244**, 466.
74. Ballard, D. W., Bohnlein, E., Lowenthal, J. W., Wano, Y., Franza, B. R., and Greene, W. C. (1988) HTLV-1 tax induces cellular proteins that activate the kappa B element in the IL-2 receptor alpha gene. *Science*, **241**, 1652.
75. Kieran, M., Blank, V., Logeat, F., Vandekerckhove, J., Lottspeich, F., LeBail, O., Urban, M. B., Kourilsky, P., Baeuerle, P. A., and Israel, A. (1990) The DNA binding subunit of NF-κB is identical to factor KBF1 and homologous to the *rel* oncogene product. *Cell*, **62**, 1007.
76. Ghosh, S., Gifford, A. M., Riviere, L. R., Tempst, P., Nolan, G. P., and Baltimore, D. (1990) Cloning of the p50 DNA Binding Subunit of NF-κB: Homology to *rel* and *dorsal*. *Cell*, **62**, 1019.
77. Gilmore, T. (1990) NF-κB, KBF1, dorsal and related matters. *Cell*, **62**, 841.
78. Bours, V., Villalobos, J., Burd, P. R., Kelly, K., and Siebenlist, U. (1990) Cloning of a mitogen-inducible gene encoding a κB DNA-binding protein with homology to the *rel* oncogene and to cell-cycle motifs. *Nature*, **348**, 76.
79. Molitor, J. A., Walker, W. H., Doerre, S., Ballard, D. W., and Greene, W. C. (1990) NF-κB: a family of inducible and differentially expressed enhancer-binding proteins in human T cells. *Proc. Natl Acad. Sci. USA*, **87**, 10028.
80. Ballard, D. W., Walker, W. H., Doerre, S., Sista, P., Molitor, J. A., Dixon, E. P., Peffer, N. J., Hannink, M., and Greene, W. C. (1990) The *v-rel* oncogene encodes a κB enhancer binding protein that inhibits NF-κB function. *Cell*, **63**, 803.
81. Meyer, R., Hatada, E. N., Hohmann, H.-P., Haiker, M., Bartsch, C., Rothlisberger, U., Lahm, H.-W., Schlaeger, E. J., Van Loon, A. P. G. M., and Scheidereit, C. (1991) Cloning of the DNA-binding subunit of human nuclear factor κB: the level of its mRNA is strongly regulated by phorbol ester or tumor necrosis factor a. *Proc. Natl Acad. Sci. USA*, **88**, 966.
82. Schmid, R. M., Perkins, N. D., Duckett, C. S., Andrews, P. C., and Nabel, G. J. (1991) Cloning of an NF-κB subunit which stimulates HIV transcription in synergy with p65. *Nature*, **352**, 733.
83. Neri, A., Chang, C.-C., Lombardi, L., Salina, M., Corradini, P., Maiolo, A. T., Chaganti, R. S., and Dalla-Favera, R. (1991) B cell lymphoma-associated chromosomal translocation involves candidate oncogene lyt-10, homologous to NF-κB p50. *Cell*, **67**, 1075.
84. Bours, V., Burd, P. R., Brown, K., Villalobos, J., Park, S., Ryseck, R.-P., Bravo, R., Kelly, K., and Siebenlist, U. (1992) A novel mitogen-inducible gene product related to p50/p105-NF-κB participates in transactivation through a κB site. *Mol. Cell. Biol.*, **12**, 685.
85. Liptay, S., Schmid, R. M., Perkins, N. D., Meltzer, P., Altherr, M. R., McPherson, J. D., Wasmuth, J., and Nabel, G. J. (1992) Related subunits of NF-κB map to two distinct loci associated with translocations in leukemia, NFKB1 and NFKB2. *Genomics*, **13**, 287.

86. Ten, R. M., Paya, C. V., Israel, N., LeBail, O., Mattei, M. G., Virelizier, J. L., Kourilsky, P., and Israel, A. (1992) The characterization of the promoter of the gene encoding the p50 subunit of NF-kappa B indicates that it participates in its own regulation. *Eur. J. Immunol.,* **11,** 195.
87. Perkins, N. D., Schmid, R. M., Duckett, C. S., Leung, K., Rice, N. R., and Nabel, G. J. (1992) Distinct combinations of NF-κB subunits determine the specificity of transcriptional activation. *Proc. Natl Acad. Sci. USA,* **89,** 1529.
88. Nolan, G. P., Ghosh, S., Liou, H.-C., Tempst, P., and Baltimore, D. (1991) DNA binding and IκB inhibition of the cloned p65 subunit of NF-κB, a rel-related polypeptide. *Cell,* **64,** 961.
89. Ryseck, R.-P., Bull, P., Takamiya, M., Bours, V., Siebenlist, U., Dobrzanski, P., and Bravo, R. (1992) Rel B, a new rel family transcription activator that can interact with p50-NF-κB. *Mol. Cell. Biol.,* **12,** 674.
90. Ruben, S. M., Dillon, P. J., Schreck, R., Henkel, T., Chen, C.-H., Maher, M., Baeuerle, P. A., and Rosen, C. A. (1991) Isolation of a rel-related human cDNA that potentially encodes the 65-kD subunit of NF-κB. *Science,* **251,** 1490.
91. Richardson, P. M. and Gilmore, T. D. (1991) vRel is an inactive member of the rel family of transcriptional activating proteins. *J. Virol.,* **65,** 3122.
92. Ruben, S. M., Narayanan, R., Klement, J. F., Chen, C.-H., and Rosen, C. A. (1992) Functional characterization of the NF-κB p65 transcriptional activator and an alternatively spliced derivative. *Mol. Cell. Biol.,* **12,** 444.
93. Nabel, G. J. (1991) HIV-tampering with transcription. *Nature,* **350,** 658.
94. Singh, H., LeBowitz, J. H., Baldwin, A. S., Jr., and Sharp, P. A. (1988) Molecular cloning of an enhancer binding protein: isolation by screening of an expression library with a recognition site DNA. *Cell,* **52,** 415.
95. Fan, C.-M. and Maniatis, T. (1990) A DNA-binding protein containing two widely separated zinc finger motifs that recognize the same DNA sequence. *Genes Dev.,* **4,** 29.
96. Franza, B. R., Josephs, S. F., Gilman, M. Z., Ryan, W., and Clarkson, B. (1987) Characterization of cellular proteins recognizing the HIV enhancer using a microscale DNA-affinity precipitation. *Nature,* **330,** 349.
97. Lee, J. H., Li, Y. C., Doerre, S., Sista, P., Ballard, D. W., Greene, W. C., and Franza, B. R., Jr. (1991) A member of the set of kappa B binding proteins, HIVEN86A, is a product of the human *c-rel* proto-oncogene. *Oncogene,* **6,** 665.
98. Clark, L. and Hay, R. T. (1989) Sequence requirement for specific interaction of an enhancer binding protein (EBP1) with DNA. *Nucleic Acids Res.,* **17,** 499.
99. Baeuerle, P. A. and Baltimore, D. (1988) Activation of DNA-binding activity in an apparently cytoplasmic precursor of the NF-κB transcription factor. *Cell,* **53,** 211.
100. Baeuerle, P. A. and Baltimore, D. (1988) I kappa B: a specific inhibitor of the NF-kappa B transcription factor. *Science,* **242,** 540.
101. Baeuerle, P. A. and Baltimore, D. (1989) A 65kD subunit of active NF-κB is required for inhibition of NF-κB by IκB. *Genes Dev.,* **3,** 1689.
102. Shirakawa, F., Chedid, M., Suttles, J., Pollok, B. A., and Mizel, S. B. (1989) Interleukin 1 and cyclic AMP induce kappa immunoglobulin light-chain expression via activation of an NF-kappa B-like DNA-binding protein. *Mol. Cell. Biol.,* **9,** 959.
103. Shirakawa, F. and Mizel, S. B. (1989) In vitro activation and nuclear translocation of NF-kappa B catalyzed by cyclic AMP-dependent protein kinase and protein kinase C. *Mol. Cell. Biol.,* **9,** 2424.

104. Ghosh, S. and Baltimore, D. (1990) Activation in vitro of NF-κB by phosphorylation of its inhibitor IκB. *Nature,* **344,** 678.
105. Haskill, S., Beg, A. A., Tompkins, S. M., Morris, J. S., Yurochko, A. D., Sampson-Johannes, A., Mondal, K., Ralph, P., and Baldwin, A. S., Jr. (1991) Characterization of an immediate-early gene induced in adherent monocytes that encodes IκB-like activity. *Cell,* **65,** 1281.
106. Davis, N., Ghosh, S., Simmons, D. L., Tempst, P., Liou, H.-C., Baltimore, D., and Bose, H. R., Jr. (1991) Rel-associated pp40: an inhibitor of the rel family of transcription factors. *Science,* **253,** 1268.
107. Inoue, J., Kerr, L. D., Kakizuka, A., and Verma, I. M. (1992) I kappa B gamma, a 70 kd protein identical to the C-terminal half of p110 NF-kappa B: a new member of the I kappa B family. *Cell,* **68,** 1109.
108. Roederer, M., Staal, F. J., Raju, P. A., Ela, S. W., and Herzenberg, L. A. (1990) Cytokine-stimulated human immunodeficiency virus replication is inhibited by *N*-acetyl-L-cysteine. *Proc. Natl Acad. Sci. USA,* **87,** 4884.
109. Staal, F. J., Roederer, M., and Herzenberg, L. A. (1990) Intracellular thiols regulate activation of nuclear factor kappa B and transcription of human immunodeficiency virus. *Proc. Natl Acad. Sci. USA,* **87,** 9943.
110. Schreck, R., Rieber, P., and Baeuerle, P. A. (1991) Reactive oxygen intermediates as apparently widely used messengers in the activation of the NF-kappa B transcription factor and HIV-1. *Eur. J. Immunol.,* **10,** 2247.
111. Kalebic, T., Kinter, A., Poli, G., Anderson, M. E., Meister, A., and Fauci, A. S. (1991) Suppression of human immunodeficiency virus expression in chronically infected monocytic cells by glutathione, glutathione ester, and *N*-acetylcysteine. *Proc. Natl Acad. Sci. USA,* **88,** 986.
112. Koenig, S., Gendelman, H. E., Orenstein, J. M., DalCanto, M. C., Pezeshkpour, G. H., Yungbluth, M., Janotta, F., Aksamit, A., Martin, M. A., and Fauci, A. S. (1986) Detection of AIDS virus in macrophages in brain tissue from AIDS. *Science,* **233,** 1089.
113. Gyorkey, F., Neinick, J. L., Sinkovics, J. G., and Gyorkey, P. (1985) Retrovirus resembling HTLV in macrophages of patients with AIDS. *Lancet,* **1,** 106.
114. Ho, D. D., Rota, T. R., and Hirsch, M. S. (1986) Infection of monocyte/macrophages by human T lymphotropic virus type III. *J. Clin. Invest.,* **77,** 1712.
115. Gartner, S., Markovits, P., Markovitz, D. M., Kaplan, M. H., Gallo, R. C., and Popovic, M. (1986) The role of mononuclear phagocytes in HTLV-III/LAV infection. *Science,* **233,** 215.
116. Wiley, C. A., Schrier, R. D., Nelson, J. A., Lampart, P. W., and Oldstone, M. B. (1986) Cellular localization of human immunodeficiency virus infection within the brains of acquired immune deficiency syndrome patients. *Proc. Natl Acad. Sci. USA,* **83,** 7089.
117. Gendelman, H. E., Orenstein, J. M., Martin, M. A., Ferrua, C., Mitra, R., Phipps, T., Wahl, L. A., Lane, H. C., Fauci, A. S., and Burke, D. S. (1988) Efficient isolation and propagation of human immunodeficiency virus on recombinant colony-stimulating factor 1-treated monocytes. *J. Exp. Med.,* **167,** 1428.
118. Orenstein, J. M., Meltzer, M. S., Phipps, T., and Gendelman, H. E. (1988) Cytoplasmic assembly and accumulation of human immunodeficiency virus types 1 and 2 in recombinant human colony-stimulating factor-1-treated human monocytes: an ultrastructural study. *J. Virol.,* **62,** 2578.
119. Donahue, R. E., Johnson, M. M., Zon, L. I., Clark, S. C., and Groopman, J. E. (1987)

Suppression of in vitro haematopoiesis following human immunodeficiency virus infection. *Nature,* **326,** 200.
120. Folks, T. M., Kessler, S. W., Orenstein, J. M., Justement, J. S., Jaffe, E. S., and Fauci, A. S. (1988) Infection and replication of HIV-1 in purified progenitor cells of normal human bone marrow. *Science,* **242,** 919.
121. Poli, G., Bressler, P., Kinter, A., Duh, E., Timmer, W. C., Rabson, A., Justement, J. S., Stanley, S., and Fauci, A. S. (1990) Interleukin 6 induces human immunodeficiency virus expression in infected monocytic cells alone and in synergy with tumor necrosis factor alpha by transcriptional and post-transcriptional mechanisms. *J. Exp. Med.,* **172,** 151.
122. Clavel, F., Guyader, M., Guetard, D., Salle, M., Montagnier, L., and Alizon, M. (1986) Molecular cloning and polymorphism of the human immune deficiency virus type 2. *Nature,* **324,** 691.
123. Clavel, F., Guetaro, D., Brun-Vezinet, F., Chamaret, S., Rey, M. A., Santos-Ferreira, M. O., Laurent, A. G., Dauguet, C., Katlama, C., Rouzioux, C., Klatzmann, D., Champalimaud, J. L., and Montagnier, L. (1986) Isolation of a new human retrovirus from West African patients with AIDS. *Science,* **233,** 343.
124. Clavel, F., Mansinho, K., Chamaret, S., Guetard, D., Favier, V., Nina, J., Santos-Ferreira, M. O., Champalimaud, J. L., and Montagnier, L. (1987) Human immunodeficiency virus type 2 infection associated with AIDS in West Africa. *N. Engl. J. Med.,* **316,** 1180.
125. Guyader, M., Emerman, M., Sonigo, P., Clavel, F., Montagnier, L., and Alizon, M. (1987) Genome organization and transactivation of the human immunodeficiency virus type 2. *Nature,* **326,** 662.
126. Barabe, P., Digoutte, J. P., Tristan, J. F., Peghini, M., Griffet, P., Jean, P., Seignot, P., Sarthou, J. L., Leguenno, B., Berlioz, C., Nbaye, P. S., Wade, B., Philippon, G., and Morcillo, R. (1988) Human immunodeficiency virus infections (HIV-1 and HIV-2) in Dakar: epidemiologic and clinical aspects. *Med. Trop.,* **48,** 337.
127. Courouce, A. M. (1988) A prospective study of HIV-2 prevalence in France. *AIDS,* **2,** 261.
128. Pokrovskii, V. V., Suvorova, Z. K., and Mangushev, T. N. (1988) Infection caused by the human immunodeficiency virus type 2 in the USSR. *Zh. Mikrobiol. Epidemiol. Immunobiol.,* **10,** 18.
129. Cortes, E., Detels, R., Aboulafia, D., Li, X., Moudgil, T., Alam, M., Bonecker, C., Gonzaga, A., Oyafuso, L., Tondo, M., Boite, C., Hammershlak, N., Capitani, C., Slamon, D. J., and Ho, D. D. (1989) HIV-1, HIV-2, and HTLV-I infection in high-risk groups in Brazil. *N. Engl. J. Med.,* **320,** 953.
130. Loveday, C., Pomeroy, L., Weller, I. V., Quirk, J., Hawkins, A., Williams, H., Smith, A., Williams, P., Tedder, R. S., and Adler, M. W. (1989) Human immunodeficiency viruses in patients attending a sexually transmitted disease clinic in London. *Br. Med. J.,* **298,** 419.
131. Neumann, P. W., O'Shaughnessy, M. V., Lepine, D., D'Souza, I., Major, C., and McLaughlin, B. (1989) Laboratory diagnosis of the first cases of HIV-2 infection in Canada. *Can. Med. Assoc. J.,* **140,** 125.
132. Poulsen, A. G., Kvinesdal, B., Aaby, P., Molbak, K., Frederiksen, K., Dias, F., and Lauritzen, E. (1989) Prevalence of and mortality from human immunodeficiency virus type 2 in Bissau, West Arica. *Lancet,* **1,** 827.
133. Ruef, C., Dickey, P., Schable, C. A., Griffith, B., Williams, A. E., and D'Aquila, R. T.

(1989) A second case of the acquired immunodeficiency syndrome due to human immunodeficiency virus type 2 in the United States: the clinical implications. *Am. J. Med.*, **86**, 709.
134. Kanki, P. (1989) *Clinical significance of HIV-2 infection in West Africa*. Dekker, New York, pp. 95–108.
135. Tong-Starksen, S. E., Welsh, T. M., and Peterlin, B. M. (1990) Differences in transcriptional enhancers of HIV-1 and HIV-2: Response to T cell activation signals. *J. Immunol.*, **145**, 4348.
136. Leiden, J. M., Wang, C. Y., Petryniak, B., Markovitz, D. M., Nabel, G. J., and Thompson, C. B. (1992) A novel ets-related transcription factor, ELF-1, binds to HIV-2 regulatory elements that are required for inducible transactivation in T cells. *J. Virol.* **66**, 5890.
137. Markovitz, D. M., Smith, M. J., Hilfinger, J., Hannibal, M. C., Petryniak, B., and Nabel, G. J. (1992) Activation of the human immunodeficiency virus type 2 enhancer is dependent on purine box and κB regulatory elements. *J. Virol.*, **66**, 5479.
138. Thompson, C. B., Wang, C.-Y., Ho, I.-C., Bohjanen, P. R., Petryniak, B., June, C. H., Miesfeldt, S., Zhang, L., Nabel, G. J., Karpinski, B., and Leiden, J. M. (1992) *cis*-acting sequences required for inducible Interleukin-2 enhancer function bind a novel Ets-related protein, Elf-1. *Mol. Cell. Biol.*, **12**, 1043.
139. Arya, S. K. (1990) Human immunodeficiency virus type-2 gene expression: Two enhancers and their activation by T-cell activators. *New Biol.*, **2**, 57.
140. Arya, S. K. and Sethi, A. (1990) Stimulation of the human immunodeficiency virus type 2 (HIV-2) gene expression by the cytomegalovirus and HIV-2 transactivator gene. *AIDS Res. Hum. Retroviruses*, **6**, 649.
141. Riviere, Y., Blank, V., Kourilsky, P., and Israel, A. (1991) Processing of the precursor of NF-kappa B by the HIV-1 protease during acute infection. *Nature*, **350**, 625.
142. Bachelerie, F., Alcami, J., Arenzana-Seisdedos, F., and Virelizier, J. L. (1991) HIV enhancer activity perpetuated by NF-kappa B induction on infection of monocytes. *Nature*, **350**, 709.
143. Malim, M. H., Bohnlein, S., Hauber, J., and Cullen, B. R. (1989) Functional dissection of the HIV-1 Rev *trans*-activator—derivation of a *trans*-dominant repressor of Rev function. *Cell*, **58**, 205.
144. Malim, M. H., Freimuth, W. W., Liu, J., *et al.* (1992) Transdominant rev protein inhibits HIV replication without affecting T cell function. (Unpublished.)
145. Orchard, K., Perkins, N., Chapman, C., Harris, J., Emery, V., Goodwin, G., Latchman, D., and Collins, M. (1990) A T-cell protein which recognizes a palindromic DNA sequence in the negative regulatory element of the HIV-1 long terminal repeat with homology to steroid/thyroid hormone receptor binding sites. *Biochem. Soc. Trans.*, **18**, 555.
146. Starcich, B., Ratner, L., Josephs, S. F., Okamoto, T., Gallo, R. C., and Wong-Staal, F. (1985) Characterization of long terminal repeat sequences of HTLV-III. *Science*, **227**, 538.
147. Dasgupta, P., Saikumar, P., Reddy, C. D., and Reddy, E. P. (1990) Myb protein binds to human immunodeficiency virus 1 long terminal repeat (LTR) sequences and transactivated LTR-mediated transcription. *Proc. Natl Acad. Sci. USA*, **87**, 8090.
148. Calvert, I., Peng, Z., Kung, H., and Raziuddin (1991) Cloning and characterization of a novel sequence-specific DNA-binding protein recognizing the negative regulatory element (NRE) region of the HIV-1 long terminal repeat. *Gene*, **101**, 171.

149. Maekawa, T., Sudo, T., Kurimoto, M., and Ishii, S. (1991) USF-related transcription factor, HIV-TF1, stimulates transcription of human immunodeficiency virus-1. *Nucleic Acids Res.,* **19,** 4689.
150. Giacca, M., Gutierrez, M. I., Menzo, S., d'Adda di Fagagna, F., and Falaschi, A. (1992) A human binding site for transcription factor USF/MLTF mimics the negative regulatory element of human immunodeficiency virus type 1. *Virology,* **186,** 133.
151. Smith, M. R. and Greene, W. C. (1989) The same 50-kDa cellular protein binds to the negative regulatory elements of the interleukin 2 receptor alpha-chain gene and the human immunodeficiency virus type 1 long terminal repeat. *Proc. Natl Acad. Sci. USA,* **86,** 8526.
152. Jones, K. A., Luciw, P. A., and Duchange, N. (1988) Structural arrangements of transcription control domains within the 5' untranslated leader regions of the HIV-1 and HIV-2 promoters. *Genes Dev.,* **2,** 1101.

4 | Tat *trans*-activator

B. MATIJA PETERLIN, MELANIE ADAMS, ALICIA ALONSO,
ANDREAS BAUR, SUBIR GHOSH, XIAOBIN LU, and YING LUO

1. Introduction

Several lentiviral promoters or their 5' long terminal repeats (LTRs) are unusual among eukaryotic promoters in that they contain an RNA structure in their leader sequences which is important for their function (for review see refs 1–10). This RNA element is called the *trans*-activation response element or TAR (11). Interactions between TARs and virally encoded *trans*-activators called Tats greatly increase levels of viral replication and expression of all viral proteins (12–15). Translated from multiply spliced mRNAs that appear early in the viral life cycle, Tat primarily affects rates of elongation of transcription (16); however, effects of Tat on initiation of transcription and translation have also been reported (17, 18). Since human immunodeficiency virus type 1 (HIV-1) lacking Tat or TAR replicates poorly and does not cause cytopathology (19, 20), approaches designed to interfere with Tat or TAR might be useful in the treatment of acquired immunodeficiency syndrome (AIDS).

2. Tat *trans*-activator

Tat is a small protein of 75–130 amino acids, depending on the viral strain and isolate (21). Tat of HIV-1, which contains 82 to 101 amino acids, is translated from three multiply spliced viral mRNAs (Fig. 1) (22). Another mRNA codes for the first exon of Tat fused to part of the envelope and the second exon of Rev (Fig. 1). Its protein product is called Tev or Tnv and functions like Tat (23, 24). Once expressed, Tat is found in the nucleus and nucleolus (25–27). Of known lentiviruses, equine infectious anemia virus (EIAV), simian immunodeficiency (SIV), HIV-1, and HIV-2 contain Tats that work via TAR RNA elements (14, 15, 28, 29). Comparing sequences and structures, Tats and TARs of EIAV appear the simplest and those of HIV-2 the most complex (Figs 2 and 3) (14, 15, 28–32). Functionally, whereas Tat of HIV-1 *trans*-activates HIV-1, HIV-2, and SIV LTRs, Tats of HIV-2 and SIV *trans*-activate efficiently their respective LTRs and weakly the HIV-1 LTR, but not the EIAV LTR, and Tat of EIAV *trans*-activates only the EIAV LTR (14, 15, 32).

By examining amino acid sequences from different lentiviruses, Tats can be divided into five distinct domains (Fig. 2) (32). In this chapter, they are called the

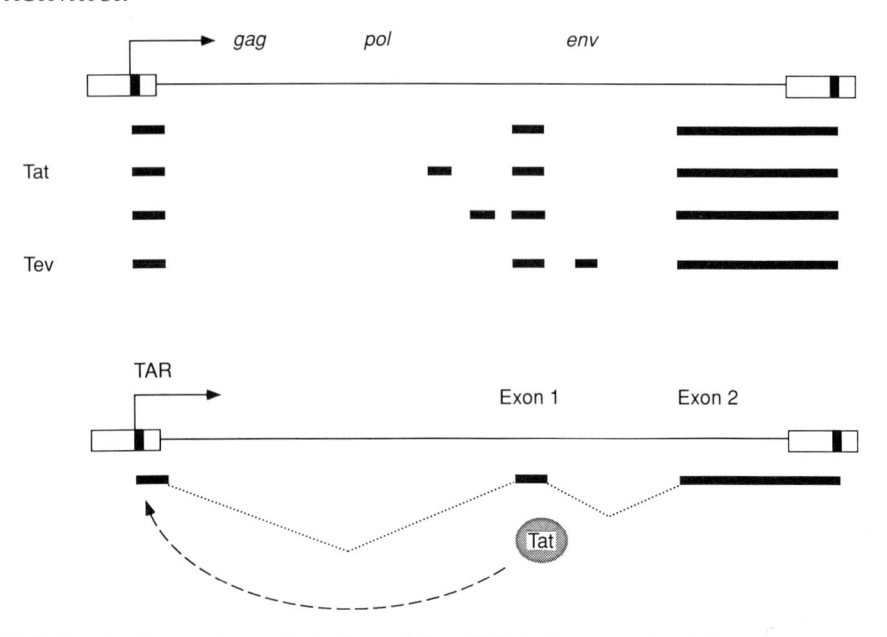

Fig. 1 Multiply spliced transcripts code for Tat and Tev of HIV-1. Once translated, Tat interacts with TAR RNA to increase levels of expression of viral RNAs and proteins. Three different multiply spliced RNAs give rise to Tat (22). Two of these transcripts contain additional sequences from *pol* and *vif* open reading frames (22). Tev or Tnv contains the first exon of Tat fused to the middle of *env* and to the second exon of *rev* (23, 24). It functions better as Tat than as Rev (23, 24). For Tat of HIV-1, exon 1 is sufficient for wild-type activity

N-terminal, cysteine-rich, core, basic, and C-terminal sequences (Fig. 2). Tats differ mostly in sequences and lengths of their N-terminal, cysteine-rich, and C-terminal regions (Fig. 2) (32). Since their core, basic, and C-terminal cysteine-rich domains are conserved, these sequences have been implicated as essential to the function of Tat (32, 33). In fact, a minimal lentiviral Tat can be constructed by fusing part of the cysteine-rich and core domains of Tat of EIAV to the basic domain of Tat of HIV-1 (Fig. 2). This protein of only 25 amino acids retains functional activity on TAR of HIV-1 (33).

To determine other residues that are important for its function, many mutations have been introduced into Tat of HIV-1 (Fig. 4) (for review see refs 3, 5–7, 9). First, neither the cysteines, with the exception of the fourth cysteine, nor lysine 41 could be substituted by other amino acids (26, 34–38). In addition, N-terminal acidic amino acids are important (26, 37, 39, 40). However, some substitutions of these residues were tolerated as long as they could form an amphiphatic α-helix (39). Second, although the basic domain is essential, its positive charge rather than precise amino acids must be preserved (35, 37, 41–43). For example, substitutions of the basic domain of Tat with those of Rev of HIV-1, Rex of HTLV-1, and even bacteriophage λ N were tolerated. However, an acidic glutamic acid at the N-terminus of the basic domain of the bacteriophage λ N protein had to be replaced by a neutral glycine for optimal activity (43). Although Tat of HIV-2 functions

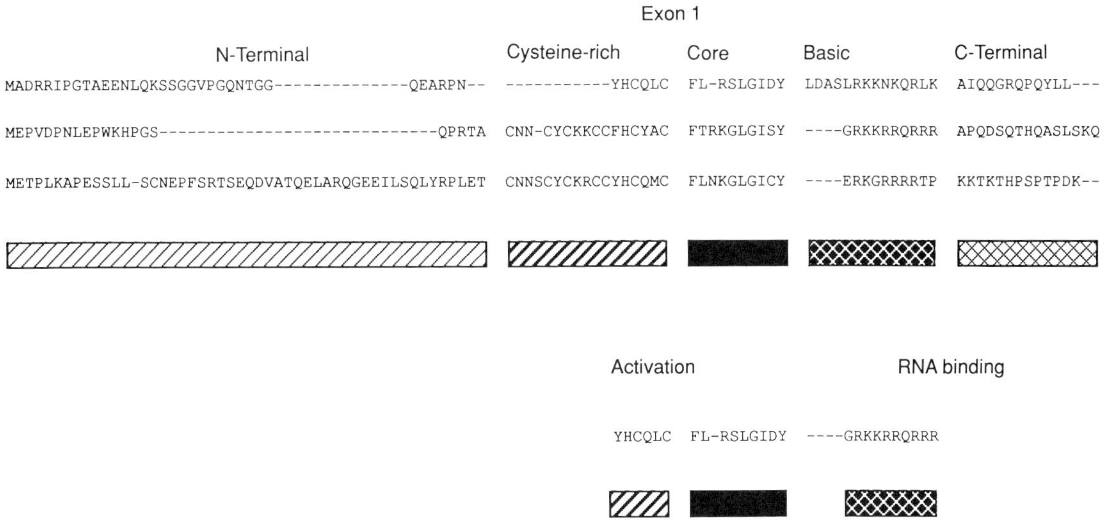

```
                                      Exon 1
                  N-Terminal              Cysteine-rich    Core        Basic            C-Terminal
MADRRIPGTAEENLQKSSGGVPGQNTGG-------------QEARPN--    ----------YHCQLC  FL-RSLGIDY  LDASLRKKNKQRLK   AIQQGRQPQYLL---

MEPVDPNLEPWKHPGS-------------------------QPRTA      CNN-CYCKKCCFHCYAC FTRKGLGISY  ----GRKKRRQRRR   APQDSQTHQASLSKQ

METPLKAPESSLL-SCNEPFSRTSEQDVATQELARQGEEILSQLYRPLET  CNNSCYCKRCCYHCQMC FLNKGLGICY  ----ERKGRRRRTP   KKTKTHPSPTPDK--
```

                                          Activation   RNA binding

                                          YHCQLC   FL-RSLGIDY   ----GRKKRRQRRR

Fig. 2 Domains of Tats. By comparing sequences from different lentiviruses, Tats can be divided into five structural domains (32). Since Tat of EIAV contains only 75 amino acids, further C-terminal extensions of Tats of HIV-1 and HIV-2/SIV have been ignored. In this chapter, these domains are called N-terminal, cysteine-rich, core, basic, and C-terminal sequences. Note that C-terminal cysteine-rich, core, and basic domains are conserved between Tats of EIAV, HIV-1, and HIV-2/SIV, whereas the N-terminal and C-terminal sequences are more divergent. Whereas activation and RNA-binding domains of Tat of EIAV contain 25 and 26 amino acids, respectively, those of Tat of HIV-1 contain 48 and 10 amino acids, respectively. Thus, activation domains of Tats of HIV-1 and HIV-2/SIV appear more complex than that of Tat of EIAV, whereas just the opposite pertains to their respective RNA-binding domains. From these two *trans*-activators, a minimal lentiviral Tat can be constructed which contains 25 cysteine-rich and core amino acids from Tat of EIAV and 10 basic amino acids from Tat of HIV-1 (33)

poorly with TAR of HIV-1, a similar substitution in the basic domain of Tat of HIV-2 rendered that *trans*-activator as potent as Tat of HIV-1 on the HIV-1 LTR (44). Finally, deletions of amino acids C-terminal to position 60 caused only a modest diminution in activity (34, 37, 45, 46). However, in addition to the basic domain, these C-terminal sequences may contribute to the nuclear localization of Tat of HIV-1 (47) and to the binding to EIAV TAR RNA of EIAV Tat (32). Although Tats contain numerous serines and threonines, no phosphorylation of Tat has been reported. However, since depleting or inhibiting protein kinase C profoundly affected Tat function, protein kinase C must target some other cellular protein that is important for *trans*-activation (48).

Heterologous DNA- and RNA-tethering mechanisms have also been used to map activation and RNA-binding domains of Tat. c-Jun, Gal-4, and Lex-A represent heterologous DNA-binding proteins that were linked to Tat (49–51) (Ghosh and Peterlin, unpublished data). Tat was also fused to Rev of HIV-1 (52) and the coat protein of bacteriophage R17 (53), two RNA-binding proteins, and appropriate target sequences were placed into the HIV-1 LTR. Using these heterologous nucleic acid tethering mechanisms, only the N-terminal 48 and 25 amino acids of Tats of HIV-1 and EIAV, respectively, were required for function (33, 53). These studies

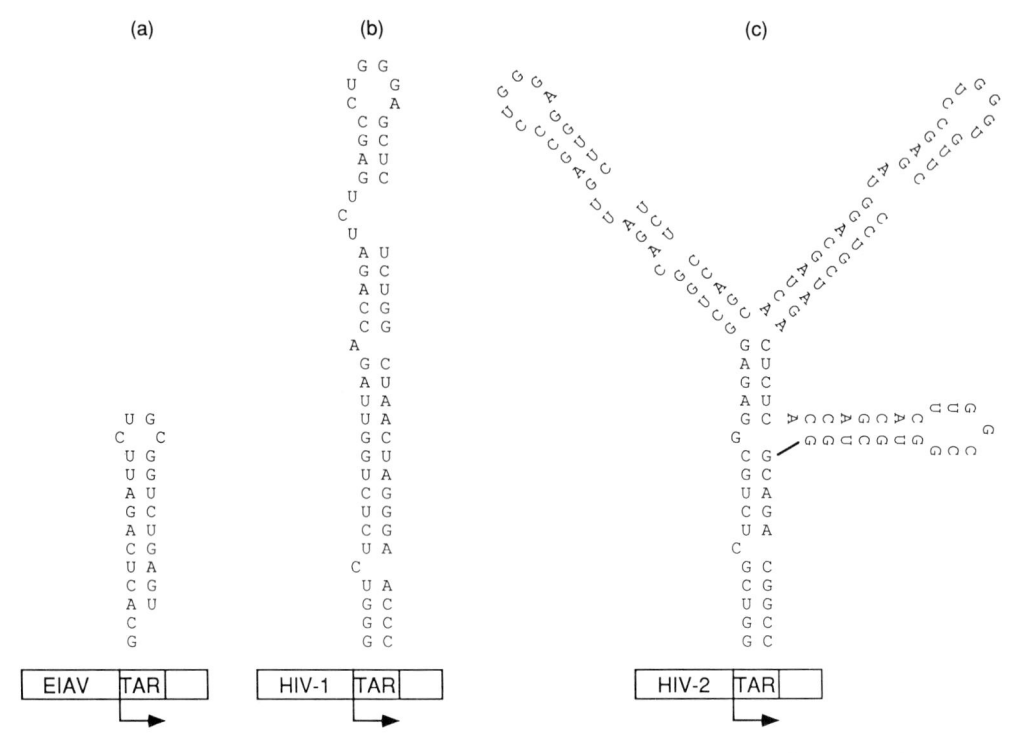

Fig. 3 Structures of (a) EIAV, (b) HIV-1, and (c) HIV-2 TARs. EIAV, HIV-1, HIV-2, and SIV have TARs that interact with Tats as nascent RNA structures (161). Whereas TAR of EIAV is only 24 nucleotides long, those of HIV-1 and HIV-2/SIV measure 59 and 123 nucleotides, respectively (14, 45, 67). TAR of EIAV has a tetranucleotide loop and no bulge. All other TARs have hexanucleotide loops and either dinucleotide or trinucleotide bulges in their 5' stems. TAR of HIV-2/SIV contains a duplication of the upper stem–loop. The 3' stem–loop of this TAR is dispensable

demonstrated that the N-terminal acidic, cysteine-rich, core domains and two cysteines followed by the core domain constitute the activation domains of Tats of HIV-1 and EIAV, respectively (33, 53). However, fusion proteins between Tat and Gal-4 or the coat protein were inactive on TAR if they did not additionally contain basic amino acids of Tat (51, 53). Since this basic domain also directs heterologous proteins to the nucleus and binds to TAR RNA *in vitro* (26, 27, 41, 54–57), the basic domain represents the RNA-binding domain of Tat.

Further structural studies of Tat have employed protease sensitivity and spectroscopic techniques. For example, Tat of HIV-1 but not of HIV-2 is a compact molecule that is protease insensitive (38, 58). However, mutations in the cysteine-rich and core domains relax this secondary structure and render Tat of HIV-1 susceptible to proteolysis (38). Although only limited spectroscopy has been performed, circular dichroism revealed that Tat of HIV-1 contains some α-helical structure, which is centred in the core domain (59, 60). Whether the N-terminal acidic domain forms an amphiphatic α-helix that is bent by several prolines and whether cysteine-rich residues form a Zn^{2+} finger or a lattice of binding pockets for

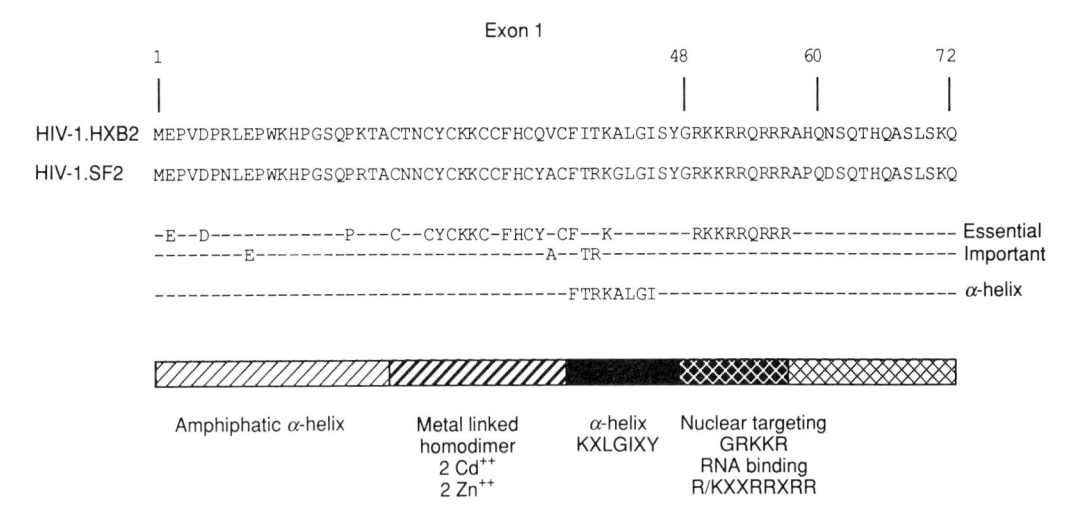

```
                                     Exon 1
            1                                    48          60          72
            |                                    |           |           |
HIV-1.HXB2  MEPVDPRLEPWKHPGSQPKTACTNCYCKKCCFHCQVCFITKALGISYGRKKRRQRRRAHQNSQTHQASLSKQ

HIV-1.SF2   MEPVDPNLEPWKHPGSQPRTACNNCYCKKCCFHCYACFTRKGLGISYGRKKRRQRRRAPQDSQTHQASLSKQ


            -E---D------------P---C--CYCKKC-FHCY-CF--K------RKKRRQRRR-------------- Essential
            --------E---------------------A--TR------------------------------------ Important

            --------------------------------FTRKALGI------------------------------- α-helix
```

| Amphiphatic α-helix | Metal linked homodimer 2 Cd⁺⁺ 2 Zn⁺⁺ | α-helix KXLGIXY | Nuclear targeting GRKKR RNA binding R/KXXRRXRR |

Fig. 4 Functionally important amino acids of Tat of HIV-1. Sequences of Tats of HIV-1.HXB2 and HIV-1.SF2 are given. Whereas activation domains extend to position +48, Tats require only additional 12 amino acids to position +60 for function. Within these sequences, N-terminal acidic, six of the seven cysteines, several amino acids in the cysteine-rich and core domains, and basic sequences are essential for *trans*-activation (for review see refs 5, 7–10). Most residues in the cysteine-rich and core domains are also required for the compact tertiary structure of Tat of HIV-1 in solution (38, 58). Several additional amino acids in the N-terminal and core domains are still important for function. Additionally, residues in the core domain form an α-helix in solution. Residues in the N-terminus might form an amphiphatic α-helix (39). The cysteine-rich domain binds divalent cations, either two Cd^{2+} or two Zn^{2+} atoms (59, 61). KXLGIXY in the core domain forms an α-helix in solution (60). Basic sequences target heterologous protein to the nucleus and also bind to TAR RNA *in vitro* (for review see refs 5, 7–10)

divalent cations are unknown. Although it was suggested that divalent cations bound by these cysteines create Tat homodimers (59, 61), no other evidence exists for Tat homodimers either free in solution or bound to TAR RNA (62).

2.1 Proteins that associate with Tat

To clarify structural features and to determine the function of Tat, proteins that interact with Tat of HIV-1 were investigated (for review see ref. 9). Two proteins have so far been identified (63, 64). The first was cloned by direct screening of expression cDNA libraries with a biotinylated Tat protein (63). A 50 kDa protein called Tat-binding protein (TBP-1) was isolated. This protein decreased levels of *trans*-activation by Tat in transient cotransfection assays (63). Recently, another protein of 36 kDa was isolated using Tat affinity chromatography followed by elution in high salt (64). This protein has been coinjected with Tat into rodent cells where it increased levels of *trans*-activation by Tat. Preliminary evidence suggested that this protein binds to the C-terminal sequences of Tat, which constitute neither the activation nor the RNA-binding domains of Tat of HIV-1 (64). However, it is possible that p36 stabilizes Tat or facilitates its interactions with TAR RNA.

Additional evidence for potential Tat-binding proteins has been obtained using mutant or hybrid Tats that function as *trans*-dominant inhibitors and squelch *trans*-activation by Tat in transient cotransfection assays (47, 65, 66). For example, Tat of HIV-1 lacking basic amino acids can compete for *trans*-activation by the wild-type Tat (65, 66). In addition, heterologous proteins that contain activation domains of Tat of HIV-1 and EIAV linked either to native or heterologous RNA-binding motifs can cross-compete for *trans*-activation (47). Thus, similar cellular proteins interact with Tats of EIAV and HIV-1, albeit with different affinities (47). Since most of these *trans*-dominant Tats are localized in the cytoplasm, these studies suggest that Tat associates with a cytoplasmic protein which it conveys to TAR and the transcription complex in the nucleus (47).

3. *Trans*-activation response element, TAR

EIAV, HIV-1, HIV-2, and SIV contain TARs which are essential for *trans*-activation (Fig. 3) (14, 30, 45, 67). TAR of EIAV with 24 nucleotides and TARs of HIV-2 and SIV with 123 nucleotides are the smallest and the largest TARs, respectively (Fig. 3) (14, 31, 67, 68). Comparing sequences and structures of these TARs, the main features are stable stems with tetranucleotide or hexanucleotide loops, which might be smaller if some of the bases in the loop can form stable base pairs in solution (Fig. 3) (45, 67, 68). However, TARs of HIV and SIV, unlike TAR of EIAV, also contain 5' dinucleotide or trinucleotide bulges (45, 54, 57, 69, 70). Several studies have revealed that whereas these bulged RNA stems are essential for *trans*-activation by Tats of HIV-1, HIV-2, and SIV (54, 57, 69, 70), the central loops are of paramount importance for all Tats (28, 67, 68, 70, 71). Also, the spacing between the bulge and the loop is important for TAR of HIV-1 (70, 72). Further deletions, mutations, and substitutions have revealed that the bottom of the stem–loop is dispensable so that the minimal functional TAR of HIV-1 extends from positions +19 to +43 (Fig. 5) (68, 73). This would make TAR of HIV-1 of similar size as TAR of EIAV and the operator of bacteriophage R17 (53). In solution, this minimal TAR might form the 5' bulge and central loop or undergo more extensive base pairing, which would result in a rod-like structure with a bend at U22 and a bulge at A34 (Fig. 5). Furthermore, the bottom stem of TAR of HIV-1 might be important for pausing of the transcription complex (74–76), for generation of short transcripts (16, 74, 77), for RNA stability (16, 68), and for possible translational effects of Tat (78–81). Finally, TAR has to be positioned very close to the site of initiation of transcription, and optimally the monomethyl CAP (7-methyl-GpppG) forms the 5' border of the TAR RNA stem–loop (68).

3.1 Interactions between Tat and TAR

Although direct binding of Tat to TAR has not been demonstrated in nuclear extracts from transfected or infected cells, recombinant Tat binds well to *in vitro*

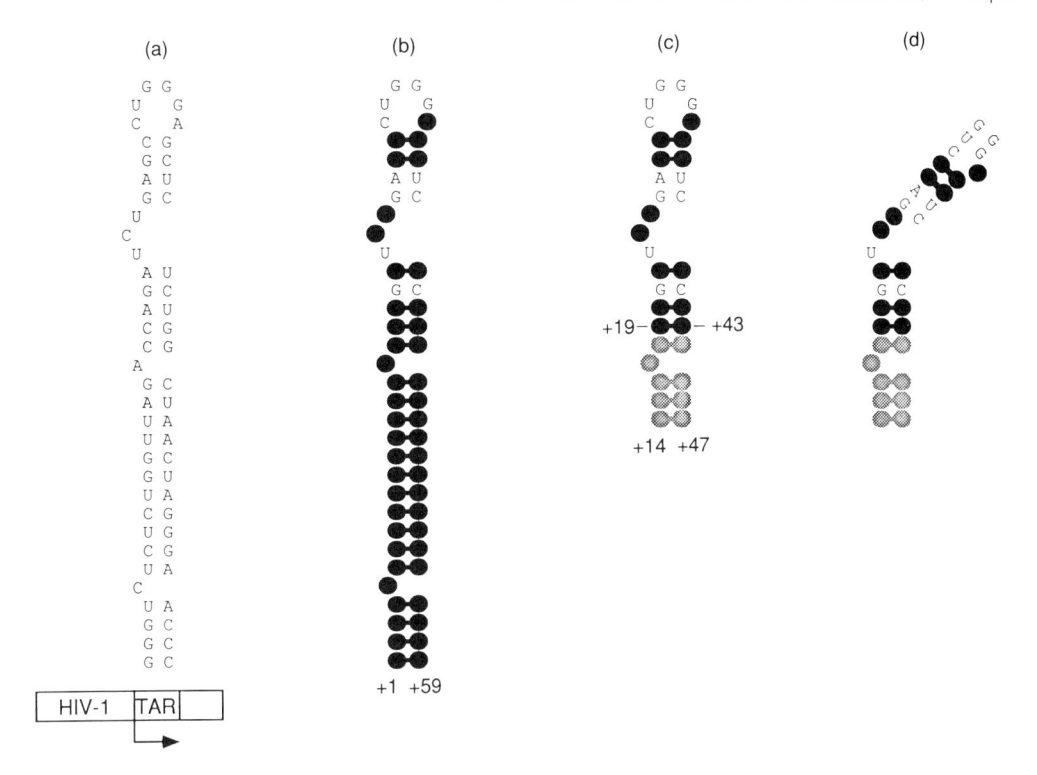

Fig. 5 Important nucleotides and possible secondary structures of TAR of HIV-1. (a) TAR of HIV-1 is positioned next to site of initiation of transcription and contains 59 nucleotides. (b) Although base-pairing in the stem is required, only 12 nucleotides in the stem, bulge, and loop cannot be substituted. These are expressed in the single letter code, whereas nucleotides required for the structure of TAR are represented as black circles. (c) Although sequences from positions +14 to +47 are important, those from positions +19 to +43 are essential to *trans*-activation (68, 73). Thus, the minimal TAR of HIV-1 has the same size as TAR of EIAV. (d) Possible secondary structure of TAR of HIV-1 in solution. Instead of a 5′ bulge, TAR might contain a bend, and instead of a loop, base-pairing of the entire stem might create a rod-like structure with an additional 3′ bulged adenosine

synthesized TAR RNA with a K_d in the nanomolar range (0.1 to 12 nM) (42, 54, 55, 82–84). Tat binds to TAR RNA at the 5′ bulge, and this binding requires specific base pairs above and below the bulge (Fig. 5) (54–56, 83–89). Furthermore, Tat might bind to the major groove in TAR RNA (84, 89). Whether shorter or longer Tat peptides bind with greater affinity than the wild-type Tat and whether certain procedures that denature and renature Tat increase its binding affinity *in vitro* remain controversial (42, 56, 84–87, 90). Surprisingly, substitutions of basic nucleotides by random combinations of arginines and lysines and a single arginine in the context of eight lysines bind equally well to TAR RNA as does the wild-type basic domain (57). In addition, only the bulged U22 seems to be important for this association which has led to the proposal that certain phosphates in the RNA backbone and the basic arginine form an arginine fork that allows for specific interactions between Tat and TAR (42, 57).

Several nuclear proteins also bind to TAR RNA (72, 91–96). Only three of these have been correlated with function in an *in vitro* transcription system (72, 95, 97). For example, p68 obtained from HeLa nuclear extracts binds to the loop and increases levels of *trans*-activation (93, 97). Of two other cellular proteins that also bind to TAR RNA, TRP-1 or TRP-185 and TRP-2 (TAR RNA-binding protein-1 or -185 and -2) bind to the loop and to the bulge, respectively (Fig. 6) (72, 95). TRP-1 binds to the loop as a heterodimer of two proteins of 185 kDa and 90 kDa, of which only the 185 kDa subunit contacts RNA (72, 95). Whereas the 90 kDa protein is present in all tissues and species including yeast, the 185 kDa protein is specific to primates (72). Although TRP-1 increases levels of transcription, its dependency on Tat is controversial (72, 95). On the other hand, TRP-2 is a complex of at least four proteins of 70–100 kDa that bind to the 5' bulge in TAR (Fig. 6). Whereas Tat can displace TRP-2 for binding to TAR RNA, TRP-1 might help Tat to bind to TAR RNA (72). Moreover, the existence of these cellular proteins and Tat that bind to TAR RNA with similar K_ds might suggest that Tat does not bind to TAR RNA directly, but might interact with these cellular proteins to affect *trans*-activation (72, 95). In fact, Tat has been found in ribonuclear complexes in HIV-infected cells (98).

That cellular proteins are required for optimal interactions between Tat and TAR was confirmed by studies of *trans*-activation in rodent cells (99–101). Following the obervation that Tat does not function efficiently in a variety of non-primate cells, Chinese hamster ovary (CHO) cells and CHO cells containing various human

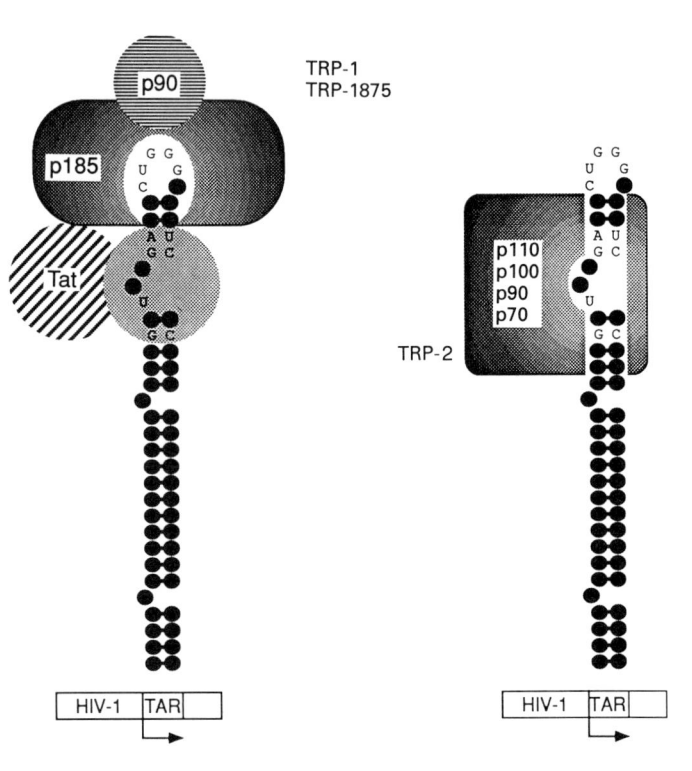

Fig. 6 Cellular proteins and Tat bind to TAR RNA. TAR RNA-binding protein-1 or 185 (TRP-1 or TRP-185) binds to the loop of TAR of HIV-1 (72, 95). Although only the 185 kDa subunit contacts RNA, an additional, ubiquitous 90 kDa subunit is required for specific binding to TAR RNA. The binding of Tat to the 5' bulge in TAR RNA requires six base-paired nucleotides and the uridine at position +22 (for review see refs 5, 7–10). RNA-binding and activation domains of Tat are pictured as interlocking circles. Another nuclear protein, TAR RNA-binding protein-2 (TRP-2) binds to the 5' bulge and stem in TAR (72). TRP-2 is composed of at least four proteins from 70 to 110 kDa in size. Although Tat and TRP-2 can displace each other for the binding to TAR, Tat and TRP-1 might be able to bind to TAR at the same time and act synergistically to activate HIV-1 transcription

chromosomes were tested in transient coexpression assays. Human chromosome 12, and to a lesser extent human chromosome 6, increased levels of *trans*-activation in CHO cells (100–102). By use of a variety of mutated and substituted HIV-1 LTRs and hybrid *trans*-activators, the defect in rodent cells was mapped to interactions between Tat and TAR and not to cellular DNA-binding proteins or factors that associate with the activation domain of Tat (103). Since heterologous RNA-tethering mechanisms negated rodent effects, cellular proteins that interact with TAR must either help Tat to bind to TAR or help to modify the structure of Tat so that the activation domain is better presented to the transcription complex (103). It is of further interest that the 185 kDa subunit of TRP-1 is missing in rodent cells (72).

3.2 TAR decoys

Since Tat binds to TAR RNA either directly or with the help of cellular TAR RNA-binding proteins, this observation has been exploited for purposes of blocking *trans*-activation by Tat. TAR decoys consisting of either multiple TARs synthesized from RNA polymerase II promoters or a single TAR synthesized from an RNA polymerase I promoter have been shown to decrease levels of *trans*-activation in transient cotransfection assays (104–107). The effect of these TAR decoys was dependent upon the sequence and structure of TAR (107). It is unknown whether these TAR decoys function at the level of mature transcripts or nascent TAR RNA in the process of transcription and whether these TAR decoys deplete TRP-1 and TRP-2 from other cellular genes. However, together with intracellular immunization, TAR decoys might become useful in the treatment of AIDS.

4. Mechanisms of *trans*-activation by Tat

Many different mechanisms of action have been proposed for Tat of HIV-1 (for review see refs 1–10). First, it was observed that Tat could increase levels of steady-state mRNA and reporter protein in transient expression assays (11–13, 108). However, soon thereafter, no changes in levels of mRNAs, despite increased levels of reporter proteins, were observed (18). Therefore, post-transcriptional, possibly translational mechanisms for Tat action were proposed. Since levels of reporter proteins were often higher than those of corresponding mRNAs, bimodal, i.e. transcriptional and translational, mechanisms for *trans*-activation by Tat were also suggested (109, 110). However, by use of more sensitive detection techniques and kinetic studies it has become apparent that Tat primarily affects transcription from the HIV-1 LTR (111, 112). Nuclear run-on transcription and RNAse protection revealed that Tat affects both initiation and elongation of HIV transcription (17).

4.1 Transcriptional effects of Tat

In a study in our laboratory, Tat of HIV-1 increased rates of elongation of transcription from the HIV-1 LTR since promoter proximal transcription did not vary in the presence or absence of Tat (16). However, in the absence of Tat, a steep gradient of transcriptional pausing or termination within and 3' to the HIV-1 LTR resulted in a paucity of full-length polyadenylated transcripts and an abundance of short, non-polyadenylated RNAs corresponding to the stem–loop in TAR (68). In the presence of Tat, this transcriptional polarity was abolished and full-length, polyadenylated transcripts predominated. Discrepancies between nuclear run-ons and amounts of short transcripts were interpreted as representing digestion by cellular RNAses, which yielded only stable TAR RNA structures that resisted further degradation. Furthermore, the short transcripts from 55 to 59 nucleotides in length that accumulate in the cytoplasm of transfected cells have also been observed in *in vitro* transcription (74), in chronically infected cell lines (113), and in cells from infected individuals (Adams and Peterlin, unpublished data). It has been suggested that TATA and LBP-1 binding sequences and their associated factors are responsible for the generation of these short transcripts (Fig. 7) (77).

Other investigators found a significant effect of Tat on the initiation of HIV transcription (17). For example, using several smaller probes corresponding to the transcribed DNA template, although promoter proximal transcription predominated in the absence of Tat, there was an increase of all transcription in the presence of Tat (17). Thus, in the basal state, contributions of Tat to the initiation and elongation of transcription were equivalent (17). However, in the presence of adenoviral E1A or with replicating vectors in COS cells, effects of Tat on the elongation of transcription predominated (114, 115). These discrepancies could be reconciled if Tat acts like the transcription factor TFIIF, which consists of two proteins of 74 kDa and 30 kDa that increase rates of initiation and elongation of RNA polymerase II transcription (116–118). Alternatively, effects of Tat on the initiation of transcription could be indirect. By clearing stalled transcription complexes 3' to the promoter, Tat might secondarily increase subsequent rounds of initiation of transcription (119).

A number of investigators demonstrated effects of Tat using *in vitro* transcription systems (76, 97). Either with preincubation with cold triphosphates or without preincubation but with the addition of citrate, Tat increased transcriptional processivity in a TAR-dependent fashion (76, 97). Although lesser effects were observed *in vitro* than *in vivo*, effects of Tat on the elongation of HIV transcription were reproduced faithfully. In these studies, *trans*-activation depended on TAR and was increased by cellular proteins that bind to TAR (97). Additionally, wild-type but not mutant Tat peptides could block this effect (97). Moreover, Tat was able to modify a paused RNA polymerase II at TAR, suggesting that interactions between Tat and nascent TAR can modify a preformed transcription complex (Fig. 8). In addition, Tat acted synergistically with TFIIS, a protein of 30 kDa that increases rates of elongation of RNA polymerase II transcription, but not with TFIIF, again suggesting that Tat works like TFIIF (76).

Two different transcription complexes, one non-processive and the other processive, are formed on the HIV-1 LTR (Fig. 7) (120). In one study, exogenously added ATP increased non-processive complexes that were responsive to Tat.

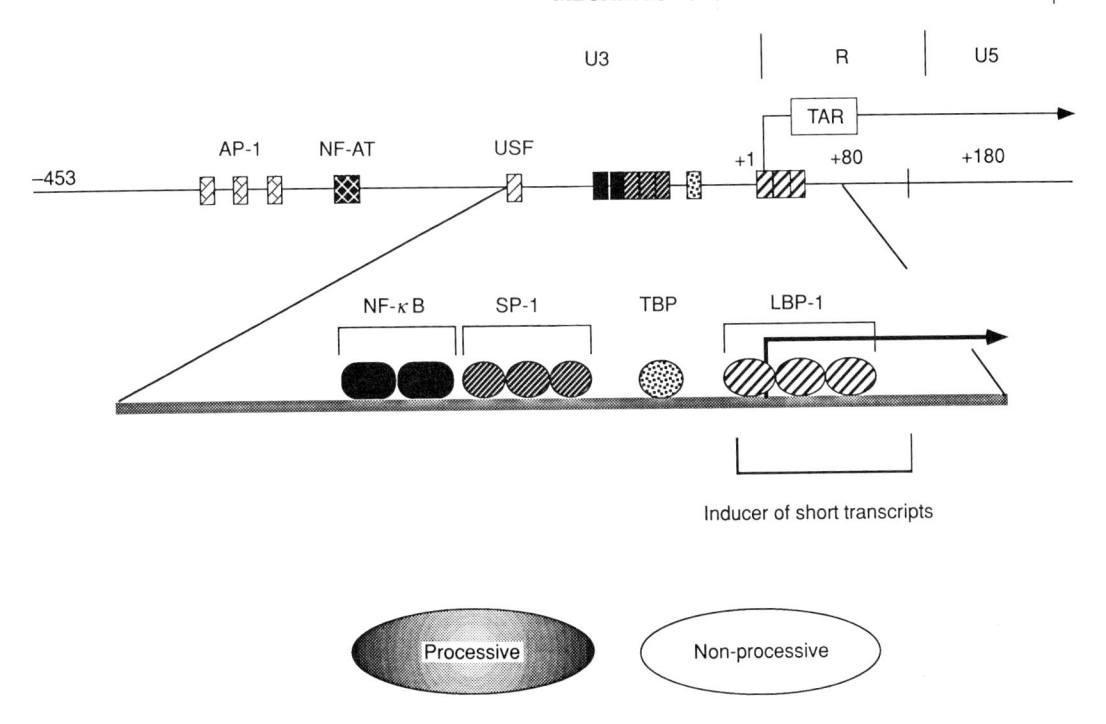

Fig. 7 HIV-1 LTR. The HIV-1 LTR, which consists of 713 nucleotides, is divided into U3, R, and U5 regions. From the 5' to the 3' direction are found functionally important binding sites for transcription factors NF-κB, SP-1, TBP (TATA-binding protein), and LBP-1 (leader binding protein-1) (for review see ref. 2, Chapter 3). Several other binding sites have also been described; however, few functional correlates exist for these proteins. Sequences from positions −5 to +80 have been implicated in the generation of short transcripts and called 'inducers of short transcripts' (77). *Trans*-activation by Tat but not effects of NF-κB absolutely require a functional TATA box (123, 162). Whereas LBP-1 is important for proper positioning of the CAP site, high levels of LBP-1 inhibit HIV-1 transcription (124, 163). Thus, whereas TATA and surrounding sequences can assemble non-processive transcription complexes that are *trans*-activated by Tat, NF-κB, and SP-1 can assemble processive transcription complexes in the absence of a functional TATA box

Conversely, the transcriptional inhibitor dichloro-1-β-D-ribofuranosylbenzimidazole (DRB), which is an analogue of ATP, did not affect processive complexes that were unresponsive to Tat, but blocked non-processive complexes and *trans*-activation by Tat (120). Since DRB inhibits cellular kinases that require ATP hydrolysis, this and previous studies suggested a role for protein kinases in *trans*-activation by Tat. Furthermore, TFIIS, TFIIF, and RNA polymerase II are all phosphoproteins that require extensive phosphorylation for function (117, 118, 121, 122). In the case of the heavy chain of RNA polymerase II, phosphorylation of the C-terminal domain is important for the transition from initiation to elongation of transcription (122). Thus, Tat could increase either the local concentration or the activity of a protein kinase.

Roles of DNA-binding proteins that interact with the HIV-1 LTR in Tat *trans*-

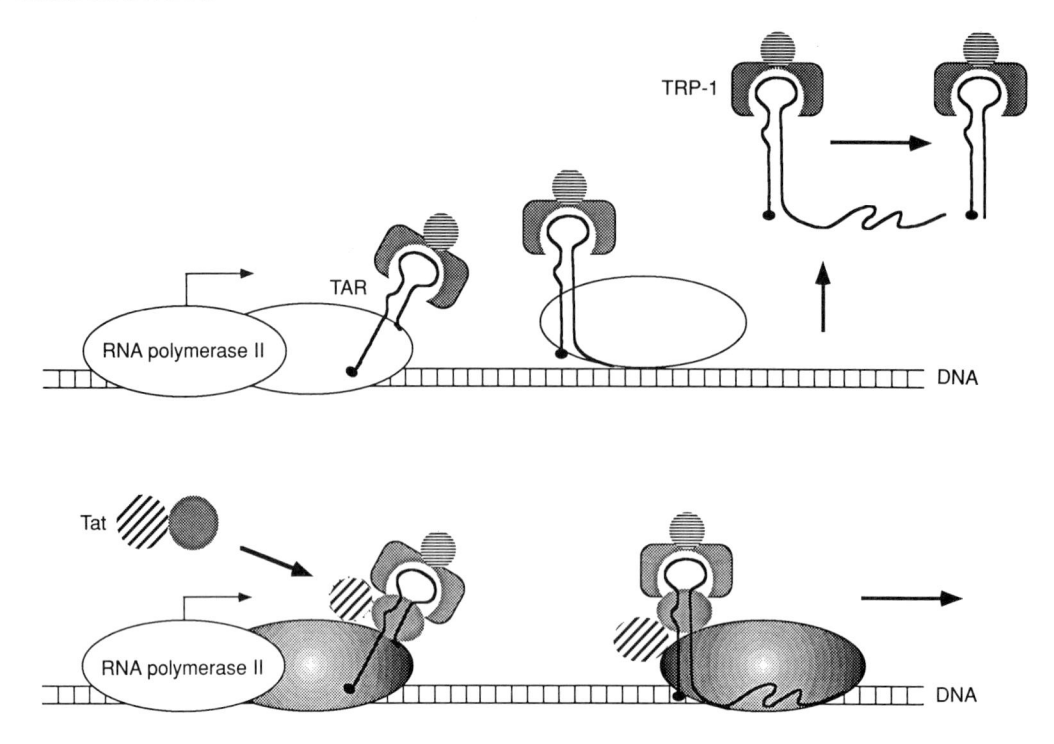

Fig. 8 A model of *trans*-activation by Tat. In the absence of Tat, non-processive transcription complexes slide along the HIV-1 LTR into viral coding sequences. Stalled, paused, and prematurely terminated transcripts dissociate from the DNA and release transcribed viral RNAs. Following RNAse digestion, stable TAR RNAs, probably complexed with cellular proteins, accumulate in the cytoplasm of transfected cells. However, in the presence of Tat, these transcription complexes are modified so that they elongate efficiently. This step also increases rates of initiation of transcription, either directly, as with TFIIF, or indirectly, by clearing the HIV-1 promoter for subsequent rounds of transcription. Whether Tats and TRPs remain associated with the transcription complexes or dissociate following *trans*-activation is unknown

activation have also been examined (Fig. 7). First, neither NF-κB nor Sp-1 are essential for *trans*-activation and could be replaced by several other transcriptional activators (50, 51, 123). However, TATA and its flanking sequences are of paramount importance since many TATA sequences, like that from the SV40 early promoter, failed to support *trans*-activation by Tat (123). In contrast, U2 snRNA promoter, which does not contain a TATA box, and a number of complete viral promoters, when fused to TAR, could be *trans*-activated by Tat (77). However, analogously modified U6 snRNA promoter, which is transcribed by RNA polymerase III, was not responsive to Tat (77). Finally, LBP-1 might play a negative role in the regulation of the HIV transcription since the addition of increased amounts of LBP-1 blocked basal transcription in an *in vitro* transcription system (124). Whether TBP, LBP-1, and their associated factors are primarily responsible for the formation of non-processive transcription complexes is unknown.

Further mechanistic details came from looking at Tat brought to the transcription complex via heterologous DNA- and RNA-binding proteins. For example, Tat linked to the DNA-binding domain of c-Jun (49), Gal-4 (50, 51), and Lex-A (Ghosh and Peterlin, unpublished data) or to Rev of HIV-1 (52) and the coat protein of bacteriophage R17 (53) was able to *trans*-activate the HIV-1 LTR. However, in all these studies RNA tethering was more efficient than DNA tethering. For example, four Gal-4 sites placed upstream of three Sp-1 sites had minimal effect (51) and six Gal-4 sites, i.e. 12 DNA-binding sites, were required to observe high levels of *trans*-activation (50), whereas a single downstream operator site could affect up to 30 and 100 per cent of wild-type activity in non-replicating and replicating systems, respectively (53). Although it was proposed that Tat acts via DNA analogously to the acidic activator VP-16 (50), VP-16 had no effect when presented to the transcription complex via RNA (40). However, it is possible that strong transcriptional activators and enhancers obviate the need for Tat since they initiate highly processive transcription complexes.

Experiments using RNA tethering also revealed that binding affinities between Tat and TAR are very important for Tat action (53). For example, when K_ds of heterologous RNA-binding mechanisms and Tat were similar, almost wild-type levels of *trans*-activation were observed with the coat protein of bacteriophage R17. However, when the binding coefficient was either reduced or increased so that Tat no longer dissociated from TAR, lower levels of *trans*-activation were observed. Thus Tat might have to be released from TAR for its activity (53).

4.2 Post-transcriptional effects of Tat

Since early experiments implicated Tat in the translational regulation of HIV (18, 109), numerous studies have investigated roles of Tat and TAR in translation of mRNAs which contain TAR. It is clear that the TAR stem–loop is inhibitory to translation both in rabbit reticulocyte lysates and in the *Xenopus* oocyte (78–81, 125–127). In eukaryotic cells, this could be due to the activation of double-stranded RNA-dependent kinase which is activated by double-stranded RNA and by the stem of TAR (79, 81, 128). However, no effect of Tat has been observed on this translational inhibition or on double-stranded RNA-dependent kinase (78, 79, 81, 128, 129). Furthermore, adding upstream leader sequences which do not base-pair with TAR reversed these deleterious effects (78). Thus, inefficient entry of polysomes on mRNAs which contain stable secondary structures at their 5' ends might explain the effect of TAR on translation of HIV transcripts *in vitro* (78, 79, 130). A further suggestion that translational effects of Tat might be minimal in cells that can support HIV replication was obtained from direct nuclear microinjections and cytoplasmic transfections of RNAs containing TAR (131). Neither was TAR inhibitory to translation of these RNAs nor could any effect of Tat on translation of these RNAs be documented in the nucleus or in the cytoplasm of these cells. However, Tat was able to transactivate the HIV-1 LTR in these cells. Thus, Tat increases only transcription from the HIV promoter in primate cells (131).

However, an intriguing model of translational control by Tat was described in the *Xenopus* oocyte (125, 126, 132). Here RNAs which contain TAR could not be translated either in the nucleus or in the cytoplasm of the *Xenopus* oocyte (125). However, after the addition of Tat, significant increase in levels of reporter proteins was observed (125, 126). Since replacing U3 sequences by the CMV promoter led to the efficient translation of TAR RNAs, the U3 region could direct these RNAs to a specific compartment where transcripts are inaccessible to translation (132). In this scenario, Tat would redirect these RNAs to a different compartment where TAR RNAs can be translated efficiently. Since effects of Tat in the *Xenopus* oocyte were blocked by DRB, this implied a direct linkage between transcription and translation of TAR RNA (132). Alternatively, Tat could facilitate enzymatic modification of TAR RNAs in the nucleus of the *Xenopus* oocyte. Indeed, the conversion of adenosine at position 27 to inosine that depends on Tat has been described (127). This inosine could destabilize base-pairing in the stem so that polysomes might load more efficiently, thereby increasing the translational efficiency of TAR RNA.

5. Tat and the HIV life cycle

Of transcribed and translated proteins, Tat is the first virally encoded protein that affects the HIV life cycle (for review see refs 1–10). In infected cells, the integrated provirus can remain transcriptionally silent or at least no full-length viral RNAs are detected (133–135). However, short transcripts might be present in most cells because promoter proximal transcription has been observed in all experimental systems. These short TAR RNAs might play an important role in viral latency since they could function like TAR decoys and remove Tat from the HIV-1 LTR (104–106). Even following cellular activation and the synthesis of more Tat, there might be sufficient TAR RNA to effect a prolonged state of viral latency. Finally, a threshold level of Tat is reached that rapidly increases rates of viral transcription. Only optimal interactions between Tat and TAR will result in high levels of viral RNA synthesis. Then, large amounts of Rev are synthesized which transport genomic RNAs into the cytoplasm where they are packaged into infectious virions (136, see Chapter 5). In this scenario, Tat could be the most important player in determining viral latency and productive replication, thus determining subsequent fates of HIV. Even in cellular models of viral latency, for example, in ACH-2 and U-1 cells that contain stably integrated HIV proviruses but do not support full levels of replication and cytopathology, Tat could be a more important determinant of viral replication than Rev (136–138). This and other subjects need to be more fully investigated.

6. Other effects of Tat

Tat has also been shown to have several other effects. For example, Tat can *trans*-activate promoters other than the HIV-1 LTR (139, 140). In microglial cells, Tat *trans*-activated the neurotropic papovavirus JCV late promoter alone or in synergy with the SV40 T-antigen, implying different mechanisms of action (141–143). Several observations have also implicated Tat in the activation of lymphokine promoters

(TNF-α, IL-6, GM-CSF) in lymphocytes, which might lead to immunoregulatory defects observed in HIV-infected individuals (144). The ability of Tat to block antigen-specific T-cell responses could be viewed in this light (145). In the absence of TAR, *cis*-acting targets of Tat in these promoters are unknown. However, if Tat interacts with a basal transcription factor like TFIIF, high levels of expression of Tat in certain cells might affect selected promoters in the absence of its TAR RNA target. Furthermore, effects of Tat in glial cells and upon neuronal tissues might play a role in the development of AIDS dementia and related central nervous system conditions.

Tat might also be responsible for skin lesions observed in HIV-infected individuals. First, transgenic mice containing Tat under the control of the HIV-1 LTR developed skin lesions resembling Kaposi's sarcoma (146). Second, Tat supported the growth of Kaposi's cells in tissue culture analogously to supernatants from HIV-infected T-cells (147–149). Third, Tat can be taken up by cells, and less convincingly, Tat can also be released from infected or transfected cells (150–152). Thus, cells that are not themselves infected by HIV could be modified by Tat. Fourth, UV light activated the HIV-1 LTR through sequences in TAR (153, 154). Thus, it is possible that an interplay between Tat and UV light might regulate the HIV-1 LTR and other promoters that are responsible for abnormal growth of dermal and vascular cells, resulting in AIDS-related skin lesions. However, although anti-Tat antibodies can block effects of Tat on Kaposi's cells in culture, epidemiological studies and low levels of circulating Tat in the serum of infected individuals do not support a direct role for Tat in the etiology of Kaposi's sarcoma.

7. Conclusion

Tat is a fascinating viral *trans*-activator whose discovery may have opened a new area in the study of eukaryotic transcription, namely that of processivity of RNA polymerase II. Both its targeting to the transcription complex via an RNA structure in the process of nascent transcription and its ability to modify stalled transcription complexes are new. However, some of these phenomena are reminiscent of bacteriophage λ, where the bacteriophage N protein interacts with cellular proteins, core RNA polymerase, and an RNA structure (nutB) to anti-terminate transcription of late bacteriophage λ genes (Fig. 9) (155, 156). There are also clear differences since Tat must be positioned near the site of initiation of HIV transcription and cannot exert its effects at a distance after TAR sequences have been transcribed (157). However, it is not surprising that this powerful regulatory mechanism would be observed in both prokaryotic and eukaryotic systems since RNA targeting offers specificity, parsimony, and rapidity to transcription, i.e. the target RNA is unique to one promoter and transcription complexes to be modified have already been assembled.

Additionally, Tat might elucidate analogous mechanisms that are required for the copying of several cellular genes, like proto-oncogenes c-*myc*, c-*myb*, c-*fos* (for

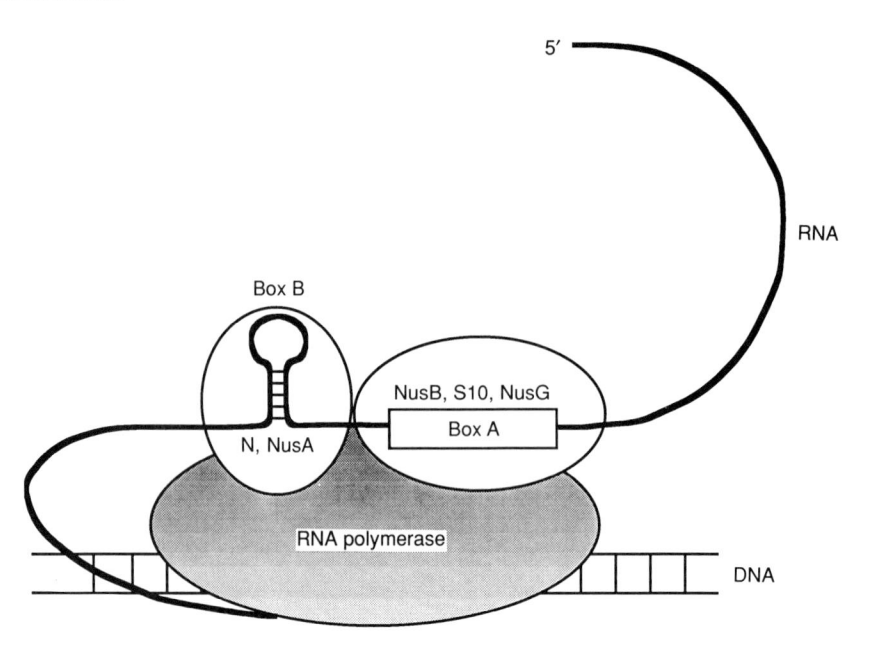

Fig. 9 Antitermination of bacteriophage λ transcription by the bacteriophage λ N protein. Being similar to Tat in that it contains a basic domain that binds to nutB RNA stem–loop, N is a small protein that is required for efficient elongation of bacteriophage λ transcription (155, 156). Furthermore, N binds to the nutB site productively only with the help of the bacterial NusA protein and core RNA polymerase (156). Optimal interactions between N and the transcription complex additionally require NusB, S10, and NusG, which interact with the nutA site (156). Appropriate and sequential interactions between these host and bacteriophage proteins modify the transcription complex so that distal termination sites are ignored and genes are expressed that profoundly affect the life cycle of bacteriophage λ

review see ref. 158) and those that span hundreds of thousands to millions of bases, like dystrophin (159). However, if Tat is unique and does not have a cellular counterpart, then strategies aimed at blocking *trans*-activation by Tat might become useful approaches in the treatment of AIDS. Presently, TAR decoys (104, 106, 107), *trans*-dominant Tats (47, 65, 66) and a drug called Ro 5-3335 (160) can interfere with *trans*-activation by Tat and might have future clinical and therapeutic importance.

References

1. Cullen, B. R. and Greene, W. C. (1989) Regulatory pathways governing HIV-1 replication. *Cell*, **58**, 423.
2. Jones, K. A. (1989) HIV trans-activation and transcription control mechanisms. *New Biol.*, **1**, 127.
3. Sharp, P. A. and Marciniak, R. A. (1989) HIV TAR: an RNA enhancer? *Cell*, **59**, 229.
4. Cullen, B. R. (1990) The HIV-1 Tat protein: an RNA sequence-specific processivity factor? *Cell*, **63**, 655.
5. Pavlakis, G. N. and Felber, B. K. (1990) Regulation of expression of human immunodeficiency virus. *New Biol.*, **2**, 20.

6. Rosen, C. A. and Pavlakis, G. N. (1990) Tat and Rev: positive regulators of HIV gene expression. *AIDS*, **4**, 499.
7. Cullen, B. R. (1991) Regulation of HIV-1 gene expression. *FASEB J.*, **5**, 2361.
8. Karn, J. (1991) Control of human immunodeficiency virus replication by the *tat, rev, nef* and *protease* genes. *Curr. Opin. Immunol.*, **3**, 526.
9. Rosen, C. A. (1991) Regulation of HIV gene expression by RNA–protein interactions. *Trends Genet.*, **7**, 9.
10. Vaishnav, Y. N. and Wong-Staal, F. (1991) The biochemistry of AIDS. *Annu. Rev. Biochem.*, **60**, 577.
11. Rosen, C. A., Sodroski, J. G., and Haseltine, W. A. (1985) The location of *cis*-acting regulatory sequences in the human T cell lymphotropic virus type III (HTLV-III/LAV) long terminal repeat. *Cell*, **41**, 813.
12. Arya, S. K., Guo, C., Josephs, S. F., and Wong-Staal, F. (1985) *Trans*-activator gene of human T-lymphotropic virus type III (HTLV-III). *Science*, **229**, 69.
13. Sodroski, J., Patarca, R., Rosen, C., Wong-Staal, F., and Haseltine, W. (1985) Location of the *trans*-activating region on the genome of human T-cell lymphotropic virus type III. *Science*, **229**, 74.
14. Emerman, M., Guyader, M., Montagnier, L., Baltimore, D., and Muesing, M. A. (1987) The specificity of the human immunodeficiency virus type 2 transactivator is different from that of human immunodeficiency virus type 1. *EMBO J.*, **6**, 3755.
15. Viglianti, G. A. and Mullins, J. I. (1988) Functional comparison of *trans*-activation by simian immunodeficiency virus from rhesus macaques and human immunodeficiency virus type 1. *J. Virol.*, **62**, 4523.
16. Kao, S. Y., Calman, A. F., Luciw, P. A., and Peterlin, B. M. (1987) Anti-termination of transcription within the long terminal repeat of HIV-1 by *tat* gene product. *Nature*, **330**, 489.
17. Laspia, M. F., Rice, A. P., and Mathews, M. B. (1989) HIV-1 Tat protein increases transcriptional initiation and stabilizes elongation. *Cell*, **59**, 283.
18. Rosen, C. A., Sodroski, J. G., Goh, W. C., Dayton, A. I., Lippke, J., and Haseltine, W. A. (1986) Post-transcriptional regulation accounts for the *trans*-activation of the human T-lymphotropic virus type III. *Nature*, **319**, 555.
19. Dayton, A. I., Sodroski, J. G., Rosen, C. A., Goh, W. C., and Haseltine, W. A. (1986) The *trans*-activator gene of the human T cell lymphotropic virus type III is required for replication. *Cell*, **44**, 941.
20. Fisher, A. G., Feinberg, M. B., Josephs, S. F., Harper, M. E., Marselle, L. M., Reyes, G., Gonda, M. A., Aldovini, A., Debouk, C., Gallo, R. C., and Wong-Staal, F. (1986) The *trans*-activator gene of HTLV-III is essential for virus replication. *Nature*, **320**, 367.
21. Myers, G., Berzofsky, J. A., Rabson, A. B., Smith, T. F., and Wong-Staal, F. (1991) *Human Retroviruses and AIDS*. Theoretical Biology and Biophysics, Los Alamos, NM.
22. Schwartz, S., Felber, B. K., Benko, D. M., Fenyo, E. M., and Pavlakis, G. N. (1990) Cloning and functional analysis of multiply spliced mRNA species of human immunodeficiency virus type 1. *J. Virol.*, **64**, 2519.
23. Benko, D. M., Schwartz, S., Pavlakis, G. N., and Felber, B. K. (1990) A novel human immunodeficiency virus type 1 protein, Tev, shares sequences with Tat, Env, and Rev proteins. *J. Virol.*, **64**, 2505.
24. Salfeld, J., Gottlinger, H. G., Sia, R. A., Park, R. E., Sodroski, J. G., and Haseltine, W. A. (1990) A tripartite HIV-1 Tat–Env–Rev fusion protein. *EMBO J.*, **9**, 965.
25. Hauber, J. and Cullen, B. R. (1988) Mutational analysis of the *trans*-activation-

responsive region of the human immunodeficiency virus type I long terminal repeat. *J. Virol.*, **62,** 673.
26. Ruben, S., Perkins, A., Purcell, R., Joung, K., Sia, R., Burghoff, R., Haseltine, W. A., and Rosen, C. A. (1989) Structural and functional characterization of human immunodeficiency virus Tat protein. *J. Virol.*, **63,** 1.
27. Siomi, H., Shida, H., Maki, M., and Hatanaka, M. (1990) Effects of a highly basic region of human immunodeficiency virus Tat protein on nucleolar localization. *J. Virol.*, **64,** 1803.
28. Feng, S. and Holland, E. C. (1988) HIV-1 Tat *trans*-activation requires the loop sequence within TAR. *Nature*, **334,** 165.
29. Dorn, P. L. and Derse, D. (1988) *Cis*- and *trans*-acting regulation of gene expression of equine infectious anemia virus. *J. Virol.*, **62,** 3522.
30. Fenrick, R., Malim, M. H., Hauber, J., Le, S. Y., Maizel, J., and Cullen, B. R. (1989) Functional analysis of the Tat *trans*-activator of human immunodeficiency virus type 2. *J. Virol.*, **63,** 5006.
31. Berkhout, B., Gatignol, A., Silver, J., and Jeang, K. T. (1990) Efficient *trans*-activation by the HIV-2 Tat protein requires a duplicated TAR RNA structure. *Nucleic Acids Res.*, **18,** 1839.
32. Carroll, R., Martarano, L., and Derse, D. (1991) Identification of lentivirus tat functional domains through generation of equine infectious anemia virus/human immunodeficiency virus type 1 *tat* gene chimeras. *J. Virol.*, **65,** 3460.
33. Derse, D., Carvalho, M., Carroll, R., and Peterlin, B. M. (1991) A minimal lentivirus Tat. *J. Virol.*, **65,** 7012.
34. Garcia, J. A., Harrich, D., Pearson, L., Mitsuyasu, R., and Gaynor, R. B. (1988) Functional domains required for Tat-induced transcriptional activation of the HIV-1 long terminal repeat. *EMBO J.*, **7,** 3143.
35. Sadaie, M. R., Rappaport, J., Benter, T., Josephs, S. F., Willis, R., and Wong-Staal, F. (1988) Missense mutations in an infectious human immunodeficiency viral genome: functional mapping of *tat* and identification of the *rev* splice acceptor. *Proc. Natl Acad. Sci. USA*, **85,** 9224.
36. Green, M., Ishino, M., and Loewenstein, P. M. (1989) Mutational analysis of HIV-1 Tat minimal domain peptides: identification of *trans*-dominant mutants that suppress HIV-LTR-driven gene expression. *Cell*, **58,** 215.
37. Kuppuswamy, M., Subramanian, T., Srinivasan, A., and Chinnadurai, G. (1989) Multiple functional domains of Tat, the *trans*-activator of HIV-1, defined by mutational analysis. *Nucleic Acids Res.*, **17,** 3551.
38. Rice, A. P. and Carlotti, F. (1990) Structural analysis of wild-type and mutant human immunodeficiency virus type 1 Tat proteins. *J. Virol.*, **64,** 6018.
39. Rappaport, J., Lee, S. J., Khalili, K., and Wong-Staal, F. (1989) The acidic amino-terminal region of the HIV-1 Tat protein constitutes an essential activating domain. *New Biol.*, **1,** 101.
40. Tiley, L. S., Brown, P. H., and Cullen, B. R. (1990) Does the human immunodeficiency virus Tat *trans*-activator contain a discrete activation domain? *Virology*, **178,** 560.
41. Hauber, J., Malim, M. H., and Cullen, B. R. (1989) Mutational analysis of the conserved basic domain of human immunodeficiency virus tat protein. *J. Virol.*, **63,** 1181.
42. Calnan, B. J., Biancalana, S., Hudson, D., and Frankel, A. D. (1991) Analysis of arginine-rich peptides from the HIV Tat protein reveals unusual features of RNA–protein recognition. *Genes Dev.*, **5,** 201.

43. Subramanian, T., Govindarajan, R., and Chinnadurai, G. (1991) Heterologous basic domain substitutions in the HIV-1 Tat protein reveal an arginine-rich motif required for *trans*-activation. *EMBO J.*, **10**, 2311.
44. Elangovan, B., Subramanian, R., and Chinnadurai, G. (1992) Functional comparison of the basic domains of the Tat proteins of human immunodeficiency virus types 1 and 2 in *trans*-activation. *J. Virol.*, **66**, 2031.
45. Muesing, M. A., Smith, D. H., and Capon, D. J. (1987) Regulation of mRNA accumulation by a human immunodeficiency virus *trans*-activator protein. *Cell*, **48**, 691.
46. Frankel, A. D., Biancalana, S., and Hudson, D. (1989) Activity of synthetic peptides from the Tat protein of human immunodeficiency virus type 1. *Proc. Natl Acad. Sci. USA*, **86**, 7397.
47. Carroll, R., Peterlin, B. M., and Derse, D. (1992) Inhibition of human immunodeficiency virus type 1 activity by co-expression of heterologous *trans*-activators. *J. Virol.*, **66**, 2000.
48. Jakobovits, A., Rosenthal, A., and Capon, D. J. (1990) *Trans*-activation of HIV-1 LTR-directed gene expression by Tat requires protein kinase C. *EMBO J.*, **9**, 1165.
49. Berkhout, B., Gatignol, A., Rabson, A. B., and Jeang, K. T. (1990) TAR-independent activation of the HIV-1 LTR: evidence that Tat requires specific regions of the promoter. *Cell*, **62**, 757.
50. Southgate, C. D. and Green, M. R. (1991) The HIV-1 Tat protein activates transcription from an upstream DNA-binding site: implications for Tat function. *Genes Dev.*, **5**, 2496.
51. Kamine, J., Subramanian, T., and Chinnadurai, G. (1991) SP-1-dependent activation of a synthetic promoter by human immunodeficiency virus type 1 Tat protein. *Proc. Natl Acad. Sci. USA*, **88**, 8510.
52. Southgate, C., Zapp, M. L., and Green, M. R. (1990) Activation of transcription by HIV-1 Tat protein tethered to nascent RNA through another protein. *Nature*, **345**, 640.
53. Selby, M. J. and Peterlin, B. M. (1990) *Trans*-activation by HIV-1 Tat via a heterologous RNA binding protein. *Cell*, **62**, 769.
54. Dingwall, C., Ernberg, I., Gait, M. J., Green, S. M., Heaphy, S., Karn, J., Lowe, A. D., Singh, M., and Skinner, M. A. (1990) HIV-1 Tat protein stimulates transcription by binding to a U-rich bulge in the stem of the TAR RNA structure. *EMBO J.*, **9**, 4145.
55. Roy, S., Delling, U., Chen, C. H., Rosen, C. A., and Sonenberg, N. (1990) A bulge structure in HIV-1 TAR RNA is required for Tat binding and Tat-mediated *trans*-activation. *Genes Dev.*, **4**, 1365.
56. Weeks, K. M., Ampe, C., Schultz, S. C., Steitz, T. A., and Crothers, D. M. (1990) Fragments of the HIV-1 Tat protein specifically bind TAR RNA. *Science*, **249**, 1281.
57. Calnan, B. J., Tidor, B., Biancalana, S., Hudson, D., and Frankel, A. D. (1991) Arginine-mediated RNA recognition: the arginine fork. *Science*, **252**, 1167.
58. Rice, A. (1991) Analysis of the structure of the HIV-1 Tat protein. In *Genetic Structure and Regulation of HIV*. Haseltine, W. A. and Wong-Staal, F. (eds). Raven Press, New York, NY, p. 143.
59. Frankel, A. D., Bredt, D. S., and Pabo, C. O. (1988) Tat protein from human immunodeficiency virus forms a metal-linked dimer. *Science*, **240**, 70.
60. Loret, E. P., Vives, E., Ho, P. S., Rochat, H., Van Rietschoten, J., and Johnson, W., Jr. (1991) Activating region of HIV-1 Tat protein: vacuum UV circular dichroism and energy minimization. *Biochemistry*, **30**, 6013.
61. Frankel, A. D., Chen, L., Cotter, R. J., and Pabo, C. O. (1988) Dimerization of the Tat protein from human immunodeficiency virus: a cysteine-rich peptide mimics the normal metal-linked dimer interface. *Proc. Natl Acad. Sci. USA*, **85**, 6297.

62. Rice, A. P. and Chan, F. (1991) Tat protein of human immunodeficiency virus type 1 is a monomer when expressed in mammalian cells. *Virology*, **185**, 451.
63. Nelbock, P., Dillon, P. J., Perkins, A., and Rosen, C. A. (1990) A cDNA for a protein that interacts with the human immunodeficiency virus Tat *trans*-activator. *Science*, **248**, 1650.
64. Desai, K., Loewenstein, P. M., and Green, M. (1991) Isolation of a cellular protein that binds to the human immunodeficiency virus Tat protein and can potentiate *trans*-activation of the viral promoter. *Proc. Natl Acad. Sci. USA*, **88**, 8875.
65. Pearson, L., Garcia, J., Wu, F., Modesti, N., Nelson, J., and Gaynor, R. (1990) A *trans*-dominant Tat mutant that inhibits Tat-induced gene expression from the human immunodeficiency virus long terminal repeat. *Proc. Natl Acad. Sci. USA*, **87**, 5079.
66. Modesti, N., Garcia, J., Debouck, C., Peterlin, M., and Gaynor, R. (1991) Trans-dominant Tat mutants with alterations in the basic domain inhibit HIV-1 gene expression. *New Biol.*, **3**, 759.
67. Carvalho, M. and Derse, D. (1991) Mutational analysis of the equine infectious anemia virus Tat-responsive element. *J. Virol.*, **65**, 3468.
68. Selby, M. J., Bain, E. S., Luciw, P. A., and Peterlin, B. M. (1989) Structure, sequence, and position of the stem–loop in TAR determine transcriptional elongation by tat through the HIV-1 long terminal repeat. *Genes Dev.*, **3**, 547.
69. Roy, S., Parkin, N. T., Rosen, C., Itovitch, J., and Sonenberg, N. (1990) Structural requirements for *trans*-activation of human immunodeficiency virus type 1 long terminal repeat-directed gene expression by Tat: importance of base pairing, loop sequence, and bulges in the Tat-responsive sequence. *J. Virol.*, **64**, 1402.
70. Berkhout, B. and Jeang, K. T. (1991) Detailed mutational analysis of TAR RNA: critical spacing between the bulge and loop recognition domains. *Nucleic Acids Res.*, **19**, 6169.
71. Garcia, J. A., Harrich, D., Soultanakis, E., Wu, F., Mitsuyasu, R., and Gaynor, R. B. (1989) Human immunodeficiency virus type 1 LTR TATA and TAR region sequences required for transcriptional regulation. *EMBO J.*, **8**, 765.
72. Sheline, C., Milocco, L., and Jones, K. A. (1991) Two distinct nuclear transcription factors recognize loop and bulge residues in the HIV-1 TAR RNA hairpin. *Genes Dev.*, **5**, 2508.
73. Jakobovits, A., Smith, D. H., Jakobovits, E. B., and Capon, D. J. (1988) A discrete element 3' of human immunodeficiency virus 1 (HIV-1) and HIV-2 mRNA initiation sites mediates transcriptional activation by an HIV *trans*-activator. *Mol. Cell. Biol.*, **8**, 2555.
74. Toohey, M. G. and Jones, K. A. (1989) *In vitro* formation of short RNA polymerase II transcripts that terminate within the HIV-1 and HIV-2 promoter-proximal downstream regions. *Genes Dev.*, **3**, 265.
75. Bengal, E. and Aloni, Y. (1991) Transcriptional elongation by purified RNA polymerase II is blocked at the *trans*-activation-responsive region of human immunodeficiency virus type 1 *in vitro*. *J. Virol.*, **65**, 4910.
76. Kato, H., Sumimoto, H., Pognonec, P., Chen, C., Rosen, C., and Roeder, B. G. (1992) HIV-1 Tat acts as a processivity factor *in vitro* in conjunction with cellular elongation factors. *Genes Dev.*, **6**, 655.
77. Ratnasabapathy, R., Sheldon, M., Johal, L., and Hernandez, N. (1990) The HIV-1 long terminal repeat contains an unusual element that induces the synthesis of short RNAs from various mRNA and snRNA promoters. *Genes Dev.*, **4**, 2061.
78. Parkin, N. T., Cohen, E. A., Darveau, A., Rosen, C., Haseltine, W., and Sonenberg,

N. (1988) Mutational analysis of the 5' non-coding region of human immunodeficiency virus type 1: effects of secondary structure on translation. *EMBO J.*, **7**, 2831.
79. Edery, I., Petryshyn, R., and Sonenberg, N. (1989) Activation of double-stranded RNA-dependent kinase (dsI) by the TAR region of HIV-1 mRNA: a novel translational control mechanism. *Cell,* **56**, 303.
80. SenGupta, D. N., Berkhout, B., Gatignol, A., Zhou, A. M., and Silverman, R. H. (1990) Direct evidence for translational regulation by leader RNA and Tat protein of human immunodeficiency virus type 1. *Proc. Natl Acad. Sci. USA,* **87**, 7492.
81. Silverman, R. H. and SenGupta, D. N. (1990) Translational regulation by HIV leader RNA, TAT, and interferon-inducible enzymes. *J. Exp. Pathol.,* **5**, 69.
82. Dingwall, C., Ernberg, I., Gait, M. J., Green, S. M., Heaphy, S., Karn, J., Lowe, A. D., Singh, M., Skinner, M. A., and Valerio, R. (1989) Human immunodeficiency virus 1 Tat protein binds *trans*-activation-responsive region (TAR) RNA *in vitro*. *Proc. Natl Acad. Sci. USA,* **86**, 6925.
83. Karn, J., Dingwall, C., Finch, J. T., Heaphy, S., and Gait, M. J. (1991) RNA binding by the Tat and Rev proteins of HIV-1. *Biochimie,* **73**, 9.
84. Weeks, K. M. and Crothers, D. M. (1991) RNA recognition by Tat-derived peptides: interaction in the major groove. *Cell,* **66**, 577.
85. Cordingley, M. G., LaFemina, R. L., Callahan, P. L., Condra, J. H., Sardana, V. V., Graham, D. J., Nguyen, T. M., LeGrow, K., Gotlib, L., Schlabach, A. J., and Colonno, R. J. (1990) Sequence-specific interaction of Tat protein and Tat peptides with the *trans*-activation-responsive sequence element of human immunodeficiency virus type 1 *in vitro*. *Proc. Natl Acad. Sci. USA,* **87**, 8985.
86. Delling, U., Roy, S., Sumner-Smith, M., Barnett, R., Reid, L., Rosen, C. A., and Sonenberg, N. (1991) The number of positively charged amino acids in the basic domain of Tat is critical for *trans*-activation and complex formation with TAR RNA. *Proc. Natl Acad. Sci. USA,* **88**, 6234.
87. Kamine, J., Loewenstein, P., and Green, M. (1991) Mapping of HIV-1 Tat protein sequences required for binding to TAR RNA. *Virology,* **182**, 570.
88. Sumner-Smith, M., Roy, S., Barnett, R., Reid, L. S., Kuperman, R., Delling, U., and Sonenberg, N. (1991) Critical chemical features in *trans*-acting-responsive RNA are required for interaction with human immunodeficiency virus type 1 Tat protein. *J. Virol.,* **65**, 5196.
89. Delling, U., Reid, L. S., Barnett, R. W., Ma, M. Y.-X., Climie, S., Sumner-Smith, M., and Sonenberg, N. (1992) Conserved nucleotides in the TAR RNA stem of human immunodeficiency virus type 1 are critical for Tat binding and *trans*-activation: model for TAR RNA tertiary structure. *J. Virol.,* **66**, 3018.
90. Harper, J. W. and Logsdon, N. J. (1991) Refolded HIV-1 Tat protein protects both bulge and loop nucleotides in TAR RNA from ribonucleolytic cleavage. *Biochemistry,* **30**, 8060.
91. Gaynor, R., Soultanakis, E., Kuwabara, M., Garcia, J., and Sigman, D. S. (1989) Specific binding of a HeLa cell nuclear protein to RNA sequences in the human immunodeficiency virus transactivating region. *Proc. Natl Acad. Sci. USA,* **86**, 4858.
92. Gatignol, A., Kumar, A., Rabson, A., and Jeang, K. T. (1989) Identification of cellular proteins that bind to the human immunodeficiency virus type 1 trans-activation-responsive TAR element RNA. *Proc. Natl Acad. Sci. USA,* **86**, 7828.
93. Marciniak, R. A., Garcia-Blanco, M. A., and Sharp, P. A. (1990) Identification and characterization of a HeLa nuclear protein that specifically binds to the *trans*-

activation-response (TAR) element of human immunodeficiency virus. *Proc. Natl Acad. Sci. USA,* **87,** 3624.
94. Gatignol, A., Buckler, W. A., Berkhout, B., and Jeang, K. T. (1991) Characterization of a human TAR RNA-binding protein that activates the HIV-1 LTR. *Science,* **251,** 1597.
95. Wu, F., Garcia, J., Sigman, D., and Gaynor, R. (1991) Tat regulates binding of the human immunodeficiency virus *trans*-activating region RNA loop-binding protein TRP-185. *Genes Dev.,* **5,** 2128.
96. Rounseville, M. P. and Kumar, A. (1992) Binding of a host cell nuclear protein to the stem region of human immunodeficiency virus type 1 *trans*-activation-responsive RNA. *J. Virol.,* **66,** 1688.
97. Marciniak, R. A., Calnan, B. J., Frankel, A. D., and Sharp, P. A. (1990) HIV-1 Tat protein *trans*-activates transcription *in vitro. Cell,* **63,** 791.
98. Pfeifer, K., Bachmann, M., Schroder, H. C., Weiler, B. E., Ugarkovic, D., Okamoto, T., and Muller, W. E. (1991) Formation of a small ribonucleoprotein particle between Tat protein and *trans*-acting response element in human immunodeficiency virus-infected cells. *J. Biol. Chem.,* **266,** 14620.
99. Barry, P. A., Pratt-Lowe, E., Unger, R. E., and Luciw, P. A. (1991) Cellular factors regulate transactivation of human immunodeficiency virus type 1. *J. Virol.,* **65,** 1392.
100. Hart, C. E., Ou, C. Y., Galphin, J. C., Moore, J., Bacheler, L. T., Wasmuth, J. J., Petteway, S. J., and Schochetman, G. (1989) Human chromosome 12 is required for elevated HIV-1 expression in human-hamster hybrid cells. *Science,* **246,** 488.
101. Newstein, M., Stanbridge, E. J., Casey, G., and Shank, P. R. (1990) Human chromosome 12 encodes a species-specific factor which increases human immunodeficiency virus type 1 Tat-mediated *trans*-activation in rodent cells. *J. Virol.,* **64,** 4565.
102. Hart, C. E., Westhafer, M. A., Galphin, J. C., Ou, C. Y., Bacheler, L. T., Petteway, S. J., Wasmuth, J. J., Chen, I. S., and Schochetman, G. (1991) Human chromosome-dependent and -independent pathways for HIV-2 *trans*-activation. *AIDS Res. Hum. Retroviruses,* **7,** 877.
103. Alonso, A., Derse, D., and Peterlin, B. M. (1992) Human chromosome 12 is required for optimal interactions between Tat and TAR of human immunodeficiency virus type 1 in rodent cells. *J. Virol.,* **66,** 4617.
104. Graham, G. J. and Maio, J. J. (1990) RNA transcripts of the human immunodeficiency virus *trans*-activation response element can inhibit action of the viral *trans*-activator. *Proc. Natl Acad. Sci. USA,* **87,** 5817.
105. Sullenger, B., Gallardo, H., Ungers, G. and Gilboa, E. (1990) Overexpression of TAR sequences renders cells resistant to human immunodeficiency virus replication. *Cell,* **63,** 601.
106. Lisziewicz, J., Rappaport, J., and Dhar, R. (1991) Tat-regulated production of multimerized TAR RNA inhibits HIV-1 gene expression. *New Biol.,* **3,** 82.
107. Sullenger, B. A., Gallardo, H. F., Ungers, G. E., and Gilboa, E. (1991) Analysis of *trans*-acting response decoy RNA-mediated inhibition of human immunodeficiency virus type 1 transactivation. *J. Virol.,* **65,** 6811.
108. Sodroski, J. G., Rosen, C. A., Wong-Staal, F., Salahuddin, S. Z., Popovic, M., Arya, S., and Gallo, R. C. (1985) *Trans*-acting transcriptional regulation of human T-cell leukemia virus type III long terminal repeat. *Science,* **227,** 171.
109. Cullen, B. R. (1986) *Trans*-activation of human immunodeficiency virus occurs via a bimodal mechanism. *Cell,* **46,** 973.

110. Wright, C. M., Felber, B. K., Paskalis, H., and Pavlakis, G. N. (1986) Expression and characterization of the *trans*-activator of HTLV-III/LAV virus. *Science*, **234**, 988.
111. Peterlin, B. M., Luciw, P. A., Barr, P. J., and Walker, M. D. (1986) Elevated levels of mRNA can account for the *trans*-activation of human immunodeficiency virus (HIV). *Proc. Natl Acad. Sci. USA*, **183**, 9734.
112. Rice, A. P. and Mathews, M. B. (1988) Transcriptional but not translational regulation of HIV-1 by the *tat* gene product. *Nature*, **332**, 551.
113. Feinberg, M. B., Baltimore, D., and Frankel, A. D. (1991) The role of Tat in the human immunodeficiency virus life cycle indicates a primary effect on transcriptional elongation. *Proc. Natl Acad. Sci. USA*, **88**, 4045.
114. Laspia, M. F., Rice, A. P., and Mathews, M. B. (1990) Synergy between HIV-1 Tat and adenovirus E1A is principally due to stabilization of transcriptional elongation. *Genes Dev.*, **4**, 2397.
115. Kessler, M. and Mathews, M. B. (1991) Tat *trans*-activation of the human immunodeficiency virus type 1 promoter is influenced by basal promoter activity and the simian virus 40 origin of DNA replication. *Proc. Natl Acad. Sci. USA*, **88**, 10018.
116. Sopta, M., Burton, Z. F., and Greenblatt, J. (1989) Structure and associated DNA-helicase activity of a general transcription initiation factor that binds to RNA polymerase II. *Nature*, **341**, 410.
117. Aso, T., Vasavada, H., Kawaguchi, T., Germino, F., Ganguly, S., Kitajima, S., Weissman, S., and Yasukochi, Y. (1992) Characterization of cDNA for the large subunit of the transcription initiation factor TFIIF. *Nature*, **355**, 461.
118. Finkelstein, A., Kostrub, C., Li, J., Chavez, D., Wang, B., Fang, S., Greenblatt, J., and Burton, Z. (1992) A cDNA encoding RAP74, a general initiation factor for transcription by RNA polymerase II. *Nature*, **355**, 464.
119. Peterlin, B. M. (1991) Transcriptional regulation of HIV. In *Genetic Structure and Regulation of HIV*. Haseltine, W. A. and Wong-Staal, F. (eds). Raven Press, New York, NY, p. 237.
120. Marciniak, R. A. and Sharp, P. A. (1991) HIV-1 Tat protein promotes formation of more-processive elongation complexes. *EMBO J.*, **10**, 4189.
121. Hirashima, S., Hirai, H., Nakanishi, Y., and Natori, S. (1988) Molecular cloning and characterization of cDNA for eukaryotic transcription factor S-II. *J. Biol. Chem.*, **263**, 3858.
122. Corden, J. F. (1990) Tails of RNA polymerase II. *Trends Biochem. Sci.*, **15**, 383.
123. Berkhout, B. and Jeang, K.-T. (1992) Functional roles for the TATA promoter and enhancers in basal and Tat-induced expression of the human immunodeficiency virus type 1 long terminal repeat. *J. Virol.*, **66**, 139.
124. Kato, H., Horikoshi, M., and Roeder, R. G. (1991) Repression of HIV-1 transcription by a cellular protein. *Science*, **251**, 1476.
125. Braddock, M., Chambers, A., Wilson, W., Esnouf, M. P., Adams, S. E., Kingsman, A. J., and Kingsman, S. M. (1989) HIV-1 TAT 'activates' presynthesized RNA in the nucleus. *Cell*, **58**, 269.
126. Braddock, M., Thorburn, A. M., Chambers, A., Elliott, G. D., Anderson, G. J., Kingsman, A. J., and Kingsman, S. M. (1990) A nuclear translational block imposed by the HIV-1 U3 region is relieved by the Tat–TAR interaction. *Cell*, **62**, 1123.
127. Sharmeen, L., Bass, B., Sonenberg, N., Weintraub, H., and Groudine, M. (1991) Tat-dependent adenosine-to-inosine modification of wild-type *trans*-activation response RNA. *Proc. Natl Acad. Sci. USA*, **88**, 8096.

128. Roy, S., Katze, M. G., Parkin, N. T., Edery, I., Hovanessian, A. G., and Sonenberg, N. (1990) Control of the interferon-induced 68-kilodalton protein kinase by the HIV-1 *tat* gene product. *Science,* **247,** 1216.
129. Roy, S., Agy, M., Hovanessian, A. G., Sonenberg, N., and Katze, M. G. (1991) The integrity of the stem structure of human immunodeficiency virus type 1 Tat-responsive sequence RNA is required for interaction with the interferon-induced 68,00-Mr protein kinase. *J. Virol.,* **65,** 632.
130. Gunnery, S., Rice, A. P., Robertson, H. D., and Mathews, M. B. (1990) Tat-responsive region RNA of human immunodeficiency virus 1 can prevent activation of the double-stranded-RNA-activated protein kinase. *Proc. Natl Acad. Sci. USA,* **87,** 8687.
131. Chin, D. J., Selby, M. J., and Peterlin, B. M. (1991) Human immunodeficiency virus type 1 Tat does not transactivate mature *trans*-acting responsive region RNA species in the nucleus or cytoplasm of primate cells. *J. Virol.,* **65,** 1758.
132. Braddock, M., Thorburn, A. M., Kingsman, A. J., and Kingsman, S. M. (1991) Blocking of Tat-dependent HIV-1 RNA modification by an inhibitor of RNA polymerase II processivity. *Nature,* **350,** 439.
133. Busch, M., Beckstead, J., Hollender, H., and Vyas, G. (1989) *In situ* hybridization: a histomolecular approach to the detection and study of HIV infection. In *HIV Detection By Genetic Engineering.* Luciw, P. and Steiner, K. S. (eds). Marcel Dekker Inc., New York, NY, p. 209.
134. Harper, M., Marselle, L., Gallo, R., and Wong-Staal, F. (1986) Detection of lymphocytes expressing Human T Lymphotropic Virus Type III in lymph nodes and peripheral blood from infected individuals by *in situ* hybridization. *Proc. Natl Acad. Sci. USA,* **83,** 772.
135. Schnittman, S., Psallidopoulos, M., Lane, H., Thompson, L., Baseler, M., Massari, F., Fox, C., Salzman, N., and Fauci, A. (1989) The reservoir for HIV-1 in human peripheral blood is a T cell that maintains expression of CD4. *Science,* **245,** 305.
136. Pomerantz, R. J., Trono, D., Feinberg, M. B., and Baltimore, D. (1990) Cells nonproductively infected with HIV-1 exhibit an aberrant pattern of viral RNA expression: a molecular model for latency. *Cell,* **61,** 1271.
137. Folks, T., Justement, J., Kinter, A., Dinarello, C. A., and Fauci, A. S. (1987) Cytokine induced expression of HIV-1 in a chronically-infected promonocyte cell line. *Science,* **238,** 800.
138. Clouse, K. A., Powell, D., Washington, I., Poli, G., Strebel, K., Farrar, W., Barstad, B., Kovacs, J., Fauci, A. S., and Folks, T. M. (1989) Monokine regulation of HIV-1 expression in a chronically-infected human T cell clone. *J. Immunol.,* **142,** 431.
139. Koka, P., Yunis, J., Passarelli, A. L., Dubey, D. P., Faller, D. V., and Ynis, E. J. (1988) Increased expression of CD4 molecules on Jurkat cells mediated by human immunodeficiency virus Tat protein. *J. Virol.,* **62,** 4353.
140. Bielinska, A., Baier, L., Hailat, N., Strahler, J. R., Nabel, G. J., and Hanash, S. (1991) Expression of a novel nuclear protein in activated and in Tat-I expressing T cells. *J. Immunol.,* **146,** 1031.
141. Chowdhury, M., Taylor, J. P., Tada, H., Rappaport, J., Wong-Staal, F., Amini, S., and Khalili, K. (1990) Regulation of the human neurotropic virus promoter by JCV-T antigen and HIV-1 Tat protein. *Oncogene,* **5,** 1737.
142. Tada, H., Rappaport, J., Lashgari, M., Amini, S., Wong-Staal, F., and Khalili, K. (1990) Trans-activation of the JC virus late promoter by the Tat protein of type 1 human immunodeficiency virus in glial cells. *Proc. Natl Acad. Sci. USA,* **87,** 3479.

143. Remenick, J., Radonovich, M. F., and Brady, J. N. (1991) Human immunodeficiency virus Tat transactivation: induction of a tissue-specific enhancer in a nonpermissive cell line. *J. Virol.,* **65,** 5641.
144. Sastry, K. J., Reddy, H. R., Pandita, R., Totpal, K., and Aggarwal, B. B. (1990) HIV-1 *tat* gene induces tumor necrosis factor-beta (lymphotoxin) in a human B-lymphoblastoid cell line. *J. Biol. Chem.,* **265,** 20091.
145. Viscidi, R. P., Mayur, K., Lederman, H. M., and Frankel, A. D. (1989) Inhibition of antigen-induced lymphocyte proliferation by Tat protein from HIV-1. *Science,* **246,** 1606.
146. Vogel, J., Hinrichs, S. H., Reynolds, R. K., Luciw, P. A., and Jay, G. (1988) The HIV *tat* gene induces dermal lesions resembling Kaposi's sarcoma in transgenic mice. *Nature,* **335,** 606.
147. Ensoli, B., Barillari, G., Salahuddin, S. Z., Gallo, R. C., and Wong-Staal, F. (1990) Tat protein of HIV-1 stimulates growth of cells derived from Kaposi's sarcoma lesions of AIDS patients. *Nature,* **345,** 84.
148. Ensoli, B., Barillari, G., and Gallo, R. C. (1991) Pathogenesis of AIDS-associated Kaposi's sarcoma. *Hematol. Oncol. Clin. North. Am.,* **5,** 281.
149. Sinkovics, J. G. (1991) Kaposi's sarcoma: its 'oncogenes' and growth factors. *Crit. Rev. Oncol. Hematol.,* **11,** 87.
150. Frankel, A. D. and Pabo, C. O. (1988) Cellular uptake of the Tat protein from human immunodeficiency virus. *Cell,* **55,** 1189.
151. Gentz, R., Chen, C. H., and Rosen, C. A. (1989) Bioassay for *trans*-activation using purified human immunodeficiency virus *tat*-encoded protein: *trans*-activation requires mRNA synthesis. *Proc. Natl Acad. Sci. USA,* **86,** 821.
152. Helland, D. E., Welles, J. L., Caputo, A., and Haseltine, W. A. (1991) Transcellular *trans*-activation by the human immunodeficiency virus type 1 Tat protein. *J. Virol.,* **65,** 4547.
153. Valerie, K., Delers, A., Bruck, C., Thiriart, C., Rosenberg, H., Debouck, C., and Rosenberg, M. (1988) Activation of human immunodeficiency virus type 1 by DNA damage in human cells. *Nature,* **333,** 78.
154. Sadaie, M. R., Tschachler, E., Valerie, K., Rosenberg, M., Felber, B. K., Pavlakis, G. N., Klotman, M. E., and Wong-Staal, F. (1990) Activation of Tat-defective human immunodeficiency virus by ultraviolet light. *New Biol.,* **2,** 479.
155. Lazinski, D., Grzadzielska, E., and Das, A. (1989) Sequence-specific recognition of RNA hairpins by bacteriophage antiterminators requires a conserved arginine-rich motif. *Cell,* **59,** 207.
156. Nodwell, J. R. and Greenblatt, J. (1991) The nut site of bacteriophage λ is made of RNA and is bound by transcription antitermination factors on the surface of RNA polymerase. *Genes Dev.,* **5,** 2141.
157. Whalen, W. A. and Das, A. (1990) Action of an RNA site at a distance: role of the nut genetic signal in transcription antitermination by phage-lambda N gene product. *New Biologist,* **2,** 975.
158. Spencer, C. A. and Groudine, M. (1990) Transcriptional elongation and eukaryotic gene regulation. *Oncogene,* **5,** 777.
159. Koenig, M., Hoffman, E. P., Bertelson, C. J., Monaco, A. P., Freener, C., and Kunkel, L. M. (1987) Complete cloning of the Duchenne muscular dystrophy (DMD) cDNA and preliminary genomic organization of the DMD gene in normal and affected individuals. *Cell,* **50,** 509.

160. Hsu, M. C., Schutt, A. D., Holly, M., Slice, L. W., Sherman, M. I., Richman, D. D., Potash, M. J., and Volsky, D. J. (1991) Inhibition of HIV replication in acute and chronic infections in vitro by a Tat antagonist. *Science,* **254,** 1799.
161. Berkhout, B., Silverman, R. H., and Jeang, K. T. (1989) Tat *trans*-activates the human immunodeficiency virus through a nascent RNA target. *Cell,* **59,** 273.
162. Bielinska, A., Krasnow, S., and Nabel, G. J. (1989) NF-κB-mediated activation of the human immunodeficiency virus enhancer: site of transcriptional initiation is independent of the TATA box. *J. Virol.,* **63,** 4097.
163. Jones, K. A., Luciw, P. A., and Duchange, N. (1988) Structural arrangements of transcription control domains within the 5'-untranslated leader regions of the HIV-1 and HIV-2 promoters. *Genes Dev.,* **2,** 1101.

5 | Post-transcriptional regulation of human retroviral gene expression

TRISTRAM G. PARSLOW

1. Introduction

The compact structure of retroviral genomes forces these viruses to rely heavily on post-transcriptional mechanisms of gene regulation. Each integrated provirus is transcribed from the single promoter in its 5' long terminal repeat (LTR) to produce a single primary transcript that spans the entire genome. Yet, from this single transcript, the virus must express a minimum of three structural proteins (Gag, Pol, and Env) and often additional regulatory and accessory proteins as well. To express all of these products, the primary transcript must first be spliced in multiple alternative ways to generate a host of different mRNAs that specify the individual proteins. The virus must then ensure that each mRNA is delivered to the cytoplasm in appropriate amounts for translation. Moreover, unlike cellular mRNAs, retroviral mRNAs often encode two or more proteins in tandem, and so pose special problems for translation. This chapter will examine the specialized post-transcriptional and translational strategies that govern retroviral gene expression, and the impact of these strategies on the viral life cycle.

2. Alternative splicing of retroviral RNAs

The primary transcripts of all known retroviruses contain RNA-splicing signals, and the survival of each virus depends absolutely upon its interactions with the cellular splicing machinery. While the basic pathway of splicing is the same for both viral and cellular RNAs (reviewed in refs 1–3), retroviruses exploit this pathway in ways seldom utilized by the cell. Alternative splicing is the rule. It is common, for example, for a viral intron to use two or more 5' (donor) or 3' (acceptor) splice sites interchangeably. Even the semantic distinction between introns and exons is blurred, as most potential introns in a retrovirus contain protein-coding information. Thus, a particular region of the primary transcript might be excised as an intron to produce one viral mRNA, but retained as an exon in another. An important corollary of this is that many of the retroviral mRNAs that ultimately find their way into the cytoplasm contain at least one complete, unexcised intron, defined by the presence of at least one unused pair of donor and acceptor splice sites. Such incompletely spliced mRNAs differ from virtually all cellular transcripts, which remain in the nucleus until all introns have been removed.

Retroviral transcripts are also notoriously inefficient substrates for RNA splicing. This is due in large part to the nucleotide sequences of the splicing signals themselves, which often differ somewhat from the optimum consensus sequences and hence are utilized poorly both *in vivo* and in cell-free splicing systems (4–6). RNA from some types of retroviruses has also been found to contain additional *cis*-acting elements that further inhibit splicing (6–11), although the mechanisms by which these elements function are not yet fully understood. Due to this inefficient processing, relatively large amounts of unspliced or incompletely spliced viral RNA tend to accumulate in the nucleus whenever a provirus is transcribed — at a given moment, as much as 80 per cent of the viral RNA may remain incompletely spliced (11). This is clearly advantageous for the virus, which must keep adequate supplies of unspliced RNA on hand to be packaged into virions. In addition, the abundance of viral nuclear transcripts at intermediate stages of processing provides a readily available source of incompletely spliced mRNAs. Both the diversity of these transcripts and their subsequent fates, however, differ significantly between simple and complex retroviruses.

2.1 Simple retroviruses

Primary transcripts of the simplest retroviruses, such as murine and avian leukemia viruses, contain only one potential intron that encompasses essentially all of the Gag and Pol coding regions (Fig. 1). Like the introns of most cellular genes, this viral intron is bounded by unique donor and acceptor splice sites. The viral transcript can, therefore, exist in only two forms — spliced and unspliced — and each of these functions as a specific mRNA. The larger, unspliced mRNA is identical to the genomic RNA found within virion particles; it includes all three of the viral structural genes, but provides an efficient template for translation of Gag and Pol only. The spliced RNA, in contrast, lacks the Gag and Pol sequences, and serves as the mRNA for Env.

A variation on this theme can be found in Rous sarcoma virus (RSV), a simple retrovirus which, in addition to *gag*, *pol*, and *env*, carries a transduced oncogene known as *v-src*. Expression of the oncogene depends upon the use of a nearby alternative splice acceptor site (11). Using its two acceptor sites interchangeably, RSV produces two different singly spliced mRNAs; one of these encodes Env, while the other specifies the oncoprotein v-Src (Fig. 1).

Owing to the inefficiency of splicing, the spliced and unspliced mRNAs accumulate in roughly equal amounts in the nucleus. An important property of simple retroviruses is that both forms of the viral RNA are constitutively released into the cytoplasm. It is not yet known why the unspliced viral mRNAs are permitted to

Fig. 1 Genomic organization and mRNA expression in simple retroviruses. The avian leukosis virus (ALV) and Rous sarcoma virus (RSV) genomes are shown schematically in proviral form, along with the mRNAs each produces. Open rectangles on the genomic maps represent the 5′ and 3′ long terminal repeat (LTR) sequences and the coding sequences for the viral Gag, Pol, Env, and Src proteins. Solid rectangles denote the R region within each LTR. Locations of splice donor (D) and splice acceptor (A) sites are also indicated. The proteins encoded by each mRNA species are indicated at right; dotted lines indicate the regions excised by splicing to generate a given mRNA. Not to scale; alignment of exons does not necessarily coincide with translational reading frames

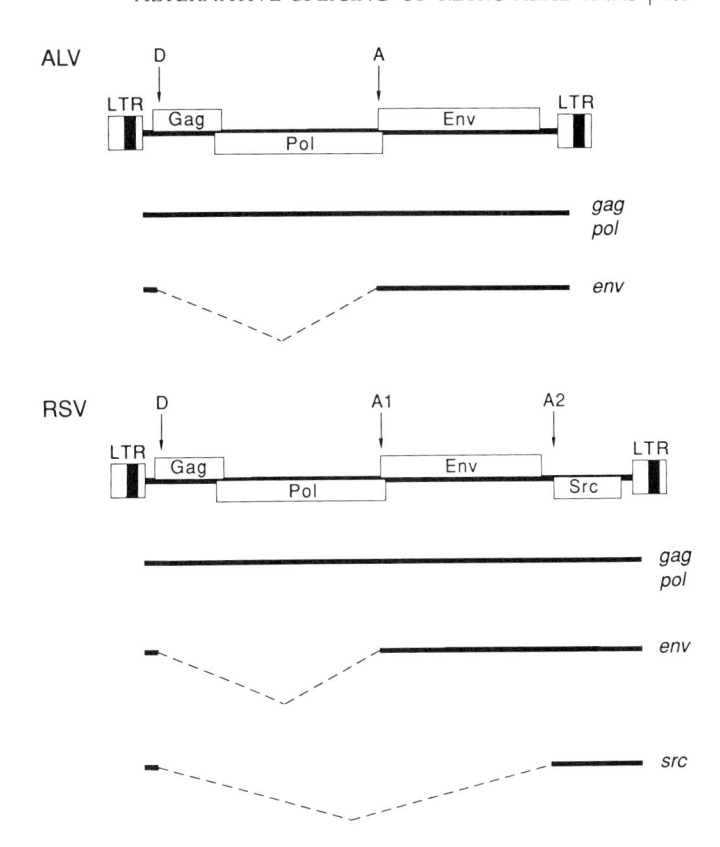

enter the cytoplasm while unspliced cellular RNAs remain inside the nucleus. No virally encoded proteins are required for this process (4, 7, 9), and indeed the ratio of expression of spliced and unspliced viral mRNAs seems not to be modulated to any significant degree by the virus. The result is that Gag, Pol, and Env proteins are synthesized in stoichiometric amounts at all times, so that complete viral particles can be produced and shed continuously throughout the life of the infected cell.

2.2 Complex retroviruses

Human immunodeficiency virus type 1 (HIV-1), human T-cell leukemia virus type I (HTLV-I), and other complex retroviruses produce not only Gag, Pol, and Env, but also additional proteins that have regulatory or accessory functions (Fig. 2). These are encoded by short exons located near the centre and/or the 3′ end of the genome, and their expression requires more elaborate use of alternative RNA splicing. Instead of a single intron, the primary transcripts of these viruses contain multiple splice donor and acceptor sites which can be used in various combinations to yield a large number of alternative mRNAs. HIV-1, for example, contains at least four splice donors and six splice acceptors, which enable it to produce more than 30

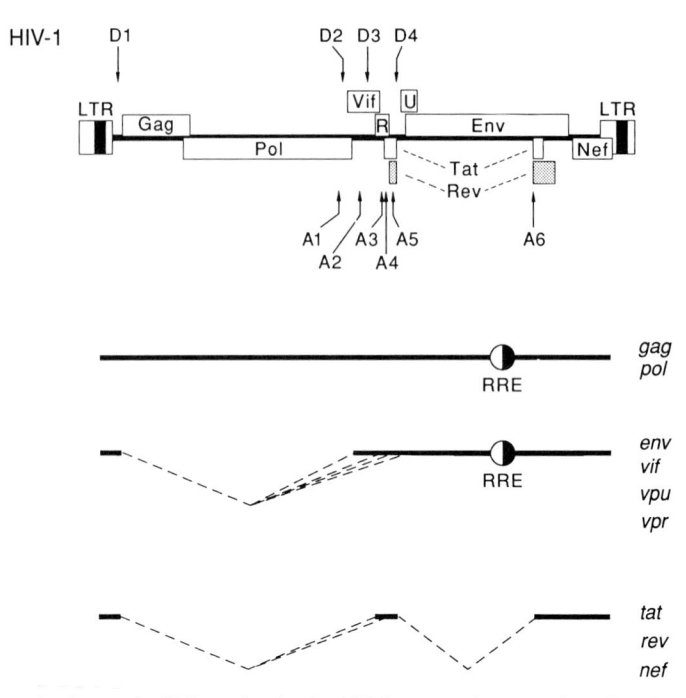

Fig. 2 Genomic organization and mRNA production by HIV-1, a complex retrovirus. The 9 kb proviral genome is shown schematically (above) with its long terminal repeat (LTR) sequences each containing an R region (solid rectangle). Coding sequences for the viral proteins are depicted by rectangles; those encoding Rev are shaded. R = Vpr; U = Vpu. Note that Tat and Rev are each specified by two coding exons separated by the same intron. Sequences encoding the four known splice donor (D1–D4) and six known splice acceptor (A1–A6) sites are indicated. Representative unspliced, singly spliced, and multiply spliced mRNAs are also depicted (below), with the proteins they encode listed at right. Dotted lines indicate the regions excised by splicing to produce a given mRNA. The use of alternative splice sites gives rise to approximately 30 different singly or multiply spliced mRNAs, only a few of which are shown here. The 240-base RRE lies within the Env coding region. Not drawn to scale; the alignment of exons as shown does not necessarily coincide with translational reading frames

different mRNAs (12–17). The pattern of splicing, in turn, determines which proteins can be expressed.

The mRNAs of complex retroviruses fall into three general classes:

1. *Unspliced mRNA.* As in simple viruses, the unspliced primary transcript serves as the mRNA for Gag and Pol, and also is the form of RNA that can be packaged into viral particles.
2. *Singly spliced mRNAs.* All singly spliced mRNAs lack the Gag/Pol coding region (which is excised as a single intron) but retain most of the 3' portion of the genome. This group always includes one or more mRNAs encoding Env, but may include mRNAs with other specificities as well. The specificities are dictated by the particular splice sites used in excising the Gag/Pol intron. For example, the choice of acceptor site determines whether a singly spliced HIV-1 mRNA will primarily yield Vif, Vpr, single-exon Tat, or Vpu and Env proteins (16, 18, 19).

3. *Multiply spliced mRNAs*. These mRNAs also lack the Gag/Pol intron, but have undergone additional splicing events which remove all or part of the Env region (12–15). Like cellular mRNAs, these are exhaustively spliced, so that no complete introns remain. As all multiply spliced mRNAs code for regulatory or accessory proteins, they are often referred to as 'regulatory' mRNAs to distinguish them from the unspliced and singly-spliced 'structural' mRNAs that code for Gag, Pol, and Env.

In contrast to the situation in simple retroviruses, only the multiply spliced regulatory mRNAs of complex retroviruses are constitutively released to the cytoplasm. The proteins produced from this class of mRNAs always include a transcriptional *trans*-activator protein, such as HIV-1 Tat, which serves as a positive feedback mechanism to increase production of the primary transcript. Inefficient processing of the primary transcript (20), in turn, leads to increased steady-state concentrations of all three classes of viral mRNAs within the nucleus. The unspliced and singly spliced mRNAs, however, remain within the nucleus unless they are acted upon by a second type of virally encoded *trans*-activator protein. Like Tat, regulatory proteins of this second type are encoded by multiply spliced viral mRNAs, and are absolutely essential for viral replication. Unlike Tat, they act after transcription. The following section will describe the properties of these post-transcriptional *trans*-activators.

3. Post-transcriptional *trans*-activation of complex retroviruses

3.1 The HIV-1 Rev protein

The post-transcriptional *trans*-activator protein from HIV-1 was the first to be discovered, and remains the best characterized protein of this type. Its existence first came to light when certain mutations immediately downstream of the Tat coding region were found to abolish HIV-1 replication (21, 22). Although proviruses containing these mutations remained transcriptionally active and were capable of producing biologically active Tat, they failed to express detectable amounts of the Gag, Pol, or Env proteins, or to generate any new infectious virions. On closer inspection, the mutations were found to disrupt a previously unrecognized protein-coding sequence in the 3' half of the genome (Fig. 2). This sequence was contained in two short exons that straddled the upstream two-thirds of the *env* region, and so, like Tat, could be expressed from multiply spliced mRNAs. Indeed, the two new exons partially overlapped (and shared a pair of splice sites with) those encoding Tat, although they occupied a different translational reading frame. These exons together specified a 116-amino-acid protein that was well conserved among HIV-1 isolates, but showed no obvious similarity to any other viral or cellular proteins known at that time.

The unprecedented regulatory effect of this protein became apparent when the

Fig. 3 Rev is essential for cytoplasmic expression of the unspliced and incompletely spliced HIV-1 mRNAs. The Northern blot shown here contrasts the expression of HIV-1 transcripts in the cytoplasm of mammalian cells infected by wild-type HIV-1 (lane 2) or by an HIV-1 mutant that lacks Rev activity (lane 3). Multiply spliced (2 kb) mRNAs are constitutively expressed, but the unspliced (9 kb) and singly spliced (4 kb) forms are detectable only when Rev is present. Lane 1 contains cytoplasmic mRNA from uninfected cells. Data provided by M. H. Malim and B. R. Cullen; reprinted with permission from ref. 38

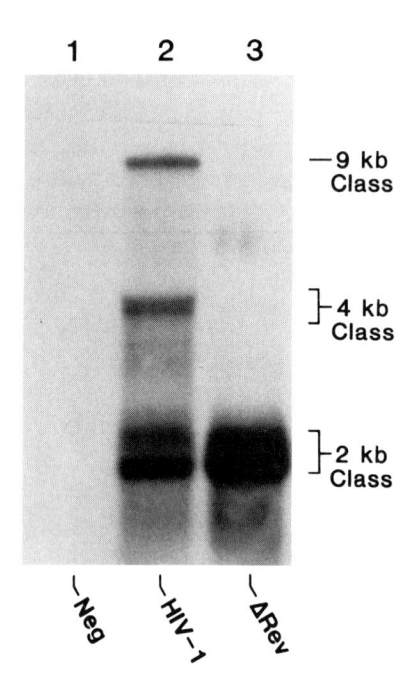

pattern of cytoplasmic mRNA expression from the mutants was compared with that of wild-type HIV-1 (21, 23). On a Northern blot, the mRNAs of HIV-1 normally migrate in three distinct size classes, with lengths of approximately 9, 4, and 2 kb; these correspond to the unspliced, singly spliced, and multiply spliced transcripts, respectively (Fig. 3). Cells containing the mutant virus were found to express normal or even increased amounts of 2 kb mRNAs, but only minimal amounts of the 9 and 4 kb forms. When an expression plasmid that encoded only the novel protein was introduced into these cells, the cytoplasmic concentrations of the larger transcripts were selectively increased. Thus, the protein was not required for HIV-1 transcription, but instead appeared to alter processing of the viral transcript in such a way as to promote cytoplasmic expression of the incompletely spliced mRNAs. By so doing, it activated synthesis of viral structural proteins, and hence the production of virion particles. Because of this ability to regulate expression of the virion, the protein became known as Rev (24).

We do not yet know precisely how Rev promotes expression of incompletely spliced HIV mRNAs. However, subsequent experiments have uncovered important clues. For example, immunofluorescence staining of cells expressing large amounts of Rev revealed that the protein is localized almost exclusively within nuclei (23, 25), suggesting that this is its primary site of action. Consistent with this view, the protein was shown to be completely inactive when anchored artificially in the cytoplasm, but to initiate its effect within minutes after entering the nucleus (26).

Further insight came from studies which compared the effects of Rev on HIV-1 mRNA expression in the nucleus and the cytoplasm (23, 27–29). Even when Rev was absent, the unspliced, singly spliced, and multiply spliced viral RNAs could all be found in significant amounts within the nucleus, although only the multiply spliced forms accumulated in the cytoplasm. Rev produced little change in the amounts of the various species in the nucleus, but selectively induced cytoplasmic expression of the larger structural mRNAs (23, 27–29). It thus became apparent that Rev acts, at least in part, by allowing intron-containing RNAs to be released from the nucleus into the cytoplasm.

It also became clear that Rev does not affect expression of all RNAs, but rather acts selectively on specific HIV-1 transcripts. By testing the effects of various deletions throughout the viral genome, a short segment of the *env* coding region was shown to be uniquely required for responsiveness to Rev (29–31). In the primary transcript, this *cis*-acting locus, known as the Rev response element (RRE), is situated within the potential intron that bisects the *tat* and *rev* genes; it is, therefore, absent from multiply spliced mRNAs, but is present in all the unspliced and singly spliced mRNAs whose expression is controlled by Rev (Fig. 2). The RRE coincides with a 240-base region of the HIV-1 transcript that is predicted to fold into a complex RNA secondary structure (29) as depicted in Fig. 4. Thermodynamic calculations suggest that this structure is highly stable; in fact, the RRE and trans-

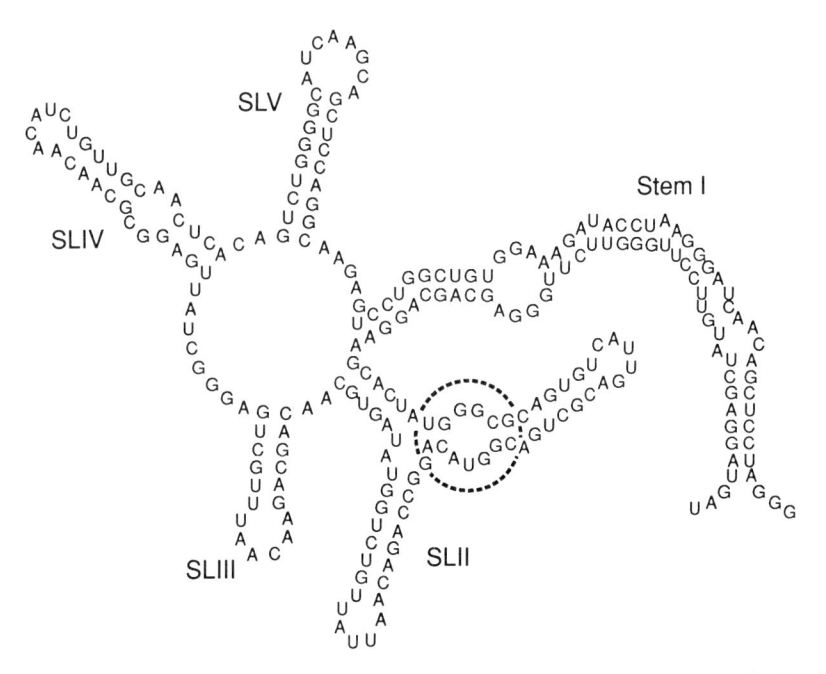

Fig. 4 Sequence and predicted secondary structure of the RRE from HIV-1. The sequence depicted corresponds to residues 7770–8011 of the viral genome, and lies within the Env coding region (see Fig. 2). This structure can be described as a major stem (Stem 1) crowned by a large loop which bears four subsidiary stem–loop domains designated SLII–SLV. Broken circle indicates the primary binding site for Rev protein, which lies in SLII. Redrawn from ref. 29, with modifications from refs 87, 121, and 128

activation response element (TAR) loci are thought to be by far the most stable regions of secondary structure in the HIV-1 transcript (29, 32).

An especially important insight was gained when it was discovered that Rev is a sequence-specific RNA-binding protein, and that its preferred binding site lies within the RRE (33, 34). Purified Rev protein was shown to bind specifically to RRE-containing transcripts *in vitro* (33–38), with multiple copies of Rev binding simultaneously on to each copy of the RRE (33, 37–39). Mutations in either the protein or the RRE that inhibited *in vitro* binding completely eliminated *trans*-activation *in vivo* (33–42), implying that this specific protein–RNA interaction is crucial for the biological response.

But the RRE alone cannot confer Rev-responsiveness on to a cellular RNA. When this element is inserted into a β-globin intron, for example, the pattern of globin RNA expression is not appreciably altered either in the presence or absence of Rev (43). To make the globin transcript responsive to Rev, it is necessary not only to insert the RRE, but also to either replace the globin splice signals with those of HIV-1, or mutate the globin donor or acceptor sites in such a way as to reduce the efficiency of splicing (43). Replacement or mutation of the splice sites alone significantly increases the steady-state level of unspliced globin transcripts in the nucleus; if these unspliced transcripts also contain the RRE, they inducibly enter the cytoplasm in response to Rev protein.

Taken together, these findings suggest that Rev acts by binding to a pre-existing pool of incompletely spliced nuclear RNAs that contain the RRE. As a result of this binding, Rev directly or indirectly induces release of these RNAs to the cytoplasm before splicing is complete. The exact mechanism by which Rev produces this effect, however, remains the subject of intense debate and research. Some of the more specific models of Rev action suggested by this research will be discussed in a later section.

3.2 Rev-like *trans*-activators in other complex retroviruses

At the time of its discovery, the finding that Rev could induce cytoplasmic expression of intron-containing RNAs was completely unprecedented. Even today, no mammalian protein with similar activity has yet been identified. It is now recognized, however, that many complex retroviruses encode proteins that are functionally equivalent to Rev. In fact, the expression of such proteins is thought to be a defining and obligatory characteristic of all complex retroviruses (44). A list of known or suspected Rev-like *trans*-activators from different retroviral families is presented in Table 1. The list includes proteins from such evolutionarily divergent groups as the primate immunodeficiency viruses HIV and simian immunodeficiency virus (SIV), the human oncogenic virus HTLV-I, and the ungulate lentivirus Visna (23, 29, 45–59).

Although they bear little obvious resemblance to one another in terms of amino acid sequence, Rev-like *trans*-activators from different viruses share a number of biological properties:

Table 1 Complex retroviruses that express Rev-like *trans*-activator proteins

Virus	Representative references
Confirmed	
Human immunodeficiency virus—Type 1	23, 29
—Type 2	45, 46
Human T-cell leukemia virus—Type I	47, 48
—Type II	49, 50
Simian immunodeficiency virus—macaque	45
—African green monkey	51
Visna virus	52
Bovine leukemia virus	53, 54
Feline immunodeficiency virus	55
Proposed	
Equine infectious anemia virus	56
Caprine arthritis-encephalitis virus	57, 58
Ovine maedi-visna virus	58, 59

1. Each is a relatively small protein (generally less than 200 amino acids long) that is encoded by one or more multiply spliced mRNAs from the parental virus.
2. Each is a nuclear protein. When expressed at sufficient levels to be detectable by immunofluorescence, the proteins are found diffusely throughout the nucleoplasm, and often associate preferentially with nucleoli (23, 25, 60).
3. In each case, the protein is required for cytoplasmic expression of unspliced and singly spliced viral structural mRNAs, and hence is essential for viral replication.
4. Each protein acts selectively on RNAs that contain a specific *cis*-acting response element. Although the sequence of the response element is unique to a given virus, it invariably coincides with a predicted region of complex RNA secondary structure approximately 200–250 bases long (29, 32, 45, 46, 48, 61, 62).
5. In every case thus far examined, the protein has sequence-specific RNA-binding activity that enables it to bind directly to its cognate response element *in vitro*.

Some of the *trans*-activators, such as HIV-2 Rev, have very limited target ranges, and can efficiently *trans*-activate only their own parental virus (45). Others, however, are more promiscuous (45, 51, 54, 61, 63–65). For example, under some conditions, the HTLV-I Rex and SIVsm Rev proteins can each substitute for Rev in *trans*-activating HIV-1. Such cross-species reactivity appears to reflect, in part, the ability of some proteins to bind response elements other than their own. Significantly, it also suggests that many (and perhaps all) *trans*-activators of this type share a common mechanism of action (44, 63).

3.3 Rev-like *trans*-activators, latency, and the biphasic life cycle of complex retroviruses

The action of a Rev-like *trans*-activator divides the life cycle of a complex retrovirus into two phases (44, 66), which are schematized for HIV-1 in Fig. 5. In the early phase, when a provirus is first transcribed, neither Tat nor Rev is present in the cell. Hence, viral transcripts are produced at a relatively low rate, and only multiply spliced (regulatory) mRNAs are released into the cytoplasm. During this phase, no virions can be produced.

Fig. 5 The biphasic life cycle of a complex retrovirus. Shown is a schematic view of viral mRNA classes that are expressed in nucleus and cytoplasm during each of the two phases. Some major proteins expressed from each class of mRNA are listed at the right. Dotted lines indicate regions excised by RNA splicing. Rev tends to inhibit cytoplasmic expression of the regulatory mRNAs during the late phase of the life cycle (18, 67), and thus down-regulates its own expression

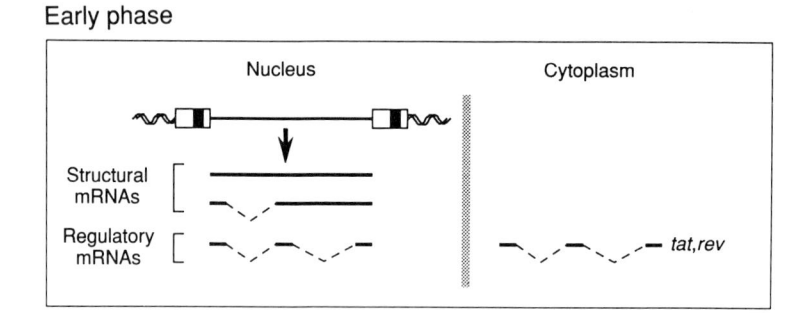

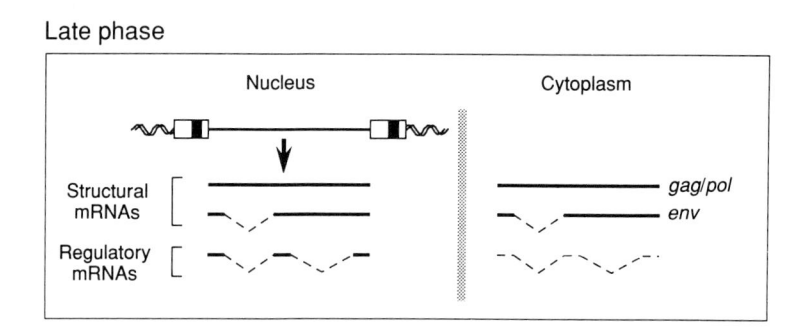

Over time, increasing amounts of Tat and Rev are synthesized and are translocated into the nucleus. When sufficient levels of these proteins are achieved, the life cycle enters its late phase, which is characterized in part by increased viral transcription in response to Tat. The main hallmark of this phase, however, is the qualitative shift in the pattern of mRNA expression that occurs as Rev permits unspliced and singly spliced viral RNAs to accumulate in the cytoplasm. This effect of Rev permits viral structural proteins to be synthesized, and leads to the production of infectious virions. At the same time, production of the regulatory mRNAs declines (18, 67) because relatively few transcripts are spliced to completion. Thus, Rev tends to down-regulate its own synthesis, and so provides the fulcrum that precisely balances regulatory and structural gene expression.

This biphasic life cycle may help to explain why infections caused by complex retroviruses typically follow a protracted and indolent course. In theory, the need to build up a threshold amount of Rev protein could relegate a provirus for extended periods to the early, non-productive phase, and thereby impose a prolonged state of latency (40, 68–71). While this notion is conceptually useful, however, it remains to be determined whether integrated human retroviruses ever actually exist in a truly latent state within the host. In the case of HIV-1, for example, continuous low levels of virus production can be detected throughout the course of natural infections (72–75), as well as in cell-culture models that otherwise appear to mimic early-phase infection (76, 77). It may be that the early and late phases are not entirely distinct, but instead represent the two ends of a continuum of viral gene expression. One central question in this regard is whether *trans*-activation takes place incrementally with rising Rev concentrations, or occurs as an all-or-none response after a threshold concentration is reached. Data from cell-culture assays can be adduced to support either position (69, 78).

Whatever its relationship to natural infections, the biphasic life cycle of complex retroviruses provides one of the most compelling rationales for studying the Rev-like *trans*-activators. It is clear from studies of HIV-1 mutants that proviruses which lack a functional *rev* gene fail to progress into the late phase, and so are unable to replicate (21, 22, 79, 80). This strongly implies that therapies aimed at blocking the activity of these *trans*-activators could provide an effective means of suppressing viral replication, and thereby slowing the progression of retroviral disease.

3.4 Experimental systems for studying Rev-like proteins

Our current understanding of the Rev-like *trans*-activators is based not only on studies of the whole viruses, but also on results obtained with simpler experimental systems.

First, a variety of cell-culture-based functional assays have been developed that can detect the biological activity of these proteins (e.g. see refs 18, 26, 29, 30, 81). Such assays generally involve transfecting cultured cells with an expression plasmid that encodes the *trans*-activator, along with a second plasmid whose expression is controlled by the *trans*-activator and thus serves as a reporter. Typical Rev reporter plasmids, for example, are derived from the HIV-1 *env* region, and produce transcripts containing an RRE flanked by a pair of HIV-1 splice sites. In response to Rev, unspliced reporter transcripts containing the RRE are permitted to enter the cytoplasm, where their presence can be detected either directly (e.g. by Northern blot) or by the expression of a marker protein (e.g. chloramphenicol acetyltransferase) encoded by the transcript. These assays are relatively rapid and sensitive, pose no infectious hazard, and are well suited for studying biological function of either the *trans*-activator or its response element. They have also made it possible to study *trans*-activation in cells that are not infectible by a particular virus. Through this approach, Rev-like *trans*-activators have been shown to function in a wide variety of cell types, and in species ranging from humans to *Drosophila melanogaster* (58, 82).

In addition, several of the Rev-like proteins have been synthesized in bacterial or baculoviral expression systems and purified to homogeneity (e.g. see refs 33–42, 49, 50, 65, 83–88). Such preparations have been extremely useful in characterizing the biophysical properties of the proteins (39, 83, 84) — most notably, their interactions with RNA targets *in vitro*. Purification of these *trans*-activators also represents an important step toward the development of cell-free systems that could mimic one or more aspects of the biological response. If such systems can be developed, they would greatly facilitate efforts to dissect the pathway of *trans*-activation using conventional biochemical techniques. An equally powerful alternative would be to reconstitute the activity of these proteins in yeast, so that components of the pathway could be identified by genetic means. Development of such genetic or cell-free functional assays remains an important goal for research.

3.5 Functional architecture of Rev-like proteins

The availability of purified preparations of HIV-1 Rev, and of other similar *trans*-activators, will eventually make it possible to determine the complete three-dimensional structures of these proteins using X-ray crystallography or nuclear magnetic resonance spectroscopy. To date, however, no detailed structural information has been reported for any Rev-like protein (83, 84).

Considerable insight has been gained, however, by studying the effects of mutations on various aspects of protein function *in vivo* and *in vitro*. The most extensive studies of this type have been performed on HIV-1 Rev (40–42, 58, 89–91). By systematically introducing mutations throughout this 116-amino-acid protein, two discrete regions have been found to be essential for its activity (Fig. 6): a multifunctional N-terminal region, and a C-terminal effector domain.

The N-terminal domain

A large region that occupies most of the N-terminal half of the protein (approximately residues 14–56) controls at least three critical aspects of Rev function: RNA binding, nuclear localization, and protein oligomerization. Since sequences re-

Fig. 6 Functional domains of the HIV-1 Rev protein. The two essential regions are indicated by shading. Amino acid sequences of the C-terminal effector domain and of an arginine-rich region within the N-terminal domain are shown. In the latter, the box denotes a hexapeptide sequence that functions as an autonomous nuclear translocation signal (90)

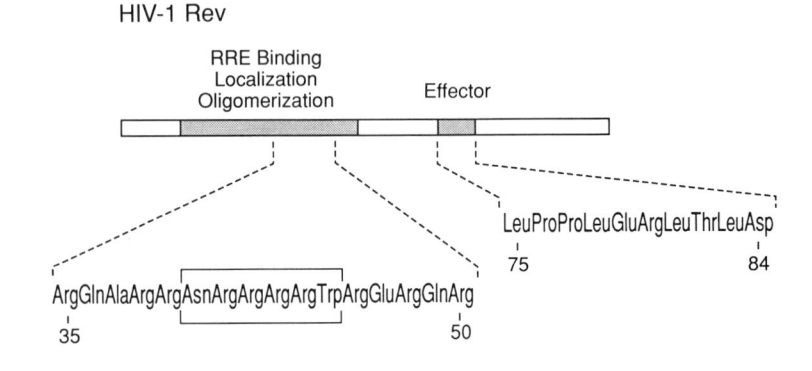

sponsible for the individual functions overlap extensively, this region is usually viewed as a single, multifunctional domain.

RNA binding

Specific binding of the RRE is primarily mediated by an arginine-rich sequence at positions 35–50. This was first demonstrated by the finding that mutations in this region of protein prevent RRE binding *in vitro* (40–42). Later, a peptide comprising only this region was reported to bind the RRE with specificity comparable to that of full-length Rev (88). Ten of the 16 amino acids in this region are arginines, and so carry a strong positive charge under physiological conditions. It is thought that these charged arginines may bind tightly to RNA by forming electrostatic contacts with the phosphodiester backbone (21, 92). However, other amino acids in this sequence also are important. For example, the tryptophan at position 45 is strictly required for interaction with the RRE (42). Based on findings in other systems (93), the aromatic sidechain of this tryptophan residue may serve to intercalate between adjacent bases of the RNA target.

Nuclear localization

A subset of residues within the RNA-binding region also provide the localization signals that target Rev protein into nuclei (89, 90). The most critical residues appear to lie at positions 40–45, in a sequence that resembles the well-characterized localization signals found in the SV40 T-antigen and certain other nuclear proteins (94). When inserted into a large cytoplasmic protein such as β-galactosidase, this hexapeptide alone is sufficient to target the protein efficiently into nuclei (90). Conversely, mutations in this sequence inhibit nuclear translocation of Rev (89–91). Because of its small size (13.5 kDa), however, Rev appears capable of diffusing passively through nuclear pores, which will admit molecules of up to 70–80 kDa (95). Thus, Rev mutants which lack this translocation signal are not completely excluded from nuclei, but rather tend to be distributed equally between the nucleoplasm and cytoplasm (89–91). The fact that such mutants are completely non-functional primarily reflects their inability to bind the RRE. Sequences in and around this same hexapeptide also are responsible for the tendency of Rev to associate with nucleoli (26, 96–98), a tendency it shares with HTLV-I Rex (60, 99, 100). Although early evidence suggested that nucleolar localization might be critical for the activity of Rev-like proteins, it now appears more likely to reflect an incidental affinity of these arginine-rich *trans*-activators for nucleolar proteins (101) or RNAs (92). Provided that RRE binding is maintained, association with this organelle is not required for Rev activity (102).

Oligomerization

Purified Rev protein has a strong tendency to oligomerize. It exists predominantly as a tetramer in solution (41, 42, 84), and can form much higher order complexes under some conditions (87), even in the absence of RNA. Evidence of substrate-independent Rev protein oligomerization has also been observed within cells (42,

78). The sequences responsible for this oligomerization lie on either side of the arginine-rich region (approximately residues 14–26 and 50–56), and may extend into it as well (40–42). Some mutations in this region block oligomerization but do not prevent RRE binding, suggesting that the monomeric form of Rev is competent for binding (39, 40). However, such mutations significantly reduce the stoichiometry of binding: whereas a single RRE can accommodate up to eight copies of wild-type Rev (37, 39), it binds only one copy of an oligomerization-defective mutant (40). This suggests that protein–protein interactions among Rev monomers may be primarily responsible for allowing more than one copy of Rev to bind the RRE. Significantly, Rev mutants that bind but fail to oligomerize are functionally inactive (40); this suggests that multiple copies of Rev must be bound to the RRE in order to trigger a response.

The effector domain

A separate, more C-terminal region also is required for Rev activity (58, 89–91, 103), but its exact function remains unknown. This second region, which will be referred to here as the effector domain (some other authors favour the terms 'activation' or 'leucine-rich' domain) has been mapped by mutagenesis to approximately residues 75–84. Mutations in the effector domain do not affect protein localization, RRE binding, or oligomerization of Rev, but completely abolish *trans*-activation. It is thought that the effector domain might mediate critical contacts with unidentified cellular proteins required for Rev activity.

Mutations in Rev that do not involve these two essential domains generally are well tolerated. In particular, residues 1–8 and 85–116 can be deleted without any loss of activity (104).

Mutational analyses suggest that Rev-like *trans*-activators from other retroviruses may be organized in a similar manner (60, 81, 100, 105–110). In several such proteins, discrete regions have been identified that mediate RNA binding, nuclear and nucleolar localization, or effector activity. (The existence and possible importance of protein oligomerization signals have not yet been investigated for most of these *trans*-activators.) The RNA-binding regions often contain short arginine-rich tracts resembling the one in HIV-1 Rev, although these may not always be sufficient for target recognition. For example, HTLV-I Rex contains an arginine-rich tract that clearly contributes to RNA binding (86), but additional sequences elsewhere in the protein are needed for discrimination among alternative RNA targets (81, 107).

Recently, the effector domains from several Rev-like proteins have been found to be functionally interchangeable (81, 108–110). This observation is particularly remarkable because the sequences of these domains have little in common, apart from their overall hydrophobic character, relatively high leucine and proline contents, and some poorly conserved features of primary sequence (58, 81). It is tempting to speculate that these dissimilar sequences might share a common secondary structure that can account for their similar activities. Whatever its basis, this interchangeability of effector domains adds credence to the view that all Rev-like *trans*-activators function through the same pathway, and it further suggests that all effector domains might serve to interact with a common cellular cofactor.

Some Rev-like *trans*-activators are known to be modified post-translationally by phosphorylation at one or more serine residues (104, 111–113). The extent to which the HTLV-I and HTLV-II Rex proteins are phosphorylated has been reported to affect their ability to bind target RNAs (113). The significance of this modification for HIV-1 Rev, on the other hand, is uncertain: there is evidence that Rev phosphorylation increases somewhat in response to agents that activate protein kinases (111), but mutations that eliminate the serines involved have no apparent effects on Rev function (89, 104).

3.6 *Trans*-dominant inhibitory mutants

Dominant negative mutations are those that give rise to a protein which not only lacks intrinsic activity, but also is capable of blocking the action of the wild-type protein in *trans* (114). In some cases, this *trans*-dominant inhibition occurs because the mutant protein competes for an essential substrate or cofactor that is present in limiting amounts. Alternatively, for proteins that multimerize, a mutant may associate with the wild-type protein to form inactive mixed multimers.

Certain mutations in Rev-like proteins give rise to a dominant negative phenotype (58, 89, 91, 98, 103, 105, 106, 108, 115, 116). The resulting mutants can block *trans*-activation by their wild-type counterparts, provided that the mutant is present in considerable excess. The first such mutations to be described were in HIV-1 Rev, where essentially all mutations that inactivate the effector domain (Fig. 6) confer dominant negativity. It was initially thought that these mutant proteins blocked *trans*-activation by competing for binding to the RRE, but later experiments suggest that they act primarily by forming non-functional oligomeric complexes with wild-type Rev (78). Mutations in the effector domains of some other Rev-like proteins are also dominant negative, but this rule does not always apply. In HTLV-I Rex, for example, the mutations that confer dominant negativity fall on either side of the effector domain; Rex variants with mutations in the effector domain itself are inactive but not inhibitory (81, 98, 105–107, 110). Interestingly, dominant negative mutants of Rex can also block HIV-1 *trans*-activation by Rev (98, 105, 106), providing additional evidence that Rev and Rex may act through the same pathway.

3.7 Properties of the response elements

The location of a response element within the viral transcript does not appear to be important. The HIV-1 RRE, for example, is normally situated within a potential intron, but works equally well when transplanted into an upstream or downstream exon (29). By contrast, the Rex response elements (RxREs) of HTLV-I (Fig. 7) and HTLV-II are encoded in the viral LTRs, and so can be found near the ends of their respective viral transcripts (46, 48, 49, 117–119). Responsiveness is lost, however,

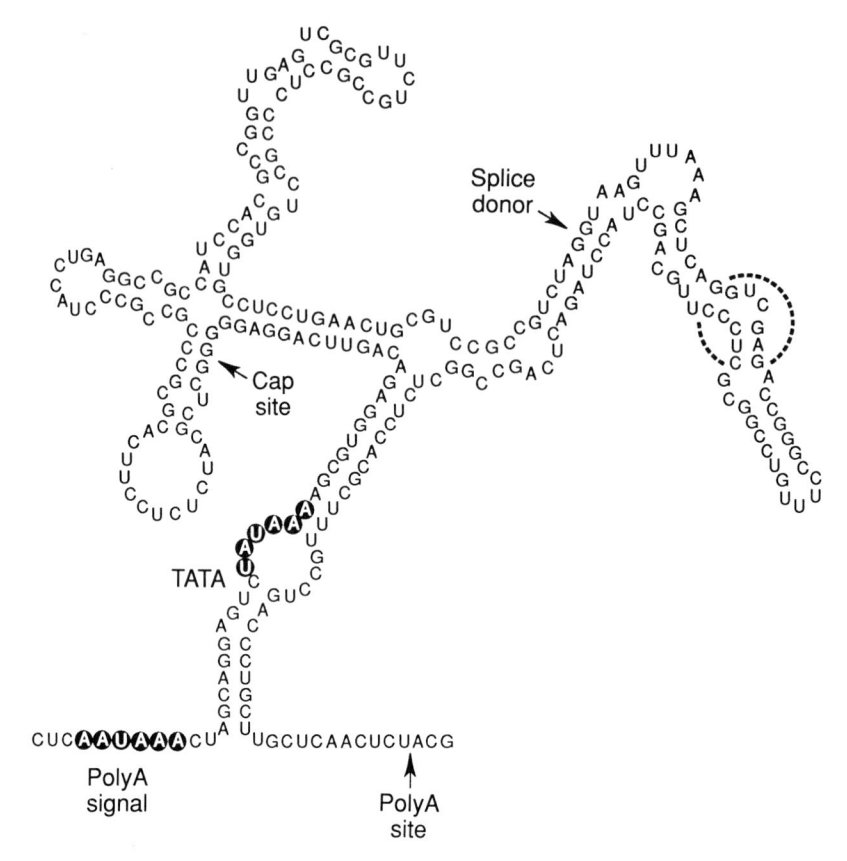

Fig. 7 Predicted structure of the Rex response element found at the 3' end of all HTLV-I mRNAs (61, 118). Locations of the cap site, a splice donor site, TATA box, polyadenylation signal (polyA signal), and the cleavage/polyadenylation site (polyA site) are indicated. A portion of this element (beginning at the cap site) is also found at the 5' end of most transcripts. The broken circle indicates the principal binding site of Rex protein

when the DNA sequence that encodes a response element is inverted or moved outside the transcription unit (29).

Apart from their overall size and potential to fold into stable stems and loops, the response elements of different retroviruses show little similarity to one another (32, 48, 53). This may, however, merely reflect our imperfect knowledge of their structures. The folding patterns usually depicted for these elements (e.g. Figs 2 and 7) are based on theoretical calculations and on inference from mutational analyses — none of these structures has been directly confirmed. Moreover, the computer algorithms used to predict these structures only considered conventional Watson–Crick interactions, with the additional possibility of G-U base-pairing. Many other types of bonding interactions can potentially occur between RNA nucleotides (120); these could significantly influence the actual folding patterns (121), and might also be important for the *trans*-activation response.

Purified forms of the HIV-1, HIV-2, HTLV-I, HTLV-II, and Visna Rev proteins have each been shown to bind directly and specifically to their cognate response elements, with binding constants in the nanomolar range (33–42, 49, 50, 65, 85–88, 98, 109, 113, 121). Binding can readily be detected using gel retention, filter-binding, and RNase- or chemical-protection assays with radiolabelled RNA substrates. HIV-1 Rev, for example, binds the RRE with a dissociation constant of approximately 1 nM, exhibits a roughly 1500-fold lower affinity for such non-specific substrates as ribosomal or transfer RNA, and shows only minimal binding to single- or double-stranded DNA (33, 37, 38). Similarly, HTLV-I Rex binds with approximately 400-fold higher affinity to its cognate RxRE from HTLV-I than to non-specific RNAs (50).

Features of the response elements that are needed for *trans*-activator binding and function have been mapped by mutagenesis (36–39, 49, 61, 62, 85–87, 121–128). The HIV-1 RRE has been the subject of particularly exhaustive study. Surprisingly, most of this 240-base element can be mutated or deleted without eliminating either Rev protein binding *in vitro* or Rev-responsiveness *in vivo* (126–128, and refs therein). The only indispensible sequences lie in a domain known as stem–loop II (SLII), where a cluster of 13 nucleotides constitutes the principal Rev-binding site of the RRE (87, 121, 128). These 13 residues are thought to form an asymmetric bulged duplex or 'bubble' structure (see Fig. 2), which may involve non-canonical (i.e. non-Watson–Crick) interactions between some of the bases. The secondary structure of this locus appears critical for Rev binding, as are the specific identities of some its constituent nucleotides (87, 121, 123, 128).

The primary binding locus in SLII is sufficient to bind a single copy of Rev (39, 128). Once this site is occupied, binding of additional Revs to the full-length RRE appears to result largely from protein–protein interactions between Rev monomers. As the additional monomers bind, some of them may contact other discrete regions in the RRE (88), but none of these secondary regions is capable of serving as an independent Rev-binding site. Thus, mutations in SLII eliminate all Rev binding to the RRE, and abolish Rev responsiveness *in vivo*. Significantly, however, such mutations do not prevent *trans*-activation by the HTLV-I Rex protein, which binds elsewhere in the RRE (61, 86, 125). This implies that no single region of the RRE is absolutely required for responsiveness, except in so far as it is needed to bind a particular *trans*-activator.

This conclusion was further extended in experiments where HIV-1 Rev and HTLV-I Rex were fused with a bacteriophage RNA-binding protein, thereby enabling them to bind a heterologous RNA sequence (102). When tethered to RNA by this phage protein, Rev and Rex each proved capable of inducing cytoplasmic expression of unspliced transcripts that contained no RRE or RxRE sequences. Like responses through the viral elements, *trans*-activation under these conditions was completely eliminated by mutations in the viral effector domains, and occurred only when multiple copies of a fusion protein were bound to each RNA. These results demonstrated that the ability to bind multiple copies of Rev or Rex protein is the only essential requirement for RRE or RxRE function.

But do the response elements really only serve as binding sites for viral *trans*-activators? To the contrary, some data suggest that these elements have other properties that help maximize the efficiency of *trans*-activation. For example, many mutations in the RRE that cause no apparent change in Rev binding nevertheless reduce the magnitude of the response (127). Similarly, responses achieved by Rev and Rex through a phage RNA-binding site were 50–65 per cent weaker than those mediated by the viral response elements (102). One possibility is that the elaborately folded structures of these elements simply help increase accessibility of the *trans*-activator binding site. Given their size and complexity, however, these elements could certainly accommodate binding of other specific factors (such as cellular proteins) that might contribute to or modulate the response. Consistent with this view, uninfected human cells express at least one protein that is capable of binding the HIV-1 RRE (129), although the possible role of this protein in *trans*-activation remains to be determined.

Given the multitude of overlapping functions in retroviral genomes, it is not surprising that the RNA making up these response elements often serves additional purposes as well. For example, the HTLV-I RxRE (Fig. 7) encompasses the TATA box and cap site of the viral promoter as well as a splice donor region (61). In so far as is known, however, none of these overlapping functions is needed for activity of a response element. Interestingly, the RxRE also lies between a polyadenylation signal sequence and the actual polyadenylation site at the 3' end of the HTLV-I transcript; it is thought that folding of the RxRE juxtaposes these two loci to yield accurate polyadenylation (118). This provides insight into the mechanism of 3' end formation in eukaryotic transcripts, and also serves as indirect evidence that the RxRE has a folded conformation *in vivo*.

3.8 The mechanism of post-transcriptional *trans*-activation

Despite intensive research, we do not yet understand how Rev and its analogues induce cytoplasmic expression of RNAs that would otherwise remain inside the nucleus. To a large extent, this is due to our lack of fundamental knowledge about the normal cellular mechanisms that regulate processing, stability, and nucleocytoplasmic export of RNA, as well as to a lack of satisfactory model systems for studying such phenomena. In fact, one compelling reason to study the Rev-like proteins lies in the insight they may provide into the inner workings of the nucleus.

The ability of Rev to function in cells from humans, mice, birds, or *Drosophila* implies that it acts through a pathway that is evolutionarily ancient and highly conserved (58, 82). Speculation about the nature of this pathway has generally focused on two types of hypothetical models (130, 131). Many variants of each model can be envisioned, but only a few will be considered here for purposes of illustration:

1. *Splicing inhibition*. Some models hold that Rev-like proteins act primarily by inhibiting splicing. According to one version, intron-containing RNAs from both cellular genes and complex retroviruses tend to be retained in the nucleus, primarily because of their interactions with cellular splicing factors. Upon binding to such

RNAs, a *trans*-activator might act to displace the splicing machinery and thereby permit the unspliced transcripts to be exported by default. These ideas are logically appealing, and are supported by some direct and indirect evidence. For example, mutations that inactivate splice junctions or certain splicing factors in yeast can lead to constitutively increased cytoplasmic expression of unspliced RNAs (132). Elegant studies in mammalian cells likewise suggest that the pathways of RNA splicing and Rev *trans*-activation may be closely intertwined (43, 133). Moreover, the addition of purified HIV-1 Rev protein to a cell-free splicing extract has been found to inhibit splicing of RRE-containing transcripts specifically *in vitro* (134). It is not yet certain, however, that splicing inhibition plays a significant role in the *trans*-activation response *in vivo*. Evidence to the contrary comes from reports that incompletely spliced viral RNAs accumulate within the nucleus even in the absence of any *trans*-activator, and that the response to a *trans*-activator is not necessarily accompanied by changes in the levels of these nuclear RNAs (23, 27–29, 53, 61, 62). Although cytoplasmic expression of the multiply spliced mRNAs does tend to decline in response to a *trans*-activator (18, 67), this may simply result from the diversion of precursor RNAs to the cytoplasm. The ability of Rev to inhibit *in vitro* splicing was ascribed to a specific interaction between the RRE and the arginine-rich region of Rev; in fact, a peptide corresponding to this region proved a much more potent inhibitor than the full-length protein (134). However, this *in vitro* phenomenon differed from *in vivo trans*-activation in that it did not require an effector domain, and occurred only if the RRE was located within an intron (134). More importantly, there is as yet no evidence that the arginine-rich region alone has any activity *in vivo*.

2. *Transport*. An alternative view is that binding of a Rev-like protein directly triggers nucleocytoplasmic transport of the bound RNA. Such models need not specify which factors ordinarily confine structural RNAs to the nucleus, but simply assume that the transport machinery can override these factors. A major drawback of the transport models is that there currently is no way to test them. We know so little about the normal mechanisms of nuclear RNA transport that it is hard to prove they exist, and even harder to distinguish transport from other intranuclear events. For example, although efforts have been made to show that Rev can control expression of transcripts that contain no functional splice sites (29), it has not been possible to rule out the possibility that cryptic sites in these transcripts interact with the splicing machinery (43).

The splicing-inhibition and transport models are not, of course, mutually exclusive. It may well be that viral *trans*-activation involves some combination of both effects. Nor are these the only models that can be envisioned. Some authors have proposed that selective changes in RNA stability or translatability might also contribute to the response (23, 135, 136). Still other data suggest that HIV-1 structural mRNAs contain specific *cis*-acting elements which cause them to be selectively degraded or retained in the nucleus, and that Rev might act, in part, by overcoming the effect of these negative elements (30, 137, 138).

One approach to resolving this issue would be to search for cellular factors that interact specifically with Rev-like proteins, and then to determine the physiological role of those factors in the cell. In one such study, where nuclear extracts were fractionated on an affinity column bearing immobilized HIV-1 Rev, this *trans*-activator was found to bind tightly and specifically to the B23 class of nucleolar proteins, which are known to shuttle continuously between nucleolus and cytoplasm (101). This interaction might well explain the nucleolar localization of Rev, but it is unlikely to contribute directly to *trans*-activation, since Rev dissociates from these proteins upon binding the RRE (101). Moreover, from a practical standpoint, the abundance of B23 proteins in the nucleus may make it difficult to detect other, less abundant Rev-binding factors. Even if such factors can be identified, it may be still more difficult to prove that they contribute to *trans*-activation. The argument might be strengthened if several *trans*-activators were found to interact with a single cellular protein, or with different components in a single cellular pathway, particularly if such interaction occurred through the effector domains. Ultimately, however, it would be necessary to demonstrate the physiological role of any such factor in a functional *trans*-activation assay. Until then, the mechanism of action of Rev-like proteins will remain one of the most tantalizing mysteries in viral gene regulation.

4. Translation of retroviral mRNAs

Most cellular mRNAs encode only one protein, and can be translated in a straightforward manner. Ribosomes initially bind to the 5' cap of an mRNA, then scan toward the 3' end of the transcript in search of an AUG start codon (139, 140). The likelihood that translation will begin at any particular AUG is determined by its location and context: start codons located nearest the 5' end, and whose flanking sequences most closely match a short consensus motif (CCPuCCAUGG), are utilized most efficiently (140–142). When a single start codon located very near the cap site satisfactorily matches the consensus, it is utilized to the exclusion of all others.

Many retroviral mRNAs, however, are multicistronic—that is, they contain two or more functional open reading frames that either overlap one another or are arranged in tandem. Although the translational machinery tends to favour the upstream reading frames, those positioned farther downstream also can be expressed. In fact, certain essential structural genes, including the *pol* genes of all retroviruses and the *env* gene of HIV-1, are always expressed from downstream reading frames. The expression of these genes illustrates two of the major strategies of translational control that are utilized by retroviruses.

4.1 Ribosomal frameshifting and the expression of Pol

The unspliced genomic transcripts of both simple and complex retroviruses provide the sole template for synthesis of Gag and Pol. Each of these proteins must be synthesized in appropriate stoichiometric amounts: in general, a given virus produces at least five times more Gag than Pol, and deviations from this ratio can severely impair virion assembly (143). The virus, therefore, confronts the need to express unequal amounts of two different translational products from a single mRNA. To make matters more complicated, no functional start codon is present at the 5′ end of the *pol* gene in any known retrovirus, so that these genes are incapable of being translated independently. Instead, *pol* genes are translated along with *gag* to produce a single Gag/Pol fusion protein. After translation, this fusion protein undergoes precise proteolytic cleavage and other modifications to produce (from the Gag sequence) the various capsid glycoproteins and (from Pol) the viral protease, integrase, and reverse transcriptase.

In some retroviruses, such as MLV, the *gag* and *pol* genes lie in the same translational reading frame separated only by an in-frame stop codon. These viruses produce the Gag/Pol fusion protein by allowing a minority of ribosomes to read through the stop codon without interrupting translation (144, 145).

In other retroviruses, however, including RSV and HIV-1, *gag* and *pol* occupy different reading frames and overlap one another slightly at their respective 3′ and 5′ ends (Fig. 8). To express the *pol* gene products, such viruses depend upon a

Fig. 8 Ribosomal frameshifting in the *gag/pol* region of HIV-1. A portion of the Gag protein is shown above the mRNA sequence that encodes it. A frameshift in the −1 direction at the heptameric shift site (highlighted in the block) transfers some ribosomes into the Pol reading frame indicated below the mRNA sequence. LTR = long terminal repeat, with internal R region (solid rectangle)

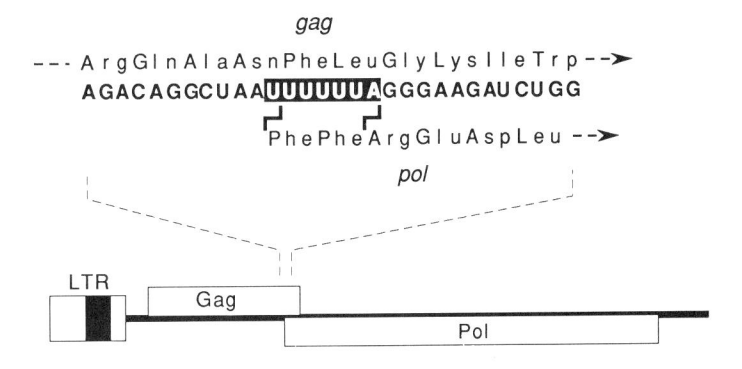

phenomenon known as ribosomal frameshifting (146, 147). In these retroviruses, as in others, all translation of the genomic mRNA initiates at the 5′ end of *gag*, and most ribosomes traverse the entire *gag* region in frame. A short distance upstream of the *gag* termination codon, however, some ribosomes shift into the *pol* reading frame without interrupting the growing polypeptide chain. These ribosomes then continue translation through the *pol* sequence to produce the Gag/Pol fusion polypeptide, which later can be cleaved to release the individual protein products.

Ribosomal frameshifting is utilized by several evolutionarily unrelated species of retroviruses to control *pol* gene expression (146–151). Some, such as the mouse mammary tumour virus (MMTV), must carry out two separate frameshift events in order to access all of the *pol* gene products (148, 149). This same phenomenon also takes place on RNAs from certain prokaryotic and yeast genetic elements (152–154), coronaviruses (155–157), and at least one bacterial gene (158, 159), but has not yet been observed in any higher eukaryotic cellular mRNAs. The frameshift events that occur in most retroviruses share at least three notable features:

1. Shifting always takes place at precisely the same location in a given mRNA.
2. Shifting invariably occurs in the −1 direction; that is, the tRNAs shift backward by one base.
3. The efficiency of shifting varies among viruses; typically, 5–20 per cent of ribosomes undergo the shift.

The effect of a ribosomal frameshift is to produce two different proteins in a fixed ratio using a single translational initiation site. Elegant studies of the mechanism of frameshifting indicate that this process depends upon two specific types of signals in a retroviral mRNA (147, 148, 156, 160). One of these is the sequence of the shift site itself; the other is a distinctive type of secondary structure that is formed by the RNA immediately downstream of this site (Fig. 9).

1. *The shift site*. Shifting occurs as the ribosome traverses a particular hepta-nucleotide in the mRNA. This site nearly always conforms to the sequence X XXY YYZ, where the two triplets as shown correspond to codons in the *gag* reading frame, and where X may be identical to Y. The significance of this sequence is that it enables the ribosome to shift backward one base (i.e. into the reading frame XXX YYY Z) without changing the identities of the first two bases in either codon. Thus,

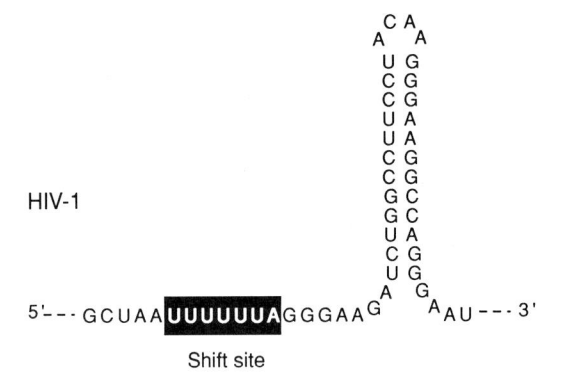

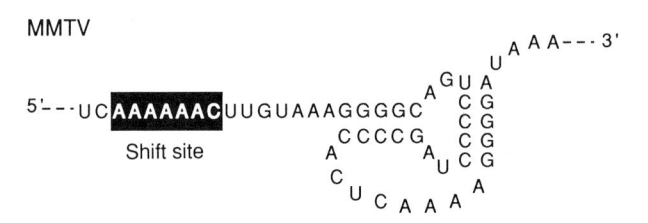

Fig. 9 Putative RNA secondary structures downstream of the ribosomal frameshift sites (highlighted by dark blocks) in *gag/pol* mRNAs from HIV-1 and from MMTV. Only one of the two frameshifting regions from the MMTV transcript is depicted. The secondary structure shown in the MMTV transcript is that of a pseudoknot

the two tRNAs associated with the ribosome at the time of the shift can each maintain at least two out of three base-pair contacts with the mRNA. Interestingly, the identity of the seventh base (Z) in the heptanucleotide appears to be crucial, as it determines which species of host tRNA will bind the second codon prior to the switch. There is some evidence that tRNAs which bind weakly to this 'wobble' base (in some cases, because they contain a modified base such as queosine at the corresponding position in the anticodon loop) may have a greater propensity to frameshift (160).

2. *The downstream secondary structure*. The second critical element lies a short distance downstream of the shift site, and coincides with a 20–40 base region of viral RNA that is predicted to fold into a stable secondary structure (Fig. 9). In some viruses, such as HIV-1, this region appears to be a simple RNA stem–loop (150, 151). In others, such as MMTV, it takes a slightly more complex form known as a pseudoknot, which consists of a short stem–loop that is folded to enable several nucleotides of the loop to interact with a complementary sequence immediately downstream (Fig. 9). The sequence of this element does not appear important, but its secondary structure is critical; mutations that destabilize the structure eliminate frameshifting. Interestingly, in viruses that normally contain a pseudoknot, substitution of a simple stem–loop at the same location yields significantly lower frameshifting efficiency (161).

According to one hypothesis, the presence of the tightly folded stem–loop or pseudoknot tends to interrupt the progress of a ribosome along the mRNA, causing it to linger at the shift site long enough for frameshifting to occur. But the story may well be more complex: certain ribosomal proteins are known to bind specifically to pseudoknots (162), and recognition of such a structure by one or more of these proteins might provoke a specific realignment of the translation complex. Neither the structured region nor the shift site alone is sufficient to trigger a frameshift *in vivo*.

4.2 Translation of HIV-1 Env and other proteins from multicistronic mRNAs

Most of the spliced mRNAs produced by a complex retrovirus are multicistronic, and so can potentially express more than one protein (17, 163). In general, the 5'-most open reading frame tends to be utilized most efficiently, as its start codon is the first to be encountered by ribosomes as they scan along the transcript (140). If the sequences flanking this start codon are suboptimal, however, many ribosomes may continue beyond it and initiate translation at a different start codon farther downstream (140, 141). Thus, depending upon their context, start sites can be inherently weak or strong. Many retroviral genes contain weak start codons, and this has important implications for the pattern of protein expression from multicistronic mRNAs.

For example, none of the mRNAs expressed by HIV-1 has *env* as its first open

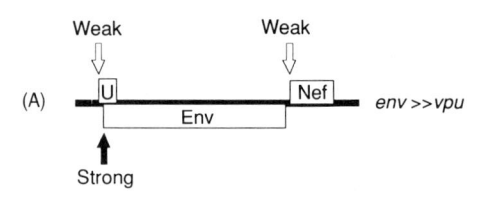

Fig. 10 Three representative multicistronic mRNAs from HIV-1. Only the complete translational reading frames present in each mRNA are shown. Arrows indicate the relative strength of the AUG initiation codon in each open reading frame; the predominant protein products of each mRNA are listed at right. U = Vpu. From data in ref. 17

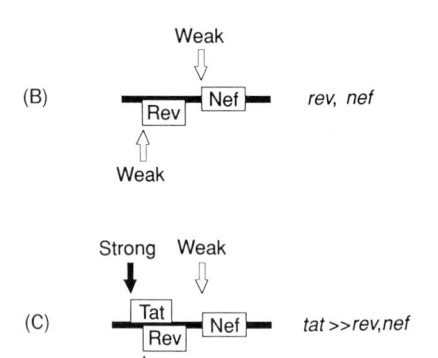

reading frame. Instead, this essential structural gene is always found downstream of *vpu*, a short accessory gene whose coding sequences overlap the *env* start site (Fig. 10A) (16). Efficient translation of *env*, therefore, depends upon the fact that *vpu* has a very weak start site, so that ribosomes frequently scan through it to reach the *env* region (17). When the *vpu* start site is deliberately altered *in vitro* to increase its efficiency, expression of the downstream *env* gene is severely reduced (17). Not surprisingly, the weakness of the *vpu* initiation site is well conserved among HIV-1 isolates; in fact, several replication-competent HIV-1 strains carry mutations that completely inactivate this site (17).

Similarly, all HIV-1 mRNAs that have *rev* as the first open reading frame also contain the *nef* sequence downstream, and, when translated, produce comparable amounts of both proteins (Fig. 10B) (15, 164). Nef expression from these mRNAs occurs because the *rev* start codon is relatively weak; engineered mutations that yield a strong *rev* start site tend to inhibit expression of the downstream *nef* gene (17). On the other hand, all HIV-1 mRNAs that code for the two-exon form of Tat also contain complete *rev* and *nef* sequences, but the latter remain translationally silent (Fig. 10C). This is primarily because the start codon of *tat* is very strong, and allows few ribosomes to pass by without initiating. Mutations that impair the *tat* start site in this mRNA lead to increased translation of both *rev* and *nef* (17).

The relative strengths of the *tat* and *rev* start codons are conserved among primate immunodeficiency viruses: in HIV-1, HIV-2, and SIV, the start site of *tat* is strong, while that of *rev* is weak. This conserved inequality may have significant biological consequences, as it tends to allow Tat to accumulate more rapidly than Rev during the early phase of infection. This may be advantageous to the virus because Rev tends to down-regulate Tat through its effect on mRNA expression (67, 165). Moreover, inefficient *rev* translation would also tend to delay the onset of the late, productive phase of the viral life cycle, and so may be in part responsible for the relatively protracted course of HIV and SIV infections.

5. Potential targets for antiretroviral therapy

The specialized post-transcriptional and translational strategies used by retroviruses are not only fascinating regulatory phenomena, but also offer potential targets for the development of new antiviral therapies. The pathways used by these viruses to accomplish ribosomal frameshifting or to express incompletely spliced RNAs in the cytoplasm are especially interesting in this regard, since both processes are indispensable for viral replication but are seldom (if ever) utilized by mammalian cells.

No useful inhibitors of ribosomal frameshifting have yet been discovered, but one can envision several ways in which such inhibitors might work. It might be possible, for example, to design pharmacologic agents that could mask RNA pseudoknots or otherwise prevent these structures from influencing the progression of ribosomes along a viral transcript. An alternative approach would be to target the specific tRNA species that seem to potentiate frameshifting (160), with the hope that the remaining species would suffice for translation of cellular mRNAs. It has been suggested that frameshifting may depend upon the presence of a modified base (such as queosine) in the tRNA anticodon loop at the position which decodes the last base of the heptameric shift site (Figs 8 and 9); if so, inhibitors might be found which could specifically block either the synthesis of such tRNAs or their interaction with the translational machinery.

The phenomenon of post-transcriptional *trans*-activation could likewise be blocked in a number of ways. For example, dominant negative mutants of the Rev-like proteins show some promise as potential antiretroviral agents: when expressed at high levels, mutant forms of HTLV-I Rex or HIV-1 Rev have each been found to reduce replication of HIV-1 in cultured cells (89, 105). It has been suggested that such mutants could be introduced into human tissues by gene therapy as a way to protect against HIV infection (166), although substantial technical hurdles would have to be overcome to make this approach practicable.

It might also be possible to prevent Rev and its counterparts from binding to viral transcripts, either by obstructing the binding domains of the proteins themselves or by disrupting the viral response elements. For example, it has been reported that short antisense oligonucleotides that are complementary to portions of the RRE can interfere with Rev binding *in vitro* (167). Antisense DNAs directed against the Rev coding sequence, when chemically modified to enhance their entry into living cells, also possess some antiviral activity under cell-culture conditions (168). In similar fashion, analogues of the RRE have been introduced into infected cells to serve as competitive inhibitors of Rev binding, and these RRE 'decoys' have been found to suppress Rev function within cultured cells (169, 170). Other promising, although still highly speculative, approaches might be to design or discover compounds that interfere with Rev protein oligomerization, or with the interaction between an effector domain and its putative cellular cofactor, since these intermolecular contacts appear critical for *trans*-activation. Eventually, through approaches such as these, the fact that the lives of retroviruses depend so completely on specialized regulatory pathways might well be used against them to our advantage.

Acknowledgements

I thank B. R. Cullen, M. B. Feinberg, T. Hope, S. Lee, and H. Madhani for helpful criticism of the manuscript. Supported in part by NIH grants AI27313 and GM37036, and by a Scholar Award from the Leukemia Society of America.

References

1. Padgett, R. A., Grabowski, P. J., Konarska, M. M., Seiler, S. R., and Sharp, P. A. (1986) Splicing of messenger RNA precursors. *Annu. Rev. Biochem.*, **55**, 1119.
2. Green, M. R. (1991) Biochemical mechanisms of constitutive and regulated pre-mRNA splicing. *Annu. Rev. Cell Biol.*, **7**, 559.
3. Maniatis, T. (1991) Mechanisms of alternative pre-mRNA splicing. *Science*, **251**, 33.
4. Katz, R. A. and Skalka, A. M. (1990) Control of retroviral RNA splicing through maintenance of suboptimal processing signals. *Mol. Cell. Biol.*, **10**, 696.
5. Berberich, S. L. and Stoltzfus, C. M. (1991) Mutations in the regions of the Rous sarcoma virus 3' splice sites: implications for regulation of alternative splicing. *J. Virol.*, **65**, 2640.
6. Fu, X.-D., Katz, R. A., Skalka, A. M., and Maniatis, T. (1991) The role of branchpoint and 3' exon sequences in the control of balanced splicing of avian retrovirus RNA. *Genes Dev.*, **5**, 211.
7. Arrigo, S. and Beemon, K. (1988) Regulation of Rous sarcoma virus RNA splicing and stability. *Mol. Cell. Biol.*, **8**, 4858.
8. Katz, R. A., Kotler, M., and Skalka, A. M. (1988) *cis*-acting intron mutations that affect the efficiency of avian retroviral RNA splicing: implication for mechanism of control. *J. Virol.*, **62**, 2686.
9. Stoltzfus, C. M. and Fogarty, S. J. (1989) Multiple regions in the Rous sarcoma virus *src* gene intron act in *cis* to affect the accumulation of unspliced RNA. *J. Virol.*, **63**, 1669.
10. McNally, M. T., Gontarek, R. R., and Beemon, K. (1991) Characterization of Rous sarcoma virus intronic sequences that negatively regulate splicing. *Virology*, **185**, 99.
11. McNally, M. T. and Beemon, K. (1992) Intronic sequences and 3' splice sites control Rous sarcoma virus RNA splicing. *J. Virol.*, **66**, 6.
12. Muesing, M. A., Smith, D. H., Cabradilla, C. D., Benton, C. V., Lasky, L. A., and Capon, D. J. (1985) Nucleic acid structure and expression of the human AIDS/lymphadenopathy retrovirus. *Nature*, **313**, 450.
13. Guatelli, J. C., Gingeras, T. R., and Richman, D. D. (1990) Alternative splice acceptor

utilization during human immunodeficiency virus type 1 infection of cultured cells. *J. Virol.*, **64**, 4093.

14. Robert-Guroff, M., Popovic, M., Gartner, S., Markham, P., Gallo, R. C., and Reitz, M. S. (1990) Structure and expression of *tat-*, *rev-*, and *nef*-specific transcripts of human immunodeficiency virus type 1 in infected lymphocytes and macrophages. *J. Virol.*, **64**, 3391.
15. Schwartz, S., Felber, B. K., Benko, D. M., Fenyo, E. M., and Pavlakis, G. N. (1990) Cloning and functional analysis of multiply spliced mRNA species of human immunodeficiency virus type 1. *J. Virol.*, **64**, 2519.
16. Schwartz, S., Felber, B. K., Fenyo, E. M., and Pavlakis, G. N. (1990) Env and Vpu proteins of human immunodeficiency virus type 1 are produced from multiple bicistronic mRNAs. *J. Virol.*, **64**, 5448.
17. Schwartz, S., Felber, B. K., and Pavlakis, G. N. (1992) Mechanism of translation of monocistronic and multicistronic human immunodeficiency virus type 1 mRNAs. *Mol. Cell. Biol.*, **12**, 207.
18. Malim, M. H., Hauber, J., Fenrick, R., and Cullen, B. R. (1988) Immunodeficiency virus *rev trans*-activator modulates the activity of the viral regulatory genes. *Nature*, **335**, 181.
19. Schwartz, S., Felber, B. K., and Pavlakis, G. N. (1991) Expression of human immunodeficiency virus type-1 vif and vpr mRNAs is Rev-dependent and regulated by splicing. *Virology*, **183**, 677.
20. Krainer, A., Conway, G. C., and Kozak, D. (1990) Purification and characterization of pre-mRNA splicing factor SF2 from HeLa cells. *Genes Dev.*, **4**, 1158.
21. Feinberg, M. B., Jarrett, R. F., Aldovini, A., Gallo, R. C., and Wong-Staal, F. (1986) HTLV-III expression and production involve complex regulation at the levels of splicing and translation of viral RNA. *Cell*, **46**, 807.
22. Sodroski, J., Goh, W. C., Rosen, C., Dayton, A., Terwilliger, E., and Haseltine, W. (1986) A second post-transcriptional *trans*-activator gene required for HTLV-III replication. *Nature*, **321**, 412.
23. Felber, B. K., Hadzopoulou-Cladaras, M., Cladaras, C., Copeland, T., and Pavlakis, G. N. (1989) rev protein of human immunodeficiency virus type 1 affects the stability and transport of the viral mRNA. *Proc. Natl Acad. Sci. USA*, **86**, 1495.
24. Gallo, R., Wong-Staal, F., Montagnier, L., Haseltine, W. A., and Yoshida, M. (1988) HIV/HTLV gene nomenclature. *Nature*, **333**, 504.
25. Cullen, B. R., Hauber, J., Campbell, K., Sodroski, J. G., Haseltine, W. A., and Rosen, C. A. (1988) Subcellular localization of the human immunodeficiency virus *trans*-acting *art* gene product. *J. Virol.*, **62**, 2498.
26. Hope, T. J., Huang, X., McDonald, D., and Parslow, T. G. (1990) Steroid-receptor fusion of the HIV-1 Rev transactivator: Mapping cryptic functions of the arginine-rich motif. *Proc. Natl Acad. Sci. USA*, **87**, 7787.
27. Emerman, M., Vazeux, R., and Peden, K. (1989) The *rev* gene product of the human immunodeficiency virus affects envelope specific RNA localization. *Cell*, **57**, 1155.
28. Hammarskjöld, M.-L., Heimer, J., Hammarskjöld, B., Sangwan, I., Albert, L., and Rekosh, D. (1989) Regulation of human immunodeficiency virus *env* expression by the *rev* gene product. *J. Virol.*, **63**, 1959.
29. Malim, M. H., Hauber, J., Le, S.-Y., Maizel, J. V., and Cullen, B. R. (1989) The HIV-1 *rev trans*-activator acts through a structured target sequence to activate nuclear export of unspliced viral mRNA. *Nature*, **338**, 254.

30. Rosen, C. A., Terwilliger, E., Dayton, A., Sodroski, J. G., and Haseltine, W. A. (1988). Intragenic cis-acting art gene-responsive sequences of the human immunodeficiency virus. *Proc. Natl Acad. Sci. USA,* **85,** 2071.
31. Hadzopoulou-Cladaras, M., Felber, B. K., Cladaras, C., Athanassopoulos, A., Tse, A., and Pavlakis, G. N. (1989) The *rev (trs/art)* protein of human immunodeficiency virus type 1 affects viral mRNA and protein expression via a *cis*-acting sequence in the *env* region. *J. Virol.,* **63,** 1265.
32. Le, S.-Y., Malim, M. H., Cullen, B. R., and Maizel, J. V. (1990) A highly conserved RNA folding region coincident with the Rev response element of primate immunodeficiency viruses. *Nucleic Acids Res.,* **18,** 1613.
33. Daly, T. J., Cook, K. S., Gray, G. S., Maione, T. E., and Rusche, J. R. (1989) Specific binding of HIV-1 recombinant Rev protein to the Rev-responsive element in vitro. *Nature,* **342,** 816.
34. Zapp, M. L. and Green, M. R. (1989) Sequence-specific RNA binding by the HIV-1 Rev protein. *Nature,* **342,** 714.
35. Cochrane, A. W., Chen, C.-H., and Rosen, C. A. (1990) Specific interaction of the human immunodeficiency virus Rev protein with a structured region in the *env* mRNA. *Proc. Natl Acad. Sci. USA,* **87,** 1198.
36. Daefler, S., Klotman, M. E., and Wong-Staal, F. (1990) Trans-activating Rev protein of the human immunodeficiency virus 1 interacts directly and specifically with its target RNA. *Proc. Natl Acad. Sci. USA,* **87,** 4571.
37. Heaphy, S., Dingwall, C., Ernberg, I., Gait, M. J., Green, S. M., Karn, J., Lowe, A. D., Singh, M., and Skinner, M. A. (1990) HIV-1 regulator of virion expression (Rev) protein binds to an RNA stem–loop structure located within the Rev response element region. *Cell,* **60,** 685.
38. Malim, M. H., Tiley, L. S., McCarn, D. F., Rusche, J. R., Hauber, J., and Cullen, B. R. (1990) HIV-1 structural gene expression requires binding of the Rev trans-activator to its RNA target sequence. *Cell,* **60,** 675.
39. Cook, K. S., Fisk, G. J., Hauber, J., Usman, N., Daly, T. J., and Rusche, J. R. (1991) Characterization of the HIV-1 REV protein: binding stoichiometry and minimal RNA substrate. *Nucleic Acids Res.,* **19,** 1577.
40. Malim, M. H. and Cullen, B. R. (1991) HIV-1 structural gene expression requires the binding of multiple Rev monomers to the viral RRE: Implications for HIV-1 latency. *Cell,* **65,** 241.
41. Olsen, H. S., Cochrane, A. W., Dillon, P. J., Nalin, C. M., and Rosen, C. A. (1990) Interaction of the human immunodeficiency virus type 1 Rev protein with a structured region in env mRNA is dependent on multimer formation mediated through a basic stretch of amino acids. *Genes Dev.,* **4,** 1357.
42. Zapp, M. L., Hope, T. J., Parslow, T. G., and Green, M. R. (1991) Oligomerization and RNA binding domains of the HIV-1 Rev protein: A dual function for an arginine-rich binding motif. *Proc. Natl Acad. Sci. USA,* **88,** 7734.
43. Chang, D. D. and Sharp, P. A. (1989) Regulation by HIV Rev depends upon recognition of splice sites. *Cell,* **59,** 789.
44. Cullen, B. R. (1991) Human immunodeficiency virus as a prototypic complex retrovirus. *J. Virol.,* **65,** 1053.
45. Malim, M. H., Böhnlein, S., Fenrick, R., Le, S.-Y., Maizel, J. V., and Cullen, B. R. (1989) Functional comparison of the Rev *trans*-activators encoded by different primate immunodeficiency virus species. *Proc. Natl Acad. Sci. USA,* **86,** 8222.

46. Lewis, N., Williams, J., Rekosh, D., and Hammarskjöld, M.-L. (1990) Identification of a *cis*-acting element in human immunodeficiency virus type 2 (HIV-2) that is responsive to the HIV-1 *rev* and human T-cell leukemia virus *rex* proteins. *J. Virol.*, **64**, 1690.
47. Hidaka, M., Inoue, J., Yoshida, M., and Seiki, M. (1988) Post-transcriptional regulator (rex) of HTLV-I initiates expression of viral structural proteins but suppresses expression of regulatory protein. *EMBO J.*, **7**, 519.
48. Ahmed, Y. F., Hanly, S. M., Malim, M. H., Cullen, B. R., and Greene, W. C. (1990) Structure–function analyses of the HTLV-I Rex and HIV-1 Rev RNA response elements: insights into the mechanism of Rex and Rev action. *Genes Dev.*, **4**, 1014.
49. Black, A. C., Ruland, C. T., Yip, M. T., Luo, J., Tran, B., Kalsi, A., Quan, E., Aboud, M., Chen, I. S. Y., and Rosenblatt, J. D. (1991) Human T-cell leukemia virus type II Rex binding and activity require an intact splice donor site and a specific RNA secondary structure. *J. Virol.*, **65**, 6645.
50. Grassmann, R., Berchtold, S., Aepinus, C., Ballaun, C., Böehnlein, E., and Fleckenstein, B. (1991) In vitro binding of human T-cell leukemia virus rex proteins to the rex-response element of viral transcripts. *J. Virol.*, **65**, 3721.
51. Sakai, H., Shibata, R., Miura, T., Hayami, M., Ogawa, K., Kiyomasu, T., Ishimoto, A., and Adachi, A. (1990) Complementation of the *rev* gene mutation among human and simian lentiviruses. *J. Virol.*, **64**, 2202.
52. Tiley, L. S., Brown, P. H., Le, S.-Y., Maizel, J. V., Clements, J. E., and Cullen, B. R. (1990) Visna virus encodes a post-transcriptional regulator of viral structural gene expression. *Proc. Natl Acad. Sci. USA*, **87**, 7497.
53. Derse, D. (1988) *Trans*-acting regulation of bovine leukemia virus mRNA processing. *J. Virol.*, **62**, 1115.
54. Felber, B. K., Derse, D., Athanoassopoulos, A., Campbell, M., and Pavlakis, G. N. (1989) Cross-activation of the Rex proteins of HTLV-1 and BLV and of the Rev protein of HIV-1 and non-reciprocal interactions with their RNA responsive elements. *New Biol.*, **1**, 318.
55. Phillips, T. R., Lamont, C., Konings, D. A. M., Shacklett, B. L., Hamson, C. A., Luciw, P. A., and Elder, J. H. (1992) Identification of the Rev transactivation and Rev-responsive elements of feline immunodeficiency virus. *J. Virol.*, **66**, 5464.
56. Stephens, R. M., Derse, D., and Rice, N. R. (1990) Cloning and characterization of cDNAs encoding equine infectious anemia virus Tat and putative Rev proteins. *J. Virol.*, **64**, 3716.
57. Saltarelli, M., Querat, G., Konings, D. A. M., Vigne, R., and Clements, J. E. (1990) Nucleotide sequence and transcriptional analysis of molecular clones of CAEV which generate infectious virus. *Virology*, **179**, 347.
58. Malim, M. H., McCarn, D. F., Tiley, L. S., and Cullen, B. R. (1991) Mutational definition of the human immunodeficiency virus type 1 rev activation domain. *J. Virol.*, **65**, 4248.
59. Querat, G., Audoly, G., Sonigo, P., and Vigne, R. (1990) Nucleotide sequence analysis of SA-OMVV, a visna-related ovine lentivirus: phylogenetic history of lentiviruses. *Virology*, **175**, 434.
60. Siomi, H., Shida, H., Nam, S. H., Nosaka, T., Maki, M., and Hatanaka, M. (1988) Sequence requirements for nucleolar localization of human T cell leukemia virus type I pX protein, which regulates viral RNA processing. *Cell*, **55**, 197.
61. Hanly, S. M., Rimsky, L. T., Malim, M. H., Kim, J. H., Hauber, J., Duc Dodon, M., Le, S.-Y., Maizel, J. V., Cullen, B. R., and Greene, W. C. (1989) Comparative analysis

of the HTLV-I Rex and HIV-1 Rev *trans*-regulatory proteins and their RNA response elements. *Genes Dev.*, **3**, 1534.
62. Toyoshima, H., Itoh, M., Inoue, J.-I., Seiki, M., Takadu, F., and Yoshida, M. (1990) Secondary structure of the human T-cell leukemia virus type I rex-responsive element is essential for rex regulation of RNA processing and transport of unspliced RNAs. *J. Virol.*, **64**, 2825.
63. Rimsky, L., Hauber, J., Dukovich, M., Malim, M. H., Langlois, A., Cullen, B. R., and Greene, W. C. (1988) Functional replacement of the HIV-1 rev protein by the HTLV-1 rex protein. *Nature*, **335**, 738.
64. Sakai, H., Siomi, H., Shida, H., Shibata, R., Kiyomasu, T., and Adachi, A. (1990) Functional comparison of transactivation by human retrovirus rev and rex genes. *J. Virol.*, **64**, 5833.
65. Yip, M. T., Dynan, W. S., Green, P. L., Black, A. C., Arrigo, S. J., Torbati, A., Heaphy, S., Ruland, C., Rosenblatt, J. D., and Chen, I. S. Y. (1991) HTLV-II Rex protein binds specifically to RNA sequences of the HTLV LTR but poorly to the HIV-1 RRE. *J. Virol.*, **65**, 2261.
66. Kim, S. Y., Byrn, R., Groopman, J., and Baltimore, D. (1989) Temporal aspects of DNA and RNA synthesis during human immunodeficiency virus infection: evidence for differential gene expression. *J. Virol.*, **63**, 3708.
67. Felber, B. K., Drysdale, C. M., and Pavlakis, G. N. (1990) Feedback regulation of human immunodeficiency virus type 1 expression by the Rev protein. *J. Virol.*, **64**, 3734.
68. Pomerantz, R. J., Trono, D., Feinberg, M. B., and Baltimore, D. (1990) Cells non-productively infected with HIV-1 exhibit an aberrant pattern of viral RNA expression: A molecular model for latency. *Cell*, **61**, 1271.
69. Pomerantz, R. J., Seshamma, T., and Trono, D. (1992) Efficient replication of human immunodeficiency virus type 1 requires a threshold level of Rev: Potential implications for latency. *J. Virol.*, **66**, 1809.
70. Garcia-Blanco, M. A. and Cullen, B. R. (1991) Molecular basis of latency in pathogenic human viruses. *Science*, **254**, 815.
71. McCune, J. M. (1991) HIV-1: the infective process in vivo. *Cell*, **64**, 351.
72. Ho, D. D., Moudgil, T., and Alam, M. (1989) Quantitation of human immunodeficiency virus type I in the blood of infected persons. *New Engl. J. Med.*, **321**, 1621.
73. Coombs, R. W., Collier, A. C., Allain, J.-P., Nikora, B., Leuther, M., Gjerset, G. F., and Corey, L. (1989) Plasma viremia in human immunodeficiency virus infection. *New Engl. J. Med.*, **321**, 1626.
74. Baltimore, D. and Feinberg, M. B. (1989) HIV revealed: Toward a natural history of the infection. *New Engl. J. Med.*, **321**, 1673.
75. Michael, N. L., Vahey, M., Burke, D. S., and Redfield, R. R. (1992) Viral DNA and mRNA expression correlate with the stage of human immunodeficiency virus (HIV) type 1 infection in humans: Evidence for viral replication in all stages of HIV disease. *J. Virol.*, **66**, 310.
76. Folks, T. M., Justeman, J., Kinter, A., Schnittman, S., Orenstein, J. Poli, G., and Fauci, A. S. (1988) Characterization of a monocyte clone chronically infected with HIV and inducible by 13-phorbol-12-myristate acetate. *J. Immunol.*, **140**, 1117.
77. Michael, N. L., Morrow, P., Mosca, J., Vahey, M. T., Burke, D. S., and Redfield, R. R. (1991) Induction of human immunodeficiency virus type 1 expression in chronically infected cells is associated primarily with a shift in RNA splicing patterns. *J. Virol.*, **65**, 1291.

78. Hope, T. J., Klein, N. P., Elder, M. E., and Parslow, T. G. (1992) *trans*-Dominant inhibition of human immunodeficiency virus type 1 Rev occurs through formation of inactive protein complexes. *J. Virol.*, **66**, 1849.
79. Sadaie, M. R., Benter, T., and Wong-Staal, F. (1988) Site-directed mutagenesis of two *trans*-regulatory genes (*tat*-III, *trs*) of HIV-1. *Science*, **239**, 910.
80. Terwilliger, E., Burghoff, R., Sia, R., Sodroski, J., Haseltine, W., and Rosen, C. (1988) The *art* gene product of human immunodeficiency virus is required for replication. *J. Virol.*, **62**, 655.
81. Hope, T. J., Bond, B. L., McDonald, D., Klein, N. P., and Parslow, T. G. (1991) Effector domains of human immunodeficiency virus type 1 Rev and human T-cell leukemia virus type I Rex are functionally interchangeable and share an essential peptide motif. *J. Virol.*, **65**, 6001.
82. Ivey-Hoyle, M. and Rosenberg, M. (1990) Rev-dependent expression of human immunodeficiency virus type 1 gp160 in *Drosophila melanogaster* cells. *Mol. Cell. Biol.*, **10**, 6152.
83. Daly, T. J., Rusche, J. R., Maione, T. E., and Frankel, A. D. (1990) Circular dichroism studies of the HIV-1 Rev protein and its specific RNA binding site. *Biochemistry*, **29**, 9791.
84. Nalin, C. M., Purcell, R. D., Antelman, D., Mueller, D., Tomchak, L., Wegrzynski, D., McCarney, E., Toome, V., Kramer, R., and Hsu, M.-C. (1990) Purification and characterization of recombinant Rev protein of human immunodeficiency virus type 1. *Proc. Natl Acad. Sci. USA*, **87**, 7593.
85. Ballaun, C., Farrington, G. K., Dobrovnik, M., Rusche, J., Hauber, J., and Böhnlein, E. (1991) Functional analysis of human T-cell leukemia virus type I *rex*-response element: Direct RNA binding of Rex protein correlates with in vivo activity. *J. Virol.*, **65**, 4408.
86. Bogerd, H. P., Huckaby, G. L., Ahmed, Y., Hanly, S. M., and Greene, W. C. (1991) The type I human T-cell leukemia virus (HTLV-I) Rex trans-activator binds directly to the HTLV-I Rex and the type 1 human immunodeficiency virus Rev RNA response elements. *Proc. Natl Acad. Sci. USA*, **88**, 5704.
87. Heaphy, S., Finch, J. T., Gait, M. J., Karn, J., and Singh, M. (1991) Human immunodeficiency virus type 1 regulator of virion expression, rev, forms nucleoprotein filaments after binding to a purine-rich 'bubble' located within the rev-responsive region of viral RNAs. *Proc. Natl Acad. Sci. USA*, **88**, 7366.
88. Kjems, J., Brown, M., Chang, D. D., and Sharp, P. A. (1991) Structural analysis of the interaction between the human immunodeficiency virus Rev protein and the Rev response element. *Proc. Natl Acad. Sci. USA*, **88**, 683.
89. Malim, M. H., Böhnlein, S., Hauber, J., and Cullen, B. R. (1989) Functional dissection of the HIV-1 Rev trans-activator—derivation of a trans-dominant repressor of Rev function. *Cell*, **58**, 205.
90. Perkins, A., Cochrane, A., Ruben, S., and Rosen, C. (1989). Structural and functional characterization of the human immunodeficiency virus *rev* protein. *J. AIDS*, **2**, 256.
91. Hope, T. J., McDonald, D., Huang, X., Low, J., and Parslow, T. G. (1990) Mutational analysis of the human immunodeficiency virus type 1 Rev transactivator: Essential residues near the amino terminus. *J. Virol.*, **64**, 5360.
92. Lazinski, D., Grzadzielska, E., and Das, A. (1989) Sequence-specific recognition of RNA hairpins by bacteriophage antiterminators requires a conserved arginine-rich motif. *Cell*, **59**, 207.

93. Saikumar, P., Murali, R., and Reddy, E. R. (1990) Role of tryptophan repeats and flanking amino acids in Myb-DNA interactions. *Proc. Natl Acad. Sci. USA*, **87**, 8452.
94. Goodson, H. and Silver, P. (1989) Nuclear protein transport. *CRC Crit. Rev.*, **24**, 419.
95. Bonner, W. M. (1975) Protein migration into nuclei. I. Frog oocyte nuclei in vivo accumulate microinjected histones, allow entry to small proteins, and exclude large proteins. *J. Cell Biol.*, **64**, 421.
96. Cochrane, A. W., Perkins, A., and Rosen, C. A. (1990) Identification of sequences important in the nucleolar localization of human immunodeficiency virus Rev: Relevance of nucleolar localization to function. *J. Virol.*, **64**, 881.
97. Venkatesh, L. K., Mohammed, S., and Chinnadurai, G. (1990) Functional domains of the HIV-1 *rev* gene required for *trans*-regulation and subcellular localization. *Virology*, **176**, 39.
98. Böhnlein, E., Berger, J., and Hauber, J. (1991) Functional mapping of the human immunodeficiency virus type 1 Rev RNA binding domain: new insights into the domain structure of Rev and Rex. *J. Virol.*, **65**, 7051.
99. Kubota, S., Siomi, H., Satoh, T., Endo, S., Maki, M., and Hatanaka, M. (1989). Functional similarity of HIV-1 Rev and HTLV-1 Rex proteins: identification of a new nucleolar-targeting signal in Rev protein. *Biochem. Biophys. Res. Commun.*, **162**, 963.
100. Nosaka, T., Siomi, H., Adachi, Y., Ishibashi, M., Kubota, S., Maki, M., and Hatanaka, M. (1989) Nucleolar targeting signal of human T-cell leukemia virus type I Rex-encoded protein is essential for cytoplasmic accumulation of unspliced viral mRNA. *Proc. Natl Acad. Sci. USA*, **86**, 9798.
101. Fankhauser, C., Izaurralde, E., Adachi, Y., Wingfield, P., and Laemmli, U. K. (1991) Specific complex of human immunodeficiency virus type 1 Rev and nucleolar B23 proteins: Dissociation by the Rev response element. *Mol. Cell. Biol.*, **11**, 2567.
102. McDonald, D., Hope, T. J., and Parslow, T. G. (1992) Post-transcriptional regulation by the human immunodeficiency virus type 1 Rev and human T-cell leukemia virus type I Rex proteins through a heterologous RNA binding site. *J. Virol.*, **66**, 7232.
103. Mermer, B., Felber, B. K., Campbell, M., and Pavlakis, G. N. (1990) Identification of *trans*-dominant HIV-1 Rev protein mutants by direct transfer of bacterially produced proteins into human cells. *Nucleic Acids Res.*, **18**, 2037.
104. Cochrane, A. W., Golub, E., Volsky, D., Ruben, S., and Rosen, C. A. (1989) Functional significance of phosphorylation to the human immunodeficiency virus Rev protein. *J. Virol.*, **63**, 4438.
105. Rimsky, L., Duc Dudon, M., Dixon, E. P., and Greene, W. C. (1989) trans-Dominant inactivation of HTLV-I and HIV-1 gene expression by mutation of the HTLV-I Rex transactivator. *Nature*, **341**, 453.
106. Böhnlein, S., Pirker, F. P., Hofer, L., Zimmermann, K., Bachmayer, H., Böhnlein, E., and Hauber, J. (1991) Transdominant repressors for human T-cell leukemia virus type I Rex and human immunodeficiency virus type 1 Rev function. *J. Virol.*, **65**, 81.
107. Hofer, L., Weichselbraun, I., Quick, S., Farrington, G. K., Böhnlein, E., and Hauber, J. (1991) Mutational analysis of the human T-cell leukemia virus type I *trans*-acting *rex* gene product. *J. Virol.*, **65**, 3379.
108. Tiley, L. S., Malim, M. H., and Cullen, B. R. (1991) Conserved functional organization of the human immunodeficiency virus type 1 and visna virus Rev proteins. *J. Virol.*, **65**, 3877.
109. Garrett, E. D. and Cullen, B. R. (1992) Comparative analysis of Rev function in human immunodeficiency viruses types 1 and 2. *J. Virol.*, **66**, 4288.

110. Weichselbraun, I., Farrington, G. K., Rusche, J. R., Böhnlein, E., and Hauber, J. (1991) Definition of the human immunodeficiency virus type 1 Rev and human T-cell leukemia virus type I Rex protein activation domains by functional exchange. *J. Virol.*, **66**, 2583.

111. Hauber, J., Bouvier, M., Malim, M. H., and Cullen, B. R. (1988) Phosphorylation of the *rev* gene product of human immunodeficiency virus type 1. *J. Virol.*, **62**, 4801.

112. Green, P. L., Xie, Y., and Chen, I. S. Y. (1991) The Rex proteins of HTLV-II differ by serine phosphorylation. *J. Virol.*, **65**, 546.

113. Green, P. L., Yip, M. T., Xie, Y., and Chen, I. S. Y. (1992) Phosphorylation regulates RNA binding by the human T-cell leukemia virus Rex protein. *J. Virol.*, **66**, 4325.

114. Herskowitz, I. (1987) Functional inactivation of genes by dominant negative mutations. *Nature*, **329**, 219.

115. Venkatesh, L. K. and Chinnadurai, G. (1990) Mutants in a conserved region near the carboxy-terminus of HIV-1 Rev identify functionally important residues and exhibit a dominant negative phenotype. *Virology*, **178**, 327.

116. Feinberg, M. B. and Trono, D. (1992) Intracellular immunization: *trans*-dominant mutants of HIV gene products as tools for the study and interruption of viral replication. *AIDS Res. Hum. Retrovirol.*, **8**, 1013.

117. Seiki, M., Inoue, J.-I., Hidaka, M., and Yoshida, M. (1988) Two cis-acting elements responsible for post-transcriptional trans-regulation of gene expression of human T cell leukemia virus type I. *Proc. Natl Acad. Sci. USA*, **85**, 7124.

118. Ahmed, Y. F., Gilmartin, G. M., Hanly, S. M., Nevins, J. R., and Greene, W. C. (1991) The HTLV-I Rex response element mediates a novel form of mRNA polyadenylation. *Cell*, **64**, 727.

119. Black, A. C., Chen, I. S. Y., Arrigo, S., Ruland, C. T., Allogiamento, T., Chin, E., and Rosenblatt, J. D. (1991) Regulation of HTLV-II gene expression by rex involves positive and negative cis-acting element in the 5' long terminal repeat. *Virology*, **181**, 433.

120. Saenger, W. (1984) *Principles of Nucleic Acid Structure*. Springer-Verlag, New York.

121. Bartel, D. P., Zapp, M. L., Green, M. R., and Szostak, J. W. (1991) HIV-1 Rev regulation involves recognition of non-Watson-Crick base pairs in viral RNA. *Cell*, **67**, 529.

122. Dayton, E. T., Powell, D. M., and Dayton, A. I. (1989) Functional analysis of CAR, the target sequence for the Rev protein of HIV-1. *Science*, **246**, 1625.

123. Holland, S. M., Ahmad, N., Maitra, R. K., Wingfield, P., and Venkatesan, S. (1990) Human immunodeficiency virus Rev protein recognizes a target sequence in Rev-responsive element RNA within the context of RNA secondary structure. *J. Virol.*, **64**, 5966.

124. Olsen, H. S., Nelbock, P., Cochrane, A. W., and Rosen, C. A. (1990) Secondary structure is the major determinant for interaction of HIV rev protein with RNA. *Science*, **247**, 845.

125. Solomin, L., Felber, B. K., and Pavlakis, G. N. (1990) Different sites of interaction for Rev, Tev, and Rex proteins within the Rev-responsive element of human immunodeficiency virus type 1. *J. Virol.*, **64**, 6010.

126. Huang, X., Hope, T. J., Bond, B. L., McDonald, D., Grahl, K., and Parslow, T. G. (1991) Minimal Rev-response element for type 1 human immunodeficiency virus. *J. Virol.*, **65**, 2131.

127. Dayton, E. T., Konings, D. A. M., Powell, D. M., Shapiro, B. A., Butini, L., Maizel, J. V., and Dayton, A. I. (1992) Extensive sequence-specific information throughout the

CAR/RRE, the target sequence of the human immunodeficiency virus type 1 Rev protein. *J. Virol.,* **66,** 1139.
128. Tiley, L. S., Malim, M. H., Tewary, H. K., Stockley, P. G., and Cullen, B. R. (1992) Identification of a high-affinity RNA-binding site for the human immunodeficiency virus type 1 Rev protein. *Proc. Natl Acad. Sci. USA,* **89,** 758.
129. Vaishnav, Y. N., Vaishnav, M., and Wong-Staal, F. (1991) Identification and characterization of a nuclear factor that specifically binds to the Rev response element (RRE) of human immunodeficiency virus type 1 (HIV-1). *New Biol.,* **3,** 142.
130. Green, M. R. and Zapp, M. L. (1989) Human immunodeficiency virus: revving up gene expression. *Nature,* **338,** 200.
131. Chang, D. D. and Sharp, P. A. (1990) Messenger RNA transport and HIV rev. *Science,* **249,** 614.
132. Legraine, P. and Rosbash, M. (1989) Some *cis-* and *trans-*acting mutants for splicing target pre-mRNA to the cytoplasm. *Cell,* **57,** 573.
133. Lu, X., Heimer, J., Rekosh, D., and Hammarskjöld, M.-L. (1990) U1 small nuclear RNA plays a direct role in the formation of a Rev-regulated human immunodeficiency virus env mRNA that remains unspliced. *Proc. Natl Acad. Sci. USA,* **87,** 7598.
134. Kjems, J., Frankel, A. D., and Sharp, P. A. (1991) Specific regulation of mRNA splicing in vitro by a peptide from HIV-1 Rev. *Cell,* **67,** 169.
135. Knight, D. M., Flomerfelt, F. A., and Ghrayeb, J. (1987) Expression of the Art/Trs protein of HIV and study of its role in viral envelope synthesis. *Science,* **236,** 837.
136. Arrigo, S. J. and Chen, I. S. Y. (1991) Rev is necessary for translation but not cytoplasmic accumulation of HIV-1 *vif, vpr,* and *env/vpu* 2 RNAs. *Genes Dev.,* **5,** 808.
137. Cochrane, A. W., Jones, K. S., Beidas, S., Dillon, P. J., Skalka, A. M., and Rosen, C. A. (1991) Identification and characterization of intragenic sequences which repress human immunodeficiency virus structural gene expression. *J. Virol.,* **65,** 5305.
138. Schwartz, S., Felber, B. K., and Pavlakis, G. N. (1992) Distinct RNA sequences in the *gag* region of human immunodeficiency virus type 1 decrease RNA stability and inhibit expression in the absence of Rev protein. *J. Virol.,* **66,** 150.
139. Shatkin, A. J. (1985) mRNA cap binding proteins: essential factors for initiation of translation. *Cell,* **40,** 223.
140. Kozak, M. (1989) The scanning model for translation: an update. *J. Cell Biol.,* **108,** 229.
141. Kozak, M. (1986) Point mutations define a sequence flanking the AUG initiator codon that modulates translation by eukaryotic ribosomes. *Cell,* **44,** 283.
142. Kozak, M. (1987) An analysis of 5′-noncoding sequences from 699 vertebrate messenger RNAs. *Nucleic Acids Res.,* **15,** 8125.
143. Felsenstein, K. M. and Goff, S P. (1988) Expression of the gag–pol fusion protein of Moloney murine leukemia virus without gag protein does not induce virion formation or proteolytic processing. *J. Virol.,* **62,** 2179.
144. Yoshinaka, Y., Katoh, I., Copeland, T. D., and Oroszlan, S. (1985) Murine leukemia virus protease is encoded by the *gag–pol* gene and is synthesized through suppression of an amber termination codon. *Proc. Natl Acad. Sci. USA,* **82,** 1618.
145. Yoshinaka, Y., Katoh, I., Copeland, T. D., and Oroszlan, S. (1985) Translational readthrough of an amber termination codon during synthesis of feline leukemia virus protease. *J. Virol.,* **55,** 870.
146. Jacks, T. and Varmus, H. E. (1985) Expression of the Rous sarcoma virus *pol* gene by ribosomal frameshifting. *Science,* **230,** 1237.

147. Jacks, T., Madhani, H. D., Masiarz, F. R., and Varmus, H. E. (1988) Signals for ribosomal frameshifting in the Rous sarcoma virus *gag–pol* region. *Cell,* **55,** 447.
148. Jacks, T., Townsley, K., Varmus, H. E., and Majors, J. (1987) Two efficient ribosomal frameshift events are required for synthesis of mouse mammary tumor virus *gag*-related polypeptides. *Proc. Natl Acad. Sci. USA,* **84,** 4298.
149. Moore, R., Dixon, M., Smith, R., Peters, G., and Dickson, C. (1987) Complete nucleotide sequence of a milk-transmitted mouse mammary tumor virus: two frameshift suppression events are required for translation of *gag* and *pol. J. Virol.,* **61,** 480.
150. Jacks, T., Power, M. D., Masiarz, F. R., Luciw, P. A., Barr, P. J., and Varmus, H. E. (1988) Characterization of ribosomal frameshifting in HIV-1 *gag–pol* expression. *Nature,* **331,** 280.
151. Parkin, N. T., Chamorro, M., and Varmus, H. E. (1992) Human immunodeficiency virus type 1 *gag-pol* frameshifting is dependent on downstream mRNA secondary structure: Demonstration by expression in vivo. *J. Virol.,* **66,** 5147.
152. Sekine, Y. and Ohtsubo, E. (1989) Frameshifting is required for production of the transposase encoded by insertion sequence 1. *Proc. Natl Acad. Sci. USA,* **86,** 4609.
153. Escoubas, J. M., Prere, M. F., Fayet, O., Salvignol, I., Galas, D., Zerbib, D., and Chandler, M. (1991) Translational control of transposition activity of the bacterial insertion sequence IS1. *EMBO J.,* **10,** 705.
154. Dinman, J. D., Icho, T., and Wickner, R. B. (1991) A −1 ribosomal frameshift in a double-stranded RNA virus of yeast forms a gag-pol fusion protein. *Proc. Natl Acad. Sci. USA,* **88,** 174.
155. Brierley, I., Boursnell, M., Birns, M., Bilmoria, B., Block, V., Brown, T., and Inglis, S. (1987) An efficient ribosomal frameshifting signal in the polymerase-encoding region of the coronavirus IBV. *EMBO J.,* **6,** 3779.
156. Brierley, I., Digard, P., and Inglis, S. C. (1989) Characterization of an efficient coronavirus ribosomal frameshift signal: requirement for an RNA pseudoknot. *Cell,* **57,** 537.
157. Snijder, E., den Boon, J., Bredenbeek, P., Horzinek, M., Rijnbrand, R., and Spaan, W. (1990) The carboxyl-terminal part of the putative Berne virus polymerase is expressed by ribosomal frameshifting and contains sequence motifs which indicate that toro- and coronaviruses are evolutionarily related. *Nucleic Acids Res.,* **18,** 4535.
158. Flower, A. M. and McHenry, C. S. (1990) The gamma subunit of DNA polymerase III holoenzyme of *Escherichia coli* is produced by ribosomal frameshifting. *Proc. Natl Acad. Sci. USA,* **87,** 3713.
159. Tsuchihashi, Z. and Kornberg, A. (1990) Translational frameshifting generates the gamma subunit of DNA polymerase III holoenzyme. *Proc. Natl Acad. Sci. USA,* **87,** 2516.
160. Chamorro, M., Parkin, N., and Varmus, H. E. (1992) An RNA pseudoknot and an optimal heptameric shift site are required for highly efficient ribosomal frameshifting on a retroviral messenger RNA. *Proc. Natl Acad. Sci. USA,* **89,** 713.
161. Brierley, I., Roley, N. J., Jenner, A. J., and Inglis, S. C. (1991) Mutational analysis of the RNA pseudoknot component of a coronavirus ribosomal frameshifting signal. *J. Mol. Biol.,* **220,** 889.
162. Tang, C. K. and Draper, D. E. (1989) Unusual mRNA pseudoknot structure is recognized by a protein translational repressor. *Cell,* **57,** 531.
163. Ciminale, V., Pavlakis, G. N., Derse, D., Cunningham, C. P., and Felber, B. K. (1992) Complex splicing in the human T-cell leukemia virus (HTLV) family of retroviruses: Novel mRNAs and proteins produced by HTLV Type I. *J. Virol.,* **66,** 1737.

164. Sadaie, M. R., Rappaport, J., Benter, T., Josephs, S. F., Willis, R., and Wong-Staal, F. (1988) Missense mutations in an infectious human immunodeficiency viral genome: functional mapping of *tat* and identification of the *rev* splice acceptor. *Proc. Natl Acad. Sci. USA*, **85,** 9224.
165. Ahmad, N., Maitra, R. K., and Venkatesan, S. (1989) Rev-induced modulation of nef protein underlies temporal regulation of human immunodeficiency virus replication. *Proc. Natl Acad. Sci. USA*, **86,** 6111.
166. Baltimore, D. (1988) Intracellular immunization. *Nature*, **335,** 395.
167. Chin, D. (1992) Inhibition of human immunodeficiency virus type 1 Rev-Rev-response element complex formation by complementary oligonucleotides. *J. Virol.*, **66,** 600.
168. Matsukura, M., Zon, G., Shinozuka, K., Robert, G. M., Shimada, T., Stein, C. A., Mitsuya, H., Wong-Staal, F., Cohen, J. S., and Broder, S. (1989) Regulation of viral expression of human immunodeficiency virus in vitro by an antisense phosphorothioate oligodeoxynucleotide against *Rev* (*art/trs*) in chronically infected cells. *Proc. Natl Acad. Sci. USA*, **86,** 4244.
169. Lee, T. C., Sullenger, B. A., Gallardo, H. F., Ungers, G. E., and Gilboa, E. (1992) Overexpression of RRE-derived sequences inhibits HIV-1 replication in CEM cells. *New Biol.*, **4,** 66.
170. Zimmermann, K., Weber, S., Dobrovnik, M., Hauber, J., and Böhnlein, E. (1992) Expression of chimeric Neo-Rev response element sequences interferes with Rev-dependent HIV-1 Gag expression. *Hum. Gene Ther.*, **3,** 155.

6 | Auxiliary proteins of the primate immunodeficiency viruses

JAMES S. GIBBS and RONALD C. DESROSIERS

1. Introduction

All known primate lentiviruses possess the essential structural genes *gag*, *pol*, and *env* and the essential regulatory genes *tat* and *rev*. However, all of them possess at least three or four additional genes that are known to encode protein products and that are apparently unique to the lentivirus subfamily of retroviruses. In general, little is known about their functions (1). However, since they are likely to be important for some aspects of the virus life cycle, it is important that their functional roles be clarified. Such auxiliary genes are often referred to as 'non-essential' due to the fact that in many circumstances they may be mutated or deleted with little or no effect on virus growth *in vitro*. However, the importance of these genes for virus growth in cell culture often depends upon the particulars of the system employed. The extent to which a mutant virus is able to replicate appears to depend in many cases upon the host cell used, the particular virus being studied, the methods used for infecting the cell, and the methods used for monitoring virus production. For example, cocultivation vs. infection or transfection, or the inoculum employed may affect the rate or extent of mutant virus replication. One major caveat of such studies is that the situation in cell culture may or may not be relevant to virus growth *in vivo*.

Despite the situation *in vitro*, it seems likely that these 'non-essential' genes play an important role for the viruses *in vivo*. With few exceptions, these genes have been conserved among the various viruses. Certainly, the complement of auxiliary genes is well conserved within an individual group of primate lentiviruses, for example, among simian immunodeficiency virus (SIV) isolates from African green monkeys or among human immunodeficiency type 1 (HIV-1) isolates from humans. Such conservation, particularly over large phylogenetic distances, implies that the genes play some important role in replication, persistence, or transmission of the virus. This is especially true for retroviruses since they replicate via an error-prone reverse transcriptase. Almost certainly, if these genes were truly non-essential, they would have been lost long ago. In the only instance where an auxiliary gene has been tested for a role *in vivo*, the *nef* gene of macaque SIV (SIV$_{mac}$) has been shown to be a major determinant of virulence and pathogenic potential in rhesus monkeys (2). Thus, the Nef gene product should become a target for drug development in the fight against acquired immunodeficiency syndrome (AIDS). If other auxiliary genes are similarly shown to play a significant role in pathogenesis, they too may become potential drug targets.

Four groups of primate lentiviruses are currently recognized based on their genome organization with respect to the auxiliary genes (Figs 1 and 2). Individual members of each group display a higher degree of nucleotide sequence identity to one another than to members of the other groups (3). All four groups contain *vif* and *nef* genes, implying important roles for these genes. An important role for *vif* is supported by the fact that mutations in *vif* have been shown to have dramatic effects on virus growth in some *in vitro* situations (see below). The importance of the *nef* gene for the pathogenesis of SIV in rhesus monkeys has already been mentioned.

Recently, it has been pointed out that the Vpx and Vpr gene products have a high level of sequence similarity and, in fact, one may have arisen from the other by gene duplication (4) (Figs 3 and 4). Such a duplication would be consistent with

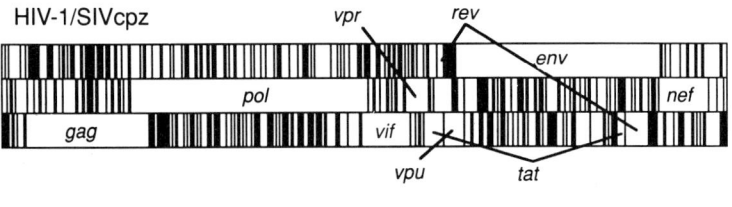

Fig. 1 Genome organization of the four groups of primate lentiviruses. Sequences from top to bottom are HIV-1BRU, SIVmac239, SIVTYO1, and SIVmnd. All stop codons in the forward three reading frames are depicted as vertical lines, making potential open reading frames stand out. This figure and the next were created with ORFwriter software for the Macintosh, written by one of the authors (J.S.G.)

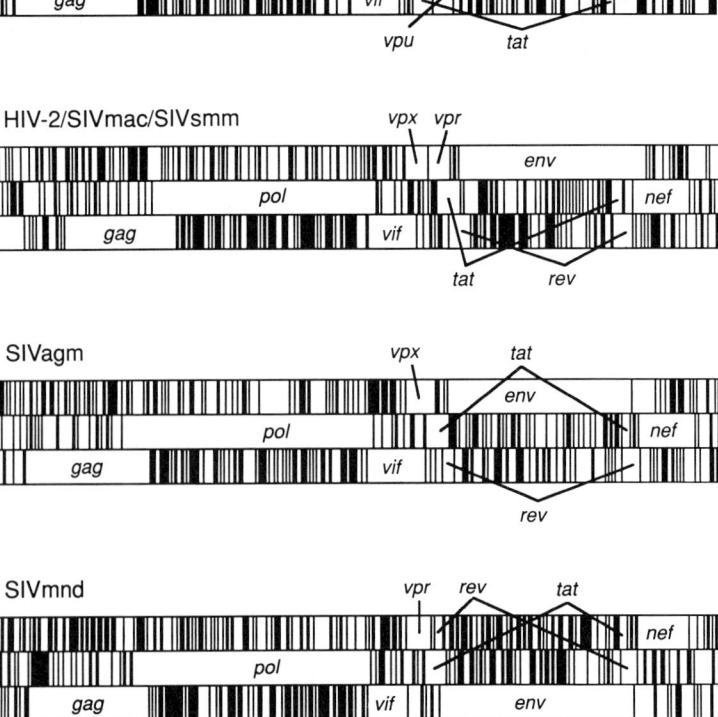

the observation that when both *vpx* and *vpr* are present, the two genes are adjacent in the genome (Figs 1 and 2). Because of the similarity of these two genes, the primate lentiviruses from African green monkeys (SIV$_{aqm}$) and mandrills (SIV$_{mnd}$) may possibly be considered to possess essentially the same genomic organization. Based on their similarity, it is possible that *vpx* and *vpr* have similar, overlapping, or identical functions and that lack of one of these gene products may be compensated for by the presence of the other in some situations. However, concrete evidence for similar function is lacking. The presence of both *vpr* and *vpx* is found only in the SIV$_{smm}$/HIV-2/SIV$_{mac}$ group of viruses. SIV$_{smm}$ is the lentivirus derived from sooty mangabey monkeys.

The *vpu* gene is found only in HIV-1 and chimpanzee SIV (SIV$_{cpz}$). However, the SIV$_{cpz}$ *vpu* gene is 64 per cent divergent from HIV-1$_{BRU}$, suggesting a possibly distinct class of lentivirus (5). It is possible that additional genome organizations may become evident as more primate lentiviruses are discovered and sequenced.

It has been suggested that the plus strand of HIV-1 contains an open reading frame that could potentially encode a gene product (6), but no evidence exists that it is either transcribed or translated. Alternate splicing has been postulated to lead

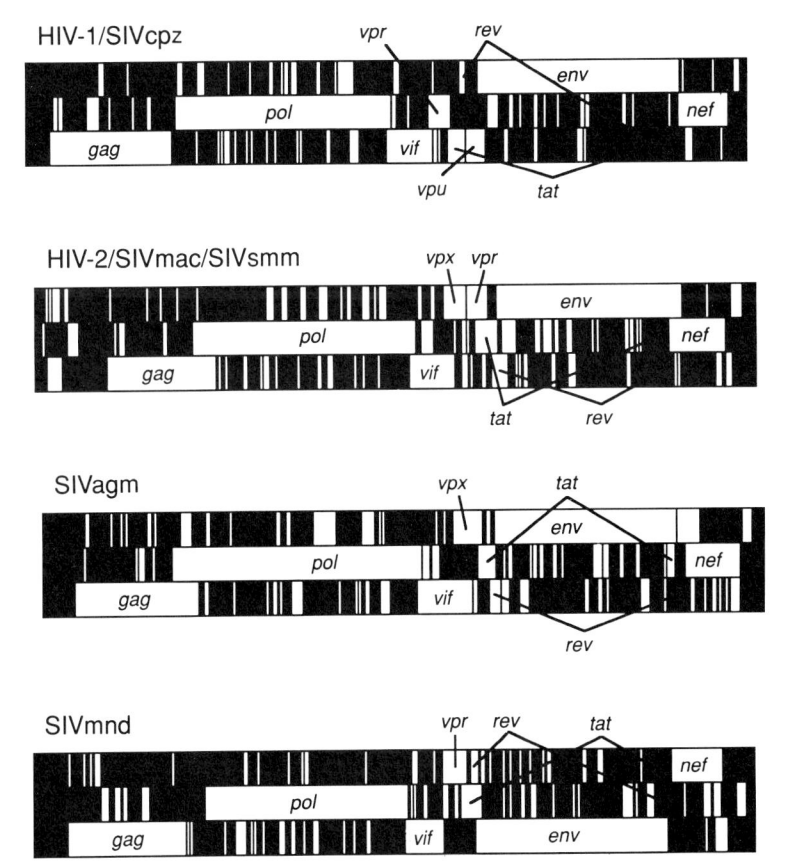

Fig. 2 Alternative view of the genome organization of the four groups of primate lentiviruses. Identical to Fig. 1, except that sequences downstream of stop codons and upstream of start codons are blacked out. All white spaces, therefore, begin with a start codon and end with a stop codon. Thus, second exons of the *rev* and *tat* genes may not show up

```
SIVMAC239  vpr  M-------EE  RPPENE--GP  QREPWDEWVV  EVLEELKEEA  IKHFDPRLLT  ALGN----HI  YNRHGDTLEG   57
HIV-2ROD   vpr  MAEAP---TE  LPPVDG--TP  LREPGDEWII  EILREIKEEA  IKHFDPRLLI  ALGK----YI  YTRHGDTLEG   61
HIV-1BRU   vpr  MEQ--------APEDQ--GP  QREPHNEWTL  ELLEELKNEA  VRHFPRIWLH  GLGQ----HI  YETYGDIWAG   56
SIVCPZ     vpr  MEQ--------APEDQ--GP  PREPYQEWAL  ETLEELKNEA  VRHFPRPWLH  QLGQ----FI  YDTYGDIWVG   56
SIVMND     vpr  MGQKR---DE  QVSEDQ--GP  PREPYNQWLA  DIMEEIKEEA  RKHFPLIILN  AVSE----YC  VQNTGSEEEA   61
SIVMAC239  vpx  MS---DPRER  IPPGNSGEET  IGEAFE-WLN  RTVEEINREA  VNHLPRELIF  QVWQRSWEYW  HDEQGMSPSY   66
HIV-2ROD   vpx  MT---DPRET  VPPGNSGEET  IGEAFA-WLN  RTVEAINREA  VNHLPRELIF  QVWQRSWRYW  HDEQGMSESY   66
SIVAGM155  vpx  MASGRDPREE  RPGGLEIWDL  SREPWDEWLR  DMVEEINNEA  KLHFGRELLY  QVWN----Y-  CQEEGERQGR   65

Consensus       M...----EE  .PPED.--GP  .REP..EWL.  ETLEEIK.EA  V.HFPR.LL.  Q.G.----YI  Y...GDT..G   70


SIVMAC239  vpr  A-----GELI  RILQRALFMH  FRGGCI----  HS-----RIG  Q-PGGNPLS   AIPPSRSMLL  102
HIV-2ROD   vpr  A-----RELI  KVLQRALFTH  FRAGCG----  HS-----RIG  Q-TRGGNPLS  AIPTPRNMQQ  106
HIV-1BRU   vpr  V-----EAII  RILQQLLFIH  FRIGCR----  HS-----RIG  V-TQ-QRRAR  N-GASRSS--   97
SIVCPZ     vpr  V-----EAII  RILQHLLFIH  FRLGCQ----  HS-----RIG  I-LP-QRRRS  N-GSNRSS--   97
SIVMND     vpr  C-----EKFI  TIMNRAIWVH  LAQGCD----  GT-----FRE  R-RP-QLPPS  GFRPRGDRLL  105
SIVMAC239  vpx  V-----KYRYL CLIQKALFMH  CKKGCR---C  LG-----EG   HGAGGWRPGP  PPPPPPGLAA  113
HIV-2ROD   vpx  T-----KYRYL CIIQKAVYMH  VRKGCT---C  LG-----RG   HGPGGWRPGP  PPPPPPGLVV  113
SIVAGM155  vpx  PIAERAYKYY  RLVQKALFVH  FRCGCRRRQP  FEPYEERRNG  QGGG--RPG-  --RVPPGLDD  120

Consensus       .-----...I  R.IQ.ALF.H  FR.GC.----  HS-----RIG  .-.G-.RP.S  ..PPPR....  130
```

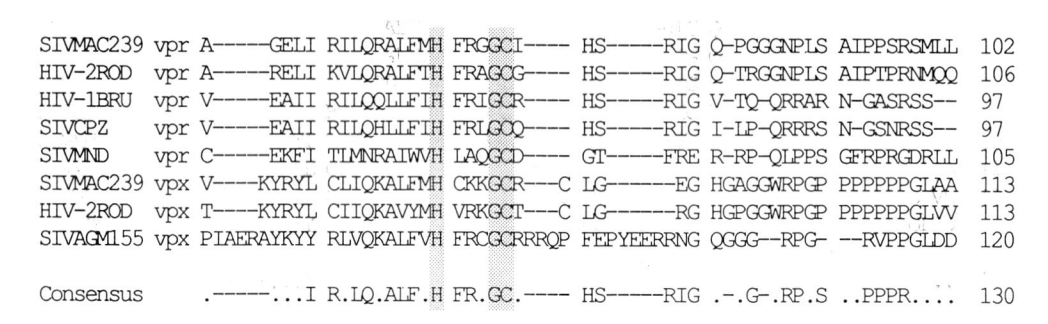

Fig. 3 Amino acid sequence alignment of primate lentivirus *vpx* and *vpr* genes. At least one *vpx* and *vpr* gene from each of the four groups of primate lentiviruses, where available, were aligned and a consensus was derived. Invariant residues are highlighted in grey

to gene product chimeras such as Tat–Env–Rev and Env–Rev fusion proteins (7). Ribosomal frameshifting has also been reported as a means of generating such chimeras (8). Whether these chimeric gene products are produced to any significant level in infected cells, or whether they have any functional role if they are produced is unclear.

Other non-primate lentiviruses, such as bovine immunodeficiency virus (BIV) (9, 10), feline immunodeficiency virus (FIV) (11–13), equine infectious anemia virus (EIAV) (14), caprine arthritis–encephalitis virus (CAEV) (15), and visna virus (16)

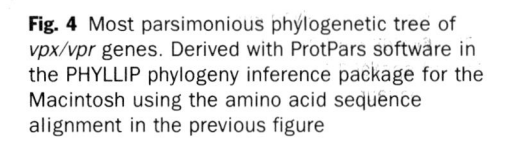

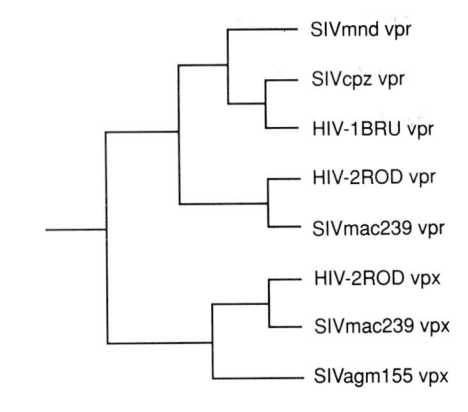

Fig. 4 Most parsimonious phylogenetic tree of *vpx/vpr* genes. Derived with ProtPars software in the PHYLLIP phylogeny inference package for the Macintosh using the amino acid sequence alignment in the previous figure

```
            ┌── SIVmnd vpr
         ┌──┤
         │  │  ┌── SIVcpz vpr
         │  └──┤
     ┌───┤     └── HIV-1BRU vpr
     │   │
     │   │  ┌── HIV-2ROD vpr
     │   └──┤
─────┤      └── SIVmac239 vpr
     │
     │   ┌── HIV-2ROD vpx
     │  ┌┤
     └──┤└── SIVmac239 vpx
        │
        └── SIVagm155 vpx
```

also contain additional auxiliary genes. *vif* and *vpr* are the only primate lentivirus 'non-essential' genes in which homologues have been unambiguously identified in these non-primate lentiviruses (9, 10, 16, 17).

2. *vif*

The *vif* gene was identified as an open reading frame between the *pol* and *env* genes of HIV-1 by the groups that first sequenced the HIV-1 genome (18–21). Later, it was found in every other primate lentivirus examined. It was originally called *sor*, which stands for short open reading frame. The current name, *vif*, is short for virion infectivity factor. The *vif* gene is phylogenetically old relative to other auxiliary genes of the primate lentiviruses, based upon significant sequence similarity between primate lentivirus *vif* genes and genes of non-primate lentiviruses, such as the Q gene of the distantly related ungulate lentivirus visna virus (16) and a gene in the bovine immunodeficiency virus (9, 10). Homologues of *vif* are present in all known lentiviruses except EIAV (17). The *vif* gene encodes an approximately 23–27 kDa protein that is expressed in cell culture and in infected individuals (22, 23). Antibodies specific to the Vif protein are present in individuals seropositive for HIV (22, 23). It does not seem to be a virion component (24) and does not seem to play any role in virion assembly (25). Expression of *vif* mRNA has been reported to be dependent on the presence of the Rev gene product (26, 27).

In addition to the sequence similarity to other lentivirus genes, there is an unexpected antigenic cross-reactivity to a surface antigen of the helminth worm *Schistosoma mansoni* (28, 29). Rat monoclonal antibodies directed against a synthetic peptide spanning amino acids 155–168 of the Vif protein selectively recognize two proteins having apparent molecular weights of 170 kDa and 65 kDa in *S. mansoni*-infected rats or humans. Such monoclonal antibodies were demonstrated to be protective against challenge in rats with *S. mansoni* cercariae. In addition, serum from HIV-1 positive individuals recognizes the 170 kDa *S. mansoni* protein. The significance of this cross-reactivity is unclear, although it has been suggested that, since the organisms are endemic in the same regions, an immune reaction against one may provide some protection against the other. This has been disputed, however. Since the original report, we know of no independent group that has confirmed or refuted these results, nor have the authors expanded upon these interesting findings.

Despite a number of studies, little is actually known about the role of *vif* in virus infection. This is especially surprising since, of all the 'non-essential' genes, *vif* can have the most dramatic effects on virus replication. Several papers have reported that it seems to play a role in virion infectivity (25, 30). Cells infected with *vif* mutants were found to produce normal amounts of virus, but cell-free virus was less infectious than wild-type virus. Virus spread in these mutants seemed to be primarily cell-to-cell. The exact extent of any decrease in virus replication in cell culture seems to depend upon several variables, including parental virus strain (either HIV-1, HIV-2, or an SIV strain), host cell, and whether virus growth assays are performed using cocultivation with uninfected cells or by straight infection/transfection (25, 30–35).

Sodroski et al. first showed in early 1986 (31) that deletion of the *vif* gene in HIV-1 results in delayed growth kinetics in the CD4$^+$ T-lymphocyte line C8166 using cocultivation with uninfected cells. Virus produced from cells infected with *vif* deletion mutants in this study was approximately 1000-fold less infectious than wild-type virus. Cells infected with *vif* deletion mutants also showed a delay in cytopathic effects. Although the deletion mutants in that study also deleted the *vpr* gene (unknown at the time) this early study demonstrated that, at least in some situations, the *vif* gene was not absolutely required for virus growth.

These findings are somewhat in line with published results using SIV$_{agm}$ (32, 33) and HIV-2 (33, 34). Studies with SIV$_{agm}$ have shown that deletions in the *vif* gene result in growth kinetics in CD4$^+$ lymphocyte lines virtually identical to wild-type virus, even in the absence of cocultivation with uninfected cells (32, 33). Work by the same group using HIV-2 have shown that deletions in the *vif* gene result in greatly decreased levels of virus replication as measured by reverse transcriptase levels relative to wild-type, and that this decrease is prevented by cocultivation with uninfected cells (33, 34).

In contrast, the studies of Fisher et al. (25) and Strebel et al. (30) indicate that HIV-1 *vif* deletion mutants are very inefficient or possibly entirely incapable of growth in at least some cell culture systems. In both studies, however, cocultivation with uninfected cells increased the rate of growth of the deletion mutants to some degree. The fact that these mutants grow better in the case of cocultivation with uninfected cells has been the basis for suggestions that cell-to-cell transmission is less affected by *vif* mutations than is virion infectivity. However, it should be pointed out that cocultivation is usually a more efficient way of transmitting even wild-type virus, so the *vif*$^-$ mutants may not be specifically deficient in the infectivity of cell-free virions.

In a highly controversial work, Guy et al. have proposed a Vif-dependent proteolytic modification of Env protein in HIV-1, specifically, proteolysis by the Vif gene product of the cytoplasmic portion of gp41 (35). Based on weak sequence similarity with cysteine proteases, they investigated the possibility that the Vif protein was a cysteine protease. Among their results, they report that Vif modifies the processing and conformation of gp41, that this modification is blocked by E64, a specific cysteine protease inhibitor, and that E64 prevents the iodoacetylation of cysteine 114 of bacterially expressed Vif protein. The authors failed, however, to observe any proteolysis of the envelope protein by the Vif protein *in vitro*. Also in conflict with these results is the observation that in HIV-1 lacking the putative cleavage site in gp41, deletion of *vif* still results in significantly decreased growth in cell culture. There has been no published work confirming these observations.

3. *vpr*

The *vpr* gene was overlooked in the original papers that reported on the complete genome sequence of HIV-1. Finer sequence analysis resulted in the identification of an open reading frame between the *vif* and *env* genes capable of encoding an approximately 78–96 amino acid protein in HIV-1. In HIV-2 and SIV, the range approximately is 95–105 amino acids. Later, it was discovered that antisera from 17–47 per cent of HIV-1 seropositive individuals specifically recognized the 11–15 kDa protein encoded by this open reading frame expressed in bacteria, proving that the gene is expressed *in vivo* (36). That study and other studies that followed (37) failed to show any pattern between expression of antibodies to the Vpr gene product and disease course. The *vpr* gene is present in all known primate lentiviruses, with the possible exception of SIV$_{agm}$, and even in that case the gene identified as *vpx* may in some cases be as related to some *vpr* genes as to other *vpx* genes (4). In any event, a *vpr* homologue is present at least once in every known primate lentivirus, suggestive of an important role in virus replication, virulence, persistence, or transmission. A homologue of *vpr* also seems to be present in the ungulate lentivirus visna virus (16). Expression of *vpr* mRNA has been reported to be dependent on the presence of the Rev gene product (26, 27).

Western blotting of subcellular fractions prepared from HIV-1-infected cells has indicated that the Vpr gene product seems to be membrane-associated, and not present in high concentrations in the cytosol or nuclear fractions of the cell (38). Yuan *et al.* (39) reported that Vpr protein could not be detected in cells infected with virus lacking the carboxy-terminal 26 amino acids, and they suggested that the carboxy terminus may be essential for Vpr expression, but it could also be that the carboxy terminus contains the predominant epitope their polyclonal antiserum was directed against. Sato *et al.* (38) also found that deletion of amino acids 64–76 resulted in loss of expression of Vpr protein in virus-infected cells. Like its homologue, Vpx, the Vpr gene product is a virion component, with at least 5–10 copies per virion (39–41). Although present in virions, Vpr and Vpx are not necessary for the formation and release of virions (42). The fact that Vpr and Vpx proteins are virion components suggests that their role in virus infection may be played out early in infection.

As with most of the other primate lentivirus auxiliary genes, there are conflicting reports as to the effects of deletion or mutation of the *Vpr* gene on virus replication in cell culture.

No phenotype was observed to be conferred to either HIV-1 or HIV-2 by the introduction of point mutations in the *vpr* gene by Dedera *et al.* (42). This group looked at viral growth curves in both primary lymphocytes and the lymphoid cell lines CEM, Jurkat, H9, Molt-3, SupT1, and U937 and found no differences in the rate of virus growth, as measured by supernatant reverse transcriptase activity beween the wild-type and mutant virus constructs. Nor did they observe any differences in cytopathic effects in infected cells. They also attempted to look for differences in growth rates at lower virus inocula, but found none. Unfortunately, according to others (43, 44), the HIV-1 IIIB strain they used contains a frameshift mutation that renders *vpr* defective.

Similarly, Shibata *et al.* (33, 34) found no difference in the ability of HIV-2 *vpr* mutants to grow in CD4$^+$ T-lymphocytes. Nor did this group find any differences in the growth rate of mutants defective in the SIV$_{agm}$ *vpx* gene relative to that of wild-type in CD4$^+$ T-lymphocyte lines (32, 33). This particular experiment has relevance because the SIV$_{agm}$ *vpx* gene seems to be more closely related to other *vpr* genes than to other *vpx* genes (4).

Ogawa *et al.* used insertions of four base pairs to mutate the *vpr* gene in HIV-1 (44). They found that, although the *vpr* mutants were infectious in all CD4$^+$ T-lymphocyte cell lines studied, the mutants grew slightly slower than wild-type virus. A distinct dependence on the size of the inoculum was noted, with less of an effect when infecting with more virus. One possible explanation for the effect on growth of the inoculum size could be that the virus inocula were based on reverse transcriptase units, not infectious dose units. However, this is true of most such studies because of the difficulties and inherent inaccuracies in trying to quantitate HIV and SIV tissue culture infectious doses.

A function for *vpr* in macrophage-tropism has been suggested by Hattori *et al.* (45) who created and analysed point mutations in the *vpr* gene of HIV-2. They found that derivatives of HIV-2-containing mutations in the *vpr* gene at both the initiation codon and three bases downstream grew like wild-type virus in peripheral blood mononuclear cells, but were severely compromised for growth in human peripheral blood monocyte/macrophages. Since virus replication or persistence in macrophages is likely to be important for the pathogenesis of AIDS, this would seem to be an important finding. However, evidence for a role for *vpr* in macrophage-tropism is complicated by the observation that the macrophage-tropic virus SIV$_{mac}$1A11 contains a truncated *vpr* gene (46). Since SIV$_{mac}$1A11 does replicate well in monocyte/macrophages, *vpr* does not appear to be universally essential for primate lentivirus replication in this cell type. Currently, portions of the *env* gene are the only finely mapped determinants of differential macrophage tropism (see Chapter 2).

Cohen *et al.* have reported that the *vpr* gene can act to *trans*-activate expression of viral genes *in vitro* (43). They oberved an approximately threefold increase in the expression of p25 Gag antigen when HIV-1 proviral DNA was cotransfected with a plasmid containing the *vpr* gene. At least one other virus, herpes simplex virus, uses a strategy of bringing a *trans*-activating protein with it in the virion. Unexpectedly, similar levels of *trans*-activation by *vpr* were also observed using heterologous promoters such as those of SV40 and cytomegalovirus. Most other *trans*-activators induce to much higher levels, and are specific for a particular class of promoter. Other investigators have observed no *trans*-activating activity of *vpr* in transfected cells (K. Partin and B. Cullen, personal communication).

4. *vpx*

The *vpx* gene is present in SIV$_{mac}$, SIV$_{smm}$, and HIV-2 viruses. A similar gene exists in SIV$_{agm}$, but based on parsimony analysis of *vpx* and *vpr* sequences, this gene may in some cases be as closely related to some *vpr* genes as to other *vpx* genes. Comparison of SIV$_{mac}$ and HIV-2 *vpx* genes reveals 86 per cent identity at the amino acid level. Such a high level of sequence identity suggests a need for a high degree of sequence conservation to retain function and an important role for this gene in the virus life cycle. The Vpx protein is predicted to be about 13 kDa and has been found to migrate as a 13–16 kDa protein in SDS-polyacrylamide gel electrophoresis. The amino terminus of the Vpx gene product is blocked to Edman degradation (47). The Vpx protein is present in virions of SIV$_{mac}$ in amounts equimolar to the Gag gene product p24, the major component of the virion (47). The Vpx protein has been shown to bind to single-stranded nucleic acid and thus could conceivably play a role in packaging (47), but like the other primate lentivirus genes, little is known about its role in virus infection.

Conflicting evidence has been presented as to whether the *vpx* gene affects the rate of virus production in cell culture. An early study by Guyader *et al.* (48) found that a mutation in the HIV-2 *vpx* gene resulted in wild-type levels of virus growth in established T-cell and monocyte lines, but severely decreased growth in primary peripheral blood lymphocytes. One of the two mutants used in this study resulted in an inadvertent additional mutation in a conserved residue of the overlapping *vif* gene, but there were no overt additional changes in the other mutant used. Kappes *et al.* reported similar results when they point-mutated the HIV-2 *vpx* gene (49). They found that *vpx*-mutated HIV-2 grew with wild-type kinetics in SupT1, H9, and CEM T-lymphocyte cell lines, but were severely impaired for growth in primary peripheral blood mononuclear cells *in vitro*. Unfortunately, these results are difficult to interpret because they, too, introduced an additional inadvertent mutation into the *vif* gene. However, Yu *et al.*, using SIV$_{mac}$, reported somewhat similar results in which *vpx* mutants grew well in T-cell lines, but slightly less well in PBLs (50). Kappes *et al.* (49) also observed sustained high levels of virus expression at later time points (9–23 months) in SupT1 cells, which also seems to be a characteristic of *vif* mutants (24, 30, J. S. Gibbs and R. C. Desrosiers, unpublished data).

In contrast to the above studies, Hu *et al.*, using HIV-2 (51), reported no differences in the growth rate between wild-type and *vpx*-mutant viruses in peripheral blood lymphocytes. Shibata *et al.* (33, 34) also failed to observe any differences between wild-type and *vpx*-mutant HIV-2, but they used CD4$^+$ T-cell lines, and not primary lymphocytes. The reasons for the conflicting results are not clear. It may be that some of the *vpx* mutants alter splicing in some situations, preventing the expression of some other gene product, as suggested by Hu *et al.* (51). Alternatively, there could be some subtle differences not accounted for in the individual experiments such as the parental virus strain, the individual host cell type, a *trans* effect of the specific mutation employed, or the particulars of the culture system. The fact that the function of *vpx* remains unknown only makes the reasons for the differences more difficult to discern.

5. *vpu*

The *vpu* gene is found only in the HIV-1/SIV$_{cpz}$ class of primate lentiviruses, distinguishing this class of lentivirus from other primate lentiviruses. It encodes an

80- to 82-amino acid protein. Approximately one-third of serum samples from AIDS patients were found to contain antibodies that reacted with Vpu protein expressed *in vitro*, proving that the gene encodes a protein *in vivo* (52–56). The presence of antibodies to Vpu may (54) or may not (55, 56) correlate with the clinical stage of disease. The Vpu protein has been shown to be phosphorylated but non-glycosylated (57). This post-translational phosphate modification probably accounts for the fact that 16 and 17 kDa proteins are sometimes observed in HIV-1-infected cells (53). The Vpu protein has been shown to insert into canine microsomal membranes during translation *in vivo* (57), and its orientation within the membrane has recently been determined. It is localized in the perinuclear region of the cytoplasm of infected cells (58). It is not believed to be a virion component (57).

Several studies have shown that lack of the *vpu* gene leads to delayed or decreased release of virion proteins from infected cells and an accumulation of virion proteins in those cells (52, 57–60). This decrease or delay of virion release results in diminished virus yields from infected cells, but the infectivity of the progeny virions is not diminished (58). The accumulation of viral proteins and the decrease in virus yield are complemented by a cell line expressing Vpu (60). This cell line has no effect relative to the parental cell line on the growth of SIV, which lacks a *vpu* gene.

Electron microscopic examination has revealed that mutations in the *vpu* gene lead to an accumulation of mutant virion particles at the surface of infected cells, many with aberrant budding structures and with significant variation in their size and shape (58). Additionally, aberrant intracytoplasmic particles can be observed, some budding at intracellular membranes, such as vacuolar membranes, which normally are restricted from virus budding.

Recent evidence has shown that coexpression of the CD4 receptor molecule and gp160 Env gene product in HeLa cells results in diminished processing of gp160 to gp120 and gp41 because of interactions between gp160 and the CD4 molecule (61). This decrease in processing was alleviated by the coexpression of Vpu, suggesting that the function of Vpu may be to destabilize intracellular gp160–CD4 interactions (Fig. 5). Such interactions would presumably occur at the Golgi apparatus, where a membrane-incorporated Vpu protein could prevent binding of unprocessed gp160 to CD4 molecules. Furthermore, the coexpression of Vpu with CD4 results in increased degradation of the CD4 receptor (61). It remains unclear whether Vpu protein acts to directly destabilize gp160–CD4 complexes, or whether such complexes are decreased due to an increase in degradation of CD4 induced by Vpu. It should be noted, however, that these effects of Vpu on destabilization of gp160–CD4 interactions were observed in transfected HeLa cells and have not yet been documented for naturally CD4-positive lymphoid cells. In HIV-1 infected cells, several bicistronic mRNAs are produced, containing both *vpu* and *env* genes (62–65). This may suggest a need for the coordinated expression of these two genes.

The above evidence suggests that Vpu allows proper virion assembly, packaging, budding, or release of infectious virion particles. Since other primate lentiviruses lack this particular gene product, it is possible that another viral or cellular

Fig. 5 Model for the action of the Vpu gene product. In the absence of Vpu, stable intracellular gp160–CD4 complexes form, preventing the processing of gp160 to gp120 and gp41. In the presence of Vpu, the amount of such complexes are decreased, allowing for processing of gp160 to occur. Model derived from ref. 61, with permission

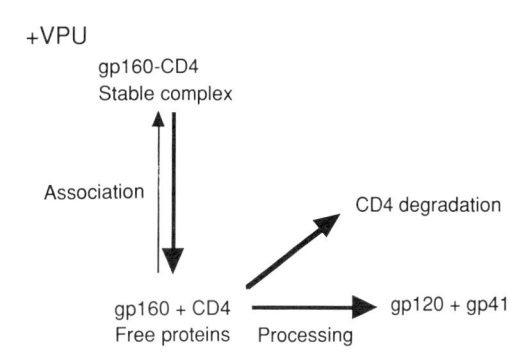

component functions in the place of Vpu, or that the Vpu activity is completely unnecessary in these other lentiviruses. Since it is believed that Vpu functions to prevent intracellular interaction between gp160 and CD4, it has been suggested that Vpu may not be needed in HIV-2 and SIV viruses because of a possibly lower affinity between the envelope glycoprotein and the CD4 receptor (61). However, few Env–CD4 affinity studies have been done, and for the most part, these have been limited to affinities for Env towards human CD4, which may not be relevant in the case of SIV where the target ligand is a monkey CD4.

A case has been made that Vpu functions as a virion release factor similar to influenza A virus M2 protein and foot-and-mouth disease virus protein p3A (58). Each of these proteins has similar hydropathicity profiles with a strongly hydrophobic amino-terminal anchor sequence and hydrophilic carboxy-terminus, each containing at least 18 charged residues. Other potentially similar proteins include the small NB protein of influenza B and the SH protein of simian virus 5. Influenza A virus M2 protein has been recently reported to possess ion-channel activities, and is thought to act to modify intracellular pH, which is critical for maturation of virion envelope proteins (66, 67). Any functional similarities between Vpu and these proteins remains to be proven, but based on the current evidence, a similarity would not be surprising.

6. *nef*

The *nef* gene is present in all primate lentiviruses sequenced to date. However, there is considerable sequence variation at the protein level at both carboxy and amino termini. *nef* consists of an open reading frame beginning within or immediately after the 3' end of the *env* gene and overlaps the U3 portion of the 3' long terminal repeat. It is expressed *in vivo*, as evidenced by the induction of antibodies directed against the Nef gene product (37, 68–70) and cytotoxic T-lymphocyte activities directed against Nef peptide-presenting cells (71). The Nef protein is 25–29 kDa in HIV-1 and somewhat larger (approximately 34 kDa) in SIV_{smm}/HIV-2/SIV_{mac} viruses. The Nef gene product is myristylated (2, 68) and its N terminus is associated with the cytoplasmic membrane (72, 73), likely by virtue of its N-myristoyl anchor.

There are several reports of the Nef gene product having sequence similarity to other proteins. In one case, Nef protein was reported to have sequence similarities with scorpion peptides that interact with potassium channels in nervous tissues (74). Two short regions of the Nef gene product were reported to have such similarities, around amino acids 91–98 and amino acids 141–149. These authors reported patch-clamp studies where the Nef gene product reversibly increased potassium flux in chick dorsal root ganglia. These authors hypothesized that *nef* gene expression could be responsible for the neuropathology observed in AIDS patients. No confirmation of this work has yet appeared.

The Nef gene products of several HIV-1, HIV-2, and SIV clones were reported to contain a conserved, if slightly degenerate, leucine zipper-like motif commencing at approximately amino acid 75–110 (75). However, instead of having the characteristic adjacent basic domain, the Nef gene products possess an acidic carboxy terminus. This was given as evidence that the Nef gene product may belong to a class of non-DNA-binding acidic transcription factors.

The *nef* gene has also been reported to have sequence similarities with signal-transducing proteins, in particular G proteins (73), and this will be discussed in more detail below.

The name *nef* was given to this gene as a reference to its purported negative regulatory effects on viral transcription and replication (76). Since then, there have been numerous conflicting reports on whether *nef* has a negative effect (76–81), has no effect (2, 82–84), or has a positive effect (81, 84, 85) on viral transcription and replication. Such discrepancies have occurred with researchers reportedly using the same virus and the same cell line. It was recently suggested that different *nef* alleles may be responsible for the differential effects of the *nef* gene on growth of HIV-1 *in vitro* (81). If there is an effect of *nef* on the growth of lentiviruses *in vitro*, it seems likely to be a small one, at least under standard cell culture conditions. Our own observations lead us to believe that differences in the rate of growth of nef^+ and nef^- viruses at very low input inocula (based on amount of Gag antigen or reverse transcriptase activity) can be due to subtle differences in the infectivity titre of different virus stocks. The absence of selective pressure to maintain open forms of the *nef* gene in standard cell culture conditions also argues against significant effects of *nef* under these culture conditions (2).

Also controversial are reports that Nef protein possesses guanine nucleotide binding, GTPase, and autophosphorylating activities. Such activities were first investigated because of the discovery of fairly good sequence similarity between short stretches of *nef* genes and a highly charged region within the intracytoplasmic phosphorylation domain of human interleukin-2 receptor and the ATP-binding site of the catalytic subunit of cAMP-dependent protein kinase and other members of the protein kinase family (86). Later, sequence similarity was reported with genes for pp60-Src, epidermal growth factor receptor, and the bovine G proteins, Gas and Gta2 (73). These observations led to work purporting to show that Nef possesses GTPase and GTP-binding activities. However, based on sequence analysis, two motifs necessary for binding the guanine base are missing (87). It remains possible, based on sequence analysis, that Nef is capable of binding gamma-phosphate residues in nucleoside triphosphates.

The original paper that purported to show GTP-binding and GTPase activities in bacterially expressed Nef protein has been criticized and contradicted by others using more highly purified preparations of Nef protein (87–90). One of these groups did confirm the autophosphorylation activity using either ATP or GTP (87), although another did not (89). Guy *et al.*, the original reporting authors, have gone so far as identifying serines 88 and 103 as the phosphorylation sites in HIV-1 Nef protein by the creation and analysis of point mutations (91).

Infection of CD4$^+$ lymphocytes by the primate lentiviruses has been shown to result in a down-regulation of membrane-surface CD4 receptor molecules (92–96). Initially, this down-regulation was thought to result from intracellular complexing of the CD4 molecule with the Env protein (96), but later it was observed that CD4 is depleted in cells infected with a vaccinia recombinant in which Nef is the only HIV-1 gene product (73). Pulse labelling studies were presented to indicate that this decrease in cell-surface CD4 receptor is caused by an almost complete block in protein translation in infected cells that is accompanied by little or no change in CD4 mRNA transcription or levels (97). However, although Garcia and Miller found a similar decrease in cell-surface CD4 receptor by FACS analysis (98), they failed to observe any change in steady-state levels of either CD4 mRNA or CD4 protein, implying that the decrease in cell-surface protein was due to a failure of nascent CD4 protein to get to or stay at the cell surface. These results on CD4 down-regulation have been completely contradicted by Cheng-Mayer *et al.* who failed to observe any change in CD4 levels in established cell lines infected with HIV-1 isolates (79). Furthermore, although Gama Sosa *et al.* have observed reductions in cell-surface CD4 after transfections with a recombinant retrovirus expressing HIV-1 *nef*, they also observed such reductions when transfecting with a recombinant lacking the HIV-1 *nef* gene, implying that any such observations were artefactual (99). However, such criticisms may be rendered moot by the recent observation that transgenic mice expressing the HIV-1 *nef* gene are severely depleted in CD4$^+$ T-lymphocytes (J. Skowronski, R. Mariani, and L. Usher, personal communication).

At this point, it is unclear what the significance of any down-regulation of the CD4 receptor may be. One possibility is that such a down-regulation may prevent the super-infection of virus-infected cells. Another possibility is that down-regulation of CD4 may be a manifestation of alterations in T-lymphocyte activation pathways. It is well known that CD4 interacts with other molecules (major histocompatibility complex class II and lck kinase) important for T-cell receptor-mediated activation, and that binding of antibodies or ligand to CD4 activates the lymphocyte (100). By this scenario, Nef could be altering activation pathways in the lymphocyte and, in so doing, could affect virus expression.

Luria et al. (101) have presented evidence that at least some Nef products can block induction of IL-2 mRNA in lymphoid cells, triggered by activating agents PMA, PHA, and/or antibodies against CD3, T-cell receptor, or CD2. Again, the significance of this blocked induction is not clear, but it could again be related to a role for Nef in altering T-lymphocyte activation pathways.

Despite the absence of a detectable effect of *nef* in standard cell culture conditions, Kestler et al. demonstrated the importance of the *nef* gene *in vivo* using cloned SIV$_{mac}$239 virus (2, 102, 103). In this model system, virus derived from cloned DNA causes AIDS in rhesus monkeys (104). Kestler et al. (2, 102) found rapid reversion of stop codon point mutations in *nef* to open forms of the reading frame *in vivo*, demonstrating selective pressure for open, presumably functional, forms of *nef in vivo*. This was in contrast to the situation observed in cell culture where such stop codons in *nef* did not reduce the ability of the virus to grow and did not revert. *nef* was found to be a positive factor in rhesus monkeys because of the requirement of *nef* for vigorous virus replication in rhesus monkeys, for maintaining normal virus loads, and for the induction of disease. Animals inoculated with *nef* deletion mutants have remained disease-free by all clinical criteria for at least two and a half years, while wild-type virus-infected animals all developed AIDS and most have died. In contrast to wild-type-infected animals, it has become increasingly difficult to isolate virus from the animals receiving the *nef*-deleted virus with time. Such a *nef*-deleted virus may potentially be of use as the basis for a multiply deleted, live-attenuated vaccine (105). Since *nef* is conserved throughout all primate lentiviruses, it is likely that the function of *nef* is the same in HIV-1 as it is in SIV$_{mac}$.

Since *nef* has been shown to be essential for the rapid pathogenesis of SIV in rhesus monkeys, it has been suggested that researchers should consider using a *nef*-deleted HIV-1, where appropriate, as a means of decreasing risk in laboratory workers (106). This idea has been criticized (107) because it is not yet known that such HIV-1 derivatives are non-pathogenic, because the function of *nef* in SIV and HIV-1 may be different, because the *nef* point mutants of Kestler et al. (2) reverted rapidly *in vivo*, and because the use of *nef*-deleted virus may result in less stringent biosafety procedures. Certainly, HIV-1 deleted in *nef* should be treated with all the respect and care afforded wild-type HIV-1. However, we feel it is highly unlikely that the *nef* genes of SIV and HIV have different functions, based on conserved genome location in all members of the primate lentivirus family, extensive amino acid sequence similarities, and similarities in the biological properties of SIV and HIV. As for reversions, it is precisely for this reason that *nef* deletions, which do not revert, were suggested for use in reducing biosafety risk. There are certainly no guarantees that human infection with an HIV-1 *nef* deletion mutant will be totally non-pathogenic, but the use of *nef*-deleted virus in the laboratory and in production facilities seems likely to be a logical safer alternative. For some applications, recombinant DNA procedures can be considered that entirely circumvent the need for live viruses.

7. Concluding remarks

The relative conservation of the primate lentivirus 'non-essential' genes, within a group and in some cases among all four primate lentivirus groups, suggests important roles for these genes in virus replication, persistence, or transmission. In the only case where a 'non-essential' gene has been examined for its effects *in vivo*, the $SIV_{mac}239$ *nef* gene was shown to be essential for vigorous virus replication, maintenance of normal virus loads, and full pathogenic potential (2, 103). Similar experiments involving other 'non-essential' genes *in vivo* are currently underway. If an important role is clearly defined for any such 'non-essential' gene, the product of the gene should immediately become a target for drug development in the fight against AIDS. Realistically, drug development will not quickly follow *in vivo* demonstration of the importance of any of these 'non-essential' genes because of the lack of specific information on what these gene products are doing. Progress has been hindered by the lack of clear-cut effects on virus replication *in vitro* and by the conflicting claims that have appeared in the literature. In the case of the *nef* studies, there is suggestive evidence that the standard cell culture systems do not accurately reflect the major modes of virus replication *in vivo* (2, 103). Hopefully, the phenotypic properties of mutant viruses in macaque monkeys will provide important clues for unravelling the functional role of each of these genes. The combination of animal model, cell culture, and biochemical studies will likely be important for eventually developing a coherent picture for the functional role of these genes.

Better understanding of these 'non-essential' genes *vif*, *nef*, *vpr*, *vpx*, and *vpu* is important not just for drug development. These genes are likely to be critical elements in the replication strategies employed by SIV/HIV and for the basic mechanisms of viral persistence, processes in which we are lacking specific fundamental insights. For example, the mechanism by which SIV and HIV are able to continue to replicate, eventually destroying their hosts, in spite of vigorous humoral and cellular immune responses, are not at all understood. The 'non-essential' genes may also be useful for creating effective vaccines against AIDS. The potential advantages of multiply deleted, live-attenuated HIV-1 strains as effective vaccines against AIDS have recently been described in detail (105).

References

1. Cullen, B. R. and Greene, W. C. (1990) Functions of the auxiliary gene products of the human immunodeficiency virus type 1. *Virology*, **178,** 1.
2. Kestler, H. W., III, Ringler, D. J., Mori, K., Panicali, D. L., Sehgal, P. K., Daniel, M. D., and Desrosiers, R. C. (1991) Importance of the *nef* gene for maintenance of high virus loads and for development of AIDS. *Cell*, **65,** 651.
3. Myers, G., MacInnes, K., and Korber, B. (1992) The emergence of simian/human immunodeficiency viruses. *AIDS Res. Hum. Retroviruses*, **8,** 373.
4. Tristem, M., Marshall, C., Karpas, A., Petrik, J., and Hill, F. (1990) Origin of *vpx* in lentiviruses. *Nature*, **347,** 341.
5. Huet, T., Cheynier, R., Meyerhans, A., Roelants, G., and Wain-Hobson, S. (1990) Genetic organization of a chimpanzee lentivirus related to HIV-1. *Nature*, **345,** 356.
6. Miller, R. H. (1988) Human immunodeficiency virus may encode a novel protein on the genomic DNA plus strand. *Science*, **239,** 1420.
7. Benko, D. M., Schwartz, S., Pavlakis, G. N., and Felber, B. K. (1990) A novel human immunodeficiency virus type 1 protein, *tev*, shares sequences with *tat*, *env*, and *rev* proteins. *J. Virol.*, **64,** 2505.
8. Cohen, E. A., Lu, Y., Göttlinger, H., Dehni, G., Jalinoos, Y., Sodroski, J. G., and Haseltine, W. A. (1990) The T open reading frame of human immunodeficiency virus type 1. *J. Acquired Immune Deficiency Syndr.*, **3,** 601.
9. Garvey, K. J., Oberste, M. S., Elser, J. E., Braun, M. J., and Gonda, M. A. (1990) Nucleotide sequence and genome organization of biologically active proviruses of the bovine immunodeficiency-like virus. *Virology*, **175,** 391.
10. Gonda, M. A., Oberste, M. S., Garvey, K. J., Pallansch, L. A., Battles, J. K., Pifat, D. Y., Bess, J. W. Jr., and Nagashima, K. (1990) Development of the bovine immuno-deficiency-like virus as a model of lentivirus disease. *Dev. Biol. Stand.*, **72,** 97.
11. Olmsted, R. A., Hirsch, V. M., Purcell, R. H., and Johnson, P. R. (1989) Nucleotide sequence analysis of feline immunodeficiency virus: genome organization and relationship to other lentiviruses. *Proc. Natl Acad. Sci. USA*, **86,** 8088.
12. Phillips, T. R., Talbott, R. L., Lamont, C., Muire, S., Lovelace, K., and Elder, J. H. (1990) Comparison of two host cell range variants of feline immunodeficiency virus. *J. Virol.*, **64,** 4605.
13. Talbott, R. L., Sparger, E. E., Lovelace, K. M., Fitch, W. M., Pedersen, N. C., Luciw, P. A., and Elder, J. H. (1989) Nucleotide sequence and genomic organization of feline immunodeficiency virus. *Proc. Natl Acad. Sci. USA*, **86,** 5743.
14. Kawakami, T., Sherman, L., Dahlberg, J., Gazit, A., Yaniv, A., Tronick, S. R., and Aaronson, S. A. (1987) Nucleotide sequence analysis of equine infectious anemia virus proviral DNA. *Virology*, **158,** 300.
15. Pyper, J. M., Clements, J. E., Gonda, M. A., and Narayan, O. (1986) Sequence homology between cloned caprine arthritis encephalitis virus and visna virus, two neurotropic lentiviruses. *J. Virol.*, **58,** 665.
16. Sonig, P., Alizon, M., Staskus, K., Klatzmann, D., Cole, S., Danos, O., Retzel, E., Tiollais, P., Haase, A., and Wain-Hobson, S. (1985) Nucleotide sequence of the visna lentivirus: relationship to the AIDS virus. *Cell*, **42,** 369.
17. Oberste, M. S. and Gonda, M. A. (1992) Conservation of amino-acid sequence motifs in lentivirus Vif proteins. *Virus Genes*, **6,** 95.

18. Muesing, M. A., Smith, D. H., Cabradilla, C. D., Benton, C. V., Lasky, L. A., and Capon, D. J. (1985) Nucleic acid structure and expression of the human AIDS/lymphadenopathy retrovirus. *Nature,* **313,** 450.
19. Ratner, L., Haseltine, W., Patarca, R., Livak, K. J., Starcich, B., Josephs, S. F., Doran, E. R., Rafalski, J. A., Whitehorn, E. A., Baumeister, K., Ivanoff, L., Petteway, S. R., Jr., Pearson, M. L., Lautenberger, J. A., Papas, T. S., Ghrayeb, J., Chang, N. T., Gallo, R. C., and Wong-Staal, F. (1985) Complete nucleotide sequence of the AIDS virus, HTLV-III. *Nature,* **313,** 277.
20. Sanchez-Pescador, R., Power, M. D., Barr, P. J., Steimer, K. S., Stempien, M. M., Brown-Shimer, S. L., Gee, W. W., Renard, A., Randolph, A., Levy, J. A., Dina, D., and Luciw, P. A. (1985) Nucleotide sequence and expression of an AIDS-associated retrovirus (ARV-2). *Science,* **227,** 484.
21. Wain-Hobson, S., Sonigo, P., Danos, O., Cole, S., and Alizon, M. (1985) Nucleotide sequence of the AIDS virus, LAV. *Cell,* **40,** 9.
22. Kan, N. C., Franchini, G., Wong-Staal, F., DuBois, G. C., Robey, W. G., Lautenberger, J. A., and Papas, T. S. (1986) Identification of HTLV-III/LAV *sor* gene product and detection of antibodies in human sera. *Science,* **231,** 1553.
23. Lee, T.-H., Coligan, J. E., Allan, J. S., McLane, M. F., Groopman, J. E., and Essex, M. (1986) A new HTLV-III/LAV protein encoded by a gene found in cytopathic retroviruses. *Science,* **231,** 1546.
24. Sakai, K., Ma, X., Gordienko, I., and Volsky, D. J. (1991) Recombinational analysis of a natural noncytopathic human immunodeficiency virus type 1 (HIV-1) isolate: role of the *vif* gene in HIV-1 infection kinetics and cytopathicity. *J. Virol.,* **65,** 5765.
25. Fisher, A. G., Ensoli, B., Ivanoff, L., Chamberlain, M., Petteway, S., Ratner, L., Gallo, R. C., and Wong-Staal, F. (1987) The *sor* gene of HIV-1 is required for efficient virus transmission in vitro. *Science,* **237,** 888.
26. Garrett, E. D., Tiley, L. S., and Cullen, B. R. (1991) Rev activates expression of the human immunodeficiency virus type 1 *vif* and *vpr* gene products. *J. Virol.,* **65,** 1653.
27. Schwartz, S., Felber, B. K., and Pavlakis, G. N. (1991) Expression of human immunodeficiency virus type 1 vif and vpr mRNAs is Rev-dependent and regulated by splicing. *Virology,* **183,** 677.
28. Cox, F. E. G. (1990) The worm and the virus. *Nature,* **347,** 618.
29. Khalife, J., Grzych, J.-M., Pierce, R., Ameisen, J.-C., Schacht, A.-M., Gras-Masse, H., Tartar, A., Lecocq, J.-P., and Capron, A. (1990) Immunological crossreactivity between the human immunodeficiency virus type 1 virion infectivity factor and a 170-kD surface antigen of *Schistosoma mansoni. J. Exp. Med.,* **172,** 1001.
30. Strebel, K., Daugherty, D., Clouse, K., Cohen, D., Folks, T., and Martin, M. A. (1987) The HIV 'A' (*sor*) gene product is essential for virus infectivity. *Nature,* **328,** 728.
31. Sodroski, J., Goh, W. C., Rosen, C., Tartar, A., Portetelle, D., Burny, A., and Haseltine, W. (1986) Replicative and cytopathic potential of HTLV-III/LAV with *sor* gene deletions. *Science,* **231,** 1549.
32. Shibata, R., Miura, T., Hayami, M., Sakai, H., Ogawa, K., Kiyomasu, T., Ishimoto, A., and Adachi, A. (1990) Construction and characterization of an infectious DNA clone and of mutants of simian immunodeficiency virus isolated from the African green monkey. *J. Virol.,* **64,** 307.
33. Shibata, R., Adachi, A., Sakai, H., Ishimoto, A., Miura, T., and Hayami, M. (1990) Mutational analysis of simian immunodeficiency virus from African green monkeys and human immunodeficiency virus type 2. *J. Med. Primatol.,* **19,** 217.

34. Shibata, R., Miura, T., Hayami, M., Ogawa, K., Sakai, H., Kiyomasu, T., Ishimoto, A., and Adachi, A. (1990) Mutational analysis of the human immunodeficiency virus type 2 (HIV-2) genome in relation to HIV-1 and simian immunodeficiency virus SIV$_{AGM}$. *J. Virol.*, **64**, 742.
35. Guy, B., Geist, M., Dott, K., Spehner, D., Kieny, M.-P., and Lecocq, J.-P. (1991) A specific inhibitor of cysteine proteases impairs a Vif-dependent modification of human immunodeficiency virus type 1 Env protein. *J. Virol.*, **65**, 1325.
36. Wong-Staal, F., Chanda, P. K., and Ghrayeb, J. (1987) Human immunodeficiency virus: The eighth gene. *AIDS Res. Hum. Retroviruses*, **3**, 33.
37. Reiss, P., de Ronde, A., Lange, J. M. A., de Wolf, F., Dekker, J., Debouck, C., and Goudsmit, J. (1989) Antibody response to the viral negative factor (*nef*) in HIV-1 infection: A correlate of levels of HIV-1 expression. *AIDS*, **3**, 227.
38. Sato, A., Igarashi, H., Adachi, A., and Hayami, M. (1990) Identification and localization of *vpr* gene product of human immunodeficiency virus type 1. *Virus Genes*, **4**, 303.
39. Yuan, X., Matsuda, Z., Matsuda, M., Essex, M., and Lee, T.-H. (1990) Human immunodeficiency virus *vpr* gene encodes a virion-associated protein. *AIDS Res. Hum. Retroviruses*, **6**, 1265.
40. Cohen, E. A., Dehni, G., Sodroski, J. G., and Haseltine, W. A. (1990) Human immunodeficiency virus *vpr* product is a virion-associated regulatory protein. *J. Virol.*, **64**, 3097.
41. Yu, X.-F., Matsuda, M., Essex, M., and Lee, T.-H. (1990) Open reading frame *vpr* of simian immunodeficiency virus encodes a virion-associated protein. *J. Virol.*, **64**, 5688.
42. Dedera, D., Hu, W., Vander Heyden, N., and Ratner, L. (1989) Viral protein R of human immunodeficiency virus types 1 and 2 is dispensable for replication and cytopathogenicity in lymphoid cells. *J. Virol.*, **63**, 3205.
43. Cohen, E. A., Terwilliger, E. F., Jalinoos, Y., Proulx, J., Sodroski, J. G., and Haseltine, W. A. (1990) Identification of HIV-1 *vpr* product and function. *J. Acquired Immune Deficiency Syndr.*, **3**, 11.
44. Ogawa, K., Shibata, R., Kiyomasu, T., Higuchi, I., Kishida, Y., Ishimoto, A., and Adachi, A. (1989) Mutational analysis of the human immunodeficiency virus *vpr* open reading frame. *J. Virol.*, **63**, 4110.
45. Hattori, N., Michaels, F., Fargnoli, K., Marcon, L., Gallo, R. C., and Franchini, G. (1990) The human immunodeficiency virus type 2 *vpr* gene is essential for productive infection of human macrophages. *Proc. Natl Acad. Sci. USA*, **87**, 8080.
46. Banapour, B., Marthas, M. L., Ramos, R. A., Lohman, B. L., Unger, R. E., Gardner, M. B., Pedersen, N. C., and Luciw, P. A. (1991) Identification of viral determinants of macrophage tropism for simian immunodeficiency virus SIV$_{mac}$. *J. Virol.*, **65**, 5798.
47. Henderson, L. E., Sowder, R. C., Copeland, T. D., Benveniste, R. E., and Oroszlan, S. (1988) Isolation and characterization of a novel protein (X-ORF product) from SIV and HIV-2. *Science*, **241**, 199.
48. Guyader, M., Emerman, M., Montagnier, L., and Peden, K. (1989) VPX mutants of HIV-2 are infectious in established cell lines but display a severe defect in peripheral blood lymphocytes. *EMBO J.*, **8**, 1169.
49. Kappes, J. C., Conway, J. A., Lee, S.-W., Shaw, G. M., and Hahn, B. H. (1991) Human immunodeficiency virus type 2 vpx protein augments viral infectivity. *Virology*, **184**, 197.
50. Yu, X.-F., Yu, Q.-C., Essex, M., and Lee, T.-H. (1991) The *vpx* gene of simian

immunodeficiency virus facilitates efficient viral replication in fresh lymphocytes and macrophages. *J. Virol.*, **65**, 5088.
51. Hu, W., Vander Heyden, N., and Ratner, L. (1989) Analysis of the function of viral protein X (VPX) of HIV-2. *Virology*, **173**, 624.
52. Strebel, K., Klimkait, T., and Martin, M. A. (1988) A novel gene of HIV-1, *vpu* and its 16-kilodalton product. *Science*, **241**, 1221.
53. Cohen, E. A., Terwilliger, E. F., Sodroski, J. G., and Haseltine, W. A. (1988) Identification of a protein encoded by the *vpu* gene of HIV-1. *Nature*, **334**, 532.
54. Schneider, T., Hildebrandt, P., Rönspeck, W., Weigelt, W., and Pauli, G. (1990) The antibody response to the HIV-1 specific 'out' (*vpu*) protein: identification of an immunodominant epitope and correlation of antibody detectability to clinical stages. *AIDS Res. Hum. Retroviruses*, **6**, 943.
55. Reiss, P., Lange, J. M. A., de Ronde, A., de Wolf, F., Dekker, J., Debouck, C., and Goudsmit, J. (1990) Speed of progression to AIDS and degree of antibody response to accessory gene products of HIV-1. *J. Med. Virol.*, **30**, 163.
56. Reiss, P., Lange, J. M. A., de Ronde, A., de Wolf, F., Dekker, J., Danner, S. A., Debouck, C., and Goudsmit, J. (1990) Antibody response to viral proteins U (*vpu*) and R (*vpr*) in HIV-1-infected individuals. *J. Acquired Immune Deficiency Syndr.*, **3**, 115.
57. Strebel, K., Klimkait, T., Maldarelli, F., and Martin, M. A. (1989) Molecular and biochemical analyses of human immunodeficiency virus type 1 *vpu* protein. *J. Virol.*, **63**, 3784.
58. Klimkait, T., Strebel, K., Hoggan, M. D., Martin, M. A., and Orenstein, J. M. (1990) The human immunodeficiency virus type 1-specific protein *vpu* is required for efficient virus maturation and release. *J. Virol.*, **64**, 621.
59. Adachi, A., Ono, N., Sakai, H., Ogawa, K., Shibata, R., Kiyomasu, T., Masuike, H., and Ueda, S. (1991) Generation and characterization of the human immunodeficiency virus type 1 mutants. *Arch. Virol.*, **117**, 45.
60. Terwilliger, E. F., Cohen, E. A., Lu, Y. C., Sodroski, J. G., and Haseltine, W. A. (1989) Functional role of human immunodeficiency virus type 1 *vpu*. *Proc. Natl Acad. Sci. USA*, **86**, 5163.
61. Willey, R. L., Maldarelli, F., Martin, M. A., and Strebel, K. (1992) Human immunodeficiency virus type 1 Vpu protein regulates the formation of intracellular gp160-CD4 complexes. *J. Virol.*, **66**, 226.
62. Schwartz, S., Felber, B. K., Fenyo, E.-M., and Pavlakis, G. N. (1990) Env and Vpu proteins of human immunodeficiency virus type 1 are produced from multiple bicistronic mRNAs. *J. Virol.*, **64**, 5448.
63. Arrigo, S. J., Weitsman, S., Zack, J. A., and Chen, I. S. Y. (1990) Characterization and expression of novel singly spliced RNA species of human immunodeficiency virus type 1. *J. Virol.*, **64**, 4585.
64. Schwartz, S., Felber, B. K., and Pavlakis, G. N. (1992) Mechanism of translation of monocistronic and multicistronic human immunodeficiency virus type 1 mRNAs. *Mol. Cell. Biol.*, **12**, 207.
65. Furtado, M. R., Balachandran, R., Gupta, P., and Wolinsky, S. M. (1991) Analysis of alternatively spliced human immunodeficiency virus type-1 mRNA species, one of which encodes a novel tat-env fusion protein. *Virology*, **185**, 258.
66. Sugrue, R. J., Bahadur, G., Zambon, M. C., Hall-Smith, M., Douglas, A. R., and Hay, A. J. (1990) Specific structural alteration of the influenza haemagglutinin by amantadine. *EMBO J.*, **9**, 3469.

67. Sugrue, R. J. and Hay, A. J. (1991) Structural characteristics of the M2 protein of influenza A viruses: evidence that it forms a tetrameric channel. *Virology*, **180**, 617.
68. Allan, J. S., Coligan, J. E., Lee, T.-H., McLane, M. F., Kanki, P. J., Groopman, J. E., and Essex, M. (1985) A new HTLV-III/LAV encoded antigen detected by antibodies from AIDS patients. *Science*, **230**, 810.
69. Arya, S. K. and Gallo, R. C. (1986) Three novel genes of human T-lymphotropic virus type III: immune reactivity of their products with sera from acquired immune deficiency syndrome patients. *Proc. Natl Acad. Sci. USA*, **83**, 2209.
70. Franchini, G., Robert-Guroff, M., Aldovini, A., Kan, N. C., and Wong-Staal, F. (1987) Spectrum of natural antibodies against five HTLV-III antigens in infected individuals: correlation of antibody prevalence with clinical status. *Blood*, **69**, 437.
71. Bahraoui, E., Yagello, M., Billaud, J.-N., Sabatier, J.-M., Guy, B., Muchmore, E., Girard, M., and Gluckman, J.-C. (1990) Immunogenicity of the human immunodeficiency virus (HIV) recombinant *nef* gene product. Mapping of T-cell and B-cell epitopes in immunized chimpanzees. *AIDS Res. Hum. Retroviruses*, **6**, 1087.
72. Franchini, G., Robert-Guroff, M., Ghrayeb, J., Chang, N. T., and Wong-Staal, F. (1986) Cytoplasmic localization of the HTLV-III 3' *orf* protein in cultured T cells. *Virology*, **155**, 593.
73. Guy, B., Kieny, M. P., Riviere, Y., Le Peuch, C., Dott, K., Girard, M., Montagnier, L., and Lecocq, J.-P. (1987) HIV F/3' *orf* encodes a phosphorylated GTP-binding protein resembling an oncogene product. *Nature*, **330**, 266.
74. Werner, T., Ferroni, S., Saermark, T., Brack-Werner, R., Banati, R. B., Mager, R., Steinaa, L., Kreutzberg, G. W., and Erfle, V. (1991) HIV-1 Nef protein exhibits structural and functional similarity to scorpion peptides interacting with K^+ channels. *AIDS*, **5**, 1301.
75. Samuel, K. P., Hodge, D. R., Chen, Y.-M. A., and Papas, T. S. (1991) Nef proteins of the human immunodeficiency viruses (HIV-1 and HIV-2) and simian immunodeficiency virus (SIV) are structurally similar to leucine zipper transcriptional activation factors. *AIDS Res. Hum. Retroviruses*, **7**, 697.
76. Terwilliger, E., Sodroski, J. G., Rosen, C. A., and Haseltine, W. A. (1986) Effects of mutations within the 3' *orf* open reading frame region of human T-cell lymphotropic virus type III (HTLV-III/LAV) on replication and cytopathogenicity. *J. Virol.*, **60**, 754.
77. Ahmad, N. and Venkatesan, S. (1988) *Nef* protein of HIV-1 is a transcriptional repressor of HIV-1 LTR. *Science*, **241**, 1481.
78. Niederman, T. M. J., Thielan, B. J., and Ratner, L. (1989) Human immunodeficiency virus type 1 negative factor is a transcriptional silencer. *Proc. Natl Acad. Sci. USA*, **86**, 1128.
79. Cheng-Mayer, C., Iannello, P., Shaw, K., Luciw, P. A., and Levy, J. A. (1989) Differential effects of *nef* on HIV replication: implications for viral pathogenesis in the host. *Science*, **246**, 1629.
80. Binninger, D., Ennen, J., Bonn, D., Norley, S. G., and Kurth, R. (1991) Mutational analysis of the simian immunodeficiency virus SIVmac *nef* gene. *J. Virol.*, **65**, 5237.
81. Terwilliger, E. F., Langhoff, E., Gabuzda, D., D. Zazopoulos, E., and Haseltine, W. A. (1991) Allelic variation in the effects of the *nef* gene on replication of human immunodeficiency virus type 1. *Proc. Natl Acad. Sci. USA*, **88**, 10971.
82. Hammes, S. R., Dixon, E. P., Malim, M. H., Cullen, B. R., and Greene, W. C. (1989) Nef protein of human immunodeficiency virus type 1: evidence against its role as a transcriptional inhibitor. *Proc. Natl Acad. Sci. USA*, **86**, 9549.

83. Bachelerie, F., Alcami, J., Hazan, U., Israël, N., Goud, B., Arenzana-Seisdedos, F., and Virelizier, J.-L. (1990). Constitutive expression of human immunodeficiency virus (HIV) *nef* protein in human astrocytes does not influence basal or induced HIV long terminal repeat activity. *J. Virol.*, **64**, 3059.
84. Kim, S., Ikeuchi, K., Byrn, R., Groopman, J., and Baltimore, D. (1989) Lack of a negative influence on viral growth by the *nef* gene of human immunodeficiency virus type 1. *Proc. Natl Acad. Sci. USA*, **86**, 9544.
85. De Ronde, A., Klaver, B., Keulen, W., Smit, L., and Goudsmit, J. (1992) Natural HIV-1 NEF accelerates virus replication in primary human lymphocytes. *Virology*, **188**, 391.
86. Samuel, K. P., Seth, A., Konopka, A., Lautenberger, J. A., and Papas, T. S. (1987) The 3'-orf protein of human immunodeficiency virus shows structural homology with the phosphorylation domain of human interleukin-2 receptor and the ATP-binding site of the protein kinase family. *FEBS Lett.*, **218**, 81.
87. Nebreda, A. R., Bryan, T., Segade, F., Wingfield, P., Venkatesan, S., and Santos, E. (1991) Biochemical and biological comparison of HIV-1 NEF and *ras* gene products. *Virology*, **183**, 151.
88. Kaminchik, J., Bashan, N., Pinchasi, D., Amit, B., Sarver, N., Johnston, M. I., Fischer, M., Yavin, Z., Gorecki, M., and Panet, A. (1990) Expression and biochemical characterization of human immunodeficiency virus type 1 *nef* gene product. *J. Virol.*, **64**, 3447.
89. Backer, J. M., Mendola, C. E., Fairhurst, J. L., and Kovesdi, I. (1991) The HIV-1 *nef* protein does not have guanine nucleotide binding, GTPase, or autophosphorylating activities. *AIDS Res. Hum. Retroviruses*, **7**, 1015.
90. Matsuura, Y., Maekawa, M., Hattori, S., Ikegami, N., Hayashi, A., Yamazaki, S., Morita, C., and Takebe, Y. (1991) Purification and characterization of human immunodeficiency virus type 1 *nef* gene product expressed by a recombinant baculovirus. *Virology*, **184**, 580.
91. Guy, B., Rivière, Y., Dott, K., Regnault, A., and Kieny, M. P. (1990) Mutational analysis of the HIV nef protein. *Virology*, **176**, 413.
92. Klatzmann, D., Barré-Sinoussi, F., Nugeyre, M. T., Dauguet, C., Vilmer, E., Griscelli, C., Brun-Vezinet, F., Rouzioux, C., Gluckman, J. C., Chermann, J.-C., and Montagnier, L. (1984) Selective tropism of lymphadenopathy associated virus (LAV) for helper-inducer T lymphocytes. *Science*, **225**, 59.
93. Dalgleish, A. G., Beverley, P. C. L., Clapham, P. R., Crawford, D. H., Greaves, M. F., and Weiss, R. A. (1984) The CD4 (T4) antigen is an essential component of the receptor for the AIDS retrovirus. *Nature*, **312**, 763.
94. Klatzmann, D., Champagne, E., Chamaret, S., Gruest, J., Guetard, D., Hercend, T., Gluckman, J.-C., and Montagnier, L. (1984) T-Lymphocyte T4 molecule behaves as the receptor for human retrovirus LAV. *Nature*, **312**, 767.
95. Hoxie, J. A., Flaherty, L. E., Haggarty, B. S., and Rackowski, J. L. (1986) Infection of T4 lymphocytes by HTLV-III does not require expression of the OKT4 epitope. *J. Immunol.*, **136**, 361.
96. Hoxie, J. A., Alpers, J. D., Rackowski, J. L., Huebner, K., Haggarty, B. S., Cedarbaum, A. J., and Reed, J. C. (1986) Alterations in T4 (CD4) protein and mRNA synthesis in cells infected with HIV. *Science*, **234**, 1123.
97. Yuille, M. A. R., Hugunin, M., John, P., Peer, L., Sacks, L. V., Poiesz, B. J., Tomar, R. H., and Silverstone, A. E. (1988) HIV-1 infection abolishes CD4 biosynthesis but not CD4 mRNA. *J. Acquir. Immune Defic. Syndr.*, **1**, 131.

98. Garcia, J. V. and Miller, A. D. (1991) Serine phosphorylation-independent down-regulation of cell-surface CD4 by *nef*. *Nature,* **350,** 508.
99. Gama Sosa, M. A., DeGasperi, R., Kim, Y.-S., Fazely, F., Sharma, P., and Ruprecht, R. M. (1991) Serine phosphorylation-independent downregulation of cell-surface CD4 by *nef. AIDS Res. Hum. Retroviruses,* **7,** 859.
100. Williams, A. F. and Beyers, A. D. (1992) T-Cell receptors: At grips with interactions. *Nature,* **356,** 746.
101. Luria, S., Chambers, I., and Berg, P. (1991) Expression of the type 1 human immunodeficiency virus Nef protein in T cells prevents antigen receptor-mediated induction of interleukin 2 mRNA. *Proc. Natl Acad. Sci. USA,* **88,** 5326.
102. Kestler, H. W., III, Naidu, Y. N., Kodama, T., King, N. W., Daniel, M. D., Li, Y., and Desrosiers, R. C. (1989) Use of infectious molecular clones of simian immunodeficiency virus for pathogenesis studies. *J. Med. Primatol.,* **18,** 305.
103. Kestler, H. W., III, Mori, K., Silva, D. P., Kodama, T., King, N. W., Daniel, M. D., and Desrosiers, R. C. (1990) *Nef* genes of SIV. *J. Med. Primatol.,* **19,** 421.
104. Kestler, H., Kodama, T., Ringler, D., Marthas, M., Pedersen, N., Lackner, A., Regier, D., Sehgal, P., Daniel, M., King, N., and Desrosiers, R. (1990) Induction of AIDS in rhesus monkeys by molecularly cloned simian immunodeficiency virus. *Science,* **248,** 1109.
105. Desrosiers, R. C. (1992) HIV with multiple gene deletions as a live attenuated vaccine for AIDS. *AIDS Res. Hum. Retroviruses,* **8,** 411.
106. Desrosiers, R. C. and Hunter, E. (1991) AIDS biosafety. *Science,* **252,** 1231.
107. Haseltine, W. A. and Levy, J. A. (1991) HIV research and *nef* alleles. *Science,* **253,** 366.

7 | The molecular biology of human T-cell leukemia viruses

SCOTT D. GITLIN, JÜRGEN DITTMER, ROBERT L. REID, and JOHN N. BRADY

1. Introduction

The human T-cell leukemia virus (HTLV) was first described in the early 1980s following the isolation of the virus from cell lines established from an American patient with a cutaneous T-cell lymphoma (1). At approximately the same time, a similar virus (adult T-cell leukemia retrovirus; ATLV) was independently isolated from a Japanese patient with adult T-cell leukemia/lymphoma (ATL) (2). Based on the homology between the viral genome and viral antigens, the name HTLV-I was proposed for viruses previously designated HTLV or ATLV. Another member of the HTLV subfamily, HTLV-II, was first isolated from the transformed T-lymphocytes of a patient with a relatively benign T-cell variant of hairy cell leukemia and later from a second patient with both a T-cell lymphoproliferative disease and hairy cell leukemia with a B-cell phenotype (3, 4). Recently, HTLV-II has also been isolated from Guaymi indians in Panama (5). An apparently distinct virus, HTLV-V, was reportedly isolated from a continuous cell line derived from the peripheral lymphocytes of a patient with $CD4^+$, IL-2Rα (Tac)-negative cutaneous T-cell lymphoma/leukemia and was subsequently identified in seven other patients, from Italy, with cutaneous T-cell lymphoma/leukemia (6).

There is little known, and in fact there is some controversy, as to the origin of the HTLVs (7–10). Epidemiologic studies have found HTLV-I to be endemic in southwestern Japan, the Caribbean basin, and parts of central and eastern Africa (7, 11). In addition, HTLV-I has been identified in populations within the southeastern and southwestern United States, South America, Philippines, Europe, and the Middle East (7, 11, 12). The geographic distribution of HTLV-II is only now being described (10) and HTLV-V has only been described in a series of patients from Italy (6).

In this chapter, we will attempt to present an overview of what has been learned about the viral, epidemiological, clinical, and cellular aspects of human T-lymphotropic virus infections from a molecular biology perspective.

2. HTLV genome structure

The HTLV-I and HTLV-II proviral genomes have been molecularly cloned and the complete nucleotide sequence has been determined (13–15). Similar to other retroviruses, open reading frames (ORFs) which code for the Gag (48 kDa), Pol (99 kDa), and Env (54 kDa) proteins are present in the 5' portion of the viral genome (Fig. 1). At the 3' end of these open reading frames is a series of open reading frames that are unique to the human retroviruses. In the HTLVs, this region is commonly referred to as *pX*. In HTLV-I, the *pX* region contains four open reading frames: X-I, X-II, X-III, and X-IV. Tax_1 is encoded primarily by the X-IV reading frame, while Rex_1 is encoded by the X-III reading frame. The AUG initiation codons for both Tax_1 and Rex_1 are located in the second exon of the doubly spliced mRNA. Two new proteins, Tof and Rof, have recently been described (16; G. Franchini, personal communication). Both Tof and Rof are encoded by an alternatively double-spliced mRNA species (16; G. Franchini, personal communication). Tof utilizes the Tax_1 AUG initiation codon and continues with the X-II open reading frame. Rof utilizes the Rex_1 AUG initiation codon and continues with the X-I open reading frame. The expression of several other alternatively spliced *pX* region mRNAs have also been detected by the reverse transcriptase-polymerase chain reaction in HTLV-I-infected cell lines, primary uncultured cells from ATL patients and from asymptomatic HTLV-I-infected carriers (17). The HTLV-II genome is similarly organized, although the equivalent alternatively spliced mRNAs have not been identified.

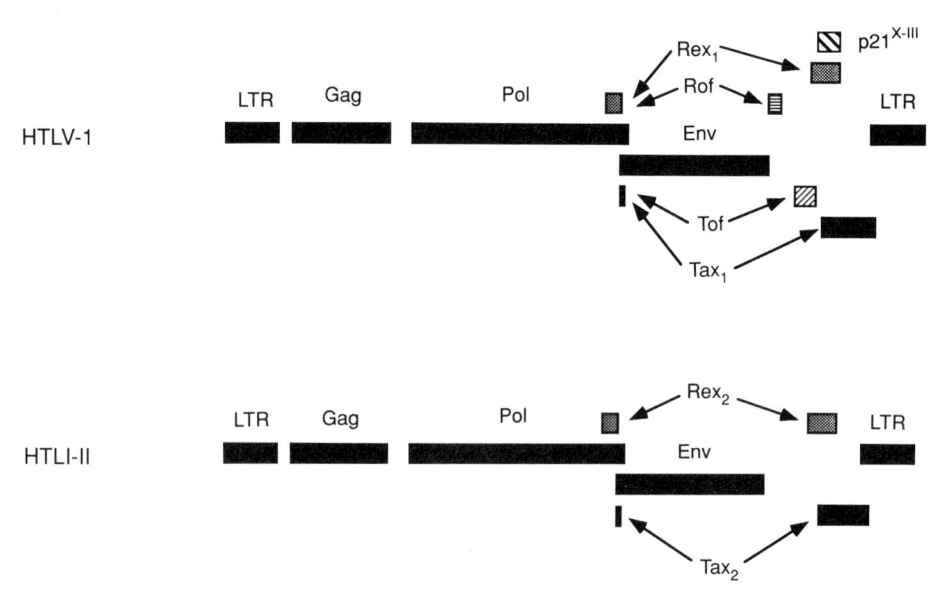

Fig. 1 The genomic organization of HTLV-I and HTLV-II. The open reading frames for the identified gene products in HTLV-I and HTLV-II are shown

Genomic sequence variability of HTLV-I and HTLV-II has been found to be dependent on geography and on the genes studied (18–20). Despite the sequence variability identified between HTLV-I isolates from ATL and tropical spastic paraparesis/HTLV-I associated myelopathy (TSP/HAM) patients, no association between sequence variation and disease has been identified (21).

3. Transformation

HTLV-I and HTLV-II have been shown to activate and immortalize human T-lymphocytes *in vitro* (22, 23). Infection of HTLV-I *in vitro* results in polyclonal proliferation of the infected cells followed by oligoclonal or monoclonal growth (23). Tumour cells derived from ATL patients are typically monoclonal. The first step in lymphocyte activation by HTLV-I appears to be mediated, at least in part, by the external envelope protein gp46 (22). Lymphocyte activation has also been reported to be mediated by the CD2 pathway (24, 25).

Tax_1 has been shown to regulate the expression of several cellular genes (Section 10), suggesting that Tax_1 may immortalize and transform cells through the induction of cellular genes and the resultant loss of the normal transcriptional control of these genes in the infected cell. For example, it has been suggested that an autocrine loop involving Tax_1-inducing IL-2Rα and IL-2 expression may lead to the immortalization and transformation of HTLV-I-infected lymphocytes (26, 27). When the *pX* coding sequences were integrated into a transformation-defective herpesvirus, it was able to transform T-cells of the same phenotype as found with HTLV-I (28). Of the three proteins encoded by the *pX* region, the Tax_1 gene is critical for transformation since specific mutations of the Tax_1 AUG initiation codon eliminated the transformation potential of the herpesvirus–HTLV-I pX recombinant virus (W. A. Haseltine, personal communication). It is not known whether the intact herpesvirus genes contributed to transformation. Tax_1, in combination with *ras*, has been shown to transform fibroblast cell lines *in vitro* (29) and to induce neurofibromas and mesenchymal and fibroblastic tumours in transgenic mice (30–32).

It seems likely that the HTLV-I virus, through expression of the viral regulatory proteins Tax_1 and Rex_1, provides some initial alteration in cell metabolism predisposing the development of ATL. Subsequently, the rearrangement or altered expression of a cellular oncogene(s) may provide the 'second hit' leading to development of ATL. Chromosomal abnormalities have been observed in patients with ATL. Trisomy 3 and trisomy 7 have been observed in patients with acute ATL (33). In addition, there appears to be a frequent abnormal rearrangement in the long arm of chromosome 6 in ATL patients from northern Kyushu, Japan (33). Critical regulatory and cellular oncogenes which are located on these three chromosomes include: chromosome 3, RAF; chromosome 6, ROS, MAS, SRC/YES-related oncogene; chromosome 7, TCR-β, TCR-γ, PDGF, ERBB, EGF receptor, ARAF2, PKS1, MET, EPH, and tumour cell invasion and metastasis gene TIM1. It has also been suggested that the rearrangement of a proto-oncogene and the T-cell antigen receptor (TCR-α), located on chromosome 14, is necessary for the development of overt ATL (34). Chromosomal analysis of human lymphocyte cultures infected with HTLV-I demonstrated that 30 per cent of the cell cultures studied contained structural abnormalities and hypodiploidy. This may be compared with 10 per cent chromosomal abnormalities in uninfected cell cultures propagated for less than 200 days. All of the immortalized cell lines infected with HTLV-I contained clonal chromosomal abnormalities characteristic for each cell line after prolonged propagation. Southern blot analysis of one T-cell line and three B-cell lines infected with HTLV-I alone or with HTLV-I and Epstein–Barr virus revealed rearrangements of the J_k gene in all B-cell lines and rearrangements of the T-cell receptor β gene in all cell lines, regardless of their lineage (35). In addition, the HTLV-I Tax_1 gene has been shown to regulate negatively the host DNA repair enzyme, β-polymerase (36), which may contribute to chromosome mutations.

4. Epidemiology

The epidemiology of HTLV infections has been extensively reviewed elsewhere (7, 12). We will briefly highlight some of the aspects which have relevance to the molecular biology of these retroviruses.

4.1 Discriminating assays for HTLV-I and HTLV-II

Distinguishing between HTLV-I and HTLV-II has been very difficult due to the inability of serologic assays to differentiate between the two viruses (37). In fact, because of a lack of discriminating assays, many of the seroepidemiologic studies of HTLV-I actually included data for both HTLV-I and HTLV-II infections. For example, in a study of intravenous drug users, 69–84 per cent of seropositive persons were found to have HTLV-II infections (39). This study suggests that the prevalence of HTLV-II is much higher than was originally believed and that much of the seroprevalence believed to be due to HTLV-I in previous epidemiologic studies was likely due to HTLV-II, at least in this US population. Additional studies have also revealed high levels of HTLV-II infections in HTLV-seropositive blood donors in New Mexico (40). Assays which can distinguish between HTLV-I and HTLV-II include the use of monoclonal antibodies directed against unique and immunogenic epitopes on one or the other virus (such as against MTA-4 found on HTLV-I and not found on HTLV-II) (38), peptide-based ELISA assays (39), peptide-based enzyme immunoassays (41), or the use of the polymerase chain reaction (40).

4.2 Transmission

HTLV-I can be transmitted by sexual intercourse, by introduction of virus-containing cells intravenously (e.g. blood transfusions and the sharing of needles by intravenous drug abusers), or by the vertical transmission from mother to child. The risk of sexual transmission of HTLV-I appears to be higher for male-to-female transmission (60.8%) than for female-to-male transmission (0.4%) (42). Interestingly, the presence of anti-Tax$_1$ antibodies in men appears to indicate a higher potential for transmission of HTLV-I to their sexual partners via heterosexual routes (43). Blood transfusion appears to be the most efficient mode of transmission, with reported seroconversion rates of 35–60 per cent following exposure to HTLV-I-contaminated cellular blood products (11). Transmission through the non-cellular fraction of blood apparently does not occur. Vertical transmission from mother to child occurs predominantly through transmission of the virus in breast milk. It has been estimated that a baby will ingest a total of approximately 10^8 HTLV-I-infected cells through breast-feeding (44, 45). The transmission of HTLV-I through breast milk can be prevented by having HTLV-I-infected mothers refrain from breast-feeding (45, 46). Reports of familial clustering, however, reveal that not all of the vertical transmission occurs through breast-feeding and raises the likelihood that intrauterine or perinatal transmission may also play a role (47), although these latter routes of transmission have been disputed (44).

5. HTLV-associated diseases
5.1 ATL

HTLV-I is the causative agent of ATL, a malignancy of mature, CD4$^+$ T-lymphocytes that was first described in 1977 in Japan (11). ATL is characterized by lymphoma/leukemia often associated with generalized lymphadenopathy, hypercalcemia, infiltrative skin lesions, hepatosplenomegaly, bone marrow involvement, and lytic bone lesions of the skull and long bones. The clinical course of ATL is progressive and can be divided into four stages: pre-ATL or 'preleukemia', smoldering ATL, chronic ATL, and acute or subacute ATL (10). The malignant cells in the peripheral blood of ATL patients are lymphocytes which have characteristic large, cleaved, or irregular nuclei, often referred to as 'flower cells'. These cells stain with CD4 monoclonal antibodies and are of the same origin as cells found infiltrating the lymph nodes in these patients (48). They contain clonally integrated HTLV-I viral sequences (48). Cytogenetic studies on ATL cells revealed these cells to be ERFC$^+$, OKT3$^+$, OKT4$^+$, OKT8$^-$, OKT10$^+$, OKIa1$^+$, and Tac (IL-2Rα)$^+$. Overexpression of IL-2Rα is commonly observed in lymphocytes from ATL patients. IL-2Rα expression on the cell surface appears to play a critical role in HTLV-I pathogenesis and has been used as a target for therapeutic agents. Treatment of ATL patients with monoclonal antibodies to the IL-2Rα has resulted in clinical improvement (49).

5.2 TSP/HAM

Tropical spastic paraparesis (TSP) and HTLV-I associated myelopathy (HAM), now considered to be the same disease (TSP/HAM), are slowly progressive myelopathies that are characterized by paraparesis and spasticity of the lower extremities, sometimes associated with posterior column and superficial sensory changes, involvement of the pyramidal tract, bladder dysfunction, and minimal sensory loss (10, 50). The incidence and prevalence of TSP/HAM correlates well with those geographic regions endemic for HTLV-I, particularly the Caribbean basin, Africa, the Seychelles Islands, Colombia, southern Japan, and the United States (11, 50–52). Lymphoid cell lines derived from the peripheral blood of HTLV-I-seropositive TSP/HAM patients have been analysed and compared with the leukemogenic prototypes. The established TSP/HAM cell lines were of T-cell lineage with $CD2^+$, $CD3^+$, $CD4^+$, $CD7^+$, $WT31^+$ with activated T-cell markers $CD25^+$, DR^+, and with a clonal rearrangement of the β and γ genes of the T-cell receptor (21). Patients with TSP/HAM revealed high levels of circulating HTLV-I-specific cytotoxic T-lymphocytes which were $CD8^+$ and HLA class I-restricted and which predominantly recognized HTLV-I Tax_1 (53).

There appears to be a state of T-cell activation that is not present in those HTLV-I-infected persons with ATL. Increased expression of IL-2 and IL-2Rα transcripts was detected in TSP/HAM and seropositive carriers that paralleled the coordinate mRNA expression of Tax_1. In addition, IL-2 and soluble IL-2Rα serum levels in TSP/HAM and seropositive carriers were elevated. In ATL, there were markedly elevated levels of soluble IL-2Rα serum levels, but transcripts for IL-2 and Tax_1 were not demonstrable in the circulating cells. These findings suggest that immune activation in TSP/HAM, in contrast to ATL, is virally driven by *trans*-activation and coordinate expression of IL-2 and IL-2Rα (54).

5.3 Other diseases

In addition to diseases involving T-lymphocytes, HTLV-I has been associated with several non-T-cell diseases. These cases have been reviewed elsewhere (12, 26).

HTLV-II was first isolated from one patient with a relatively benign T-cell variant of hairy cell leukemia (3) and from the T-cells from a patient with both a T-cell lymphoproliferative disease and a B-cell hairy cell leukemia (4). However, most cases of hairy cell leukemia are seronegative for HTLV and there was no evidence of increased risk for hairy cell leukemia, mycosis fungoides, or chronic lymphocytic leukemia in HTLV-II endemic populations (55). Despite anecdotal associations of HTLV-II infection with a variety of diseases, there has been no consistent disease association with HTLV-II.

HTLV-V has only been reported in eight patients with cutaneous T-cell lymphoma/leukemias, although the etiologic role of HTLV-V in these diseases has not yet been established (6).

5.4 Animal models

Rabbits and transgenic mice have been used most extensively in the search for a good animal model for HTLV-I infections. Rabbits have been used to examine

HTLV-I virus infection and replication (56, 57), transmission (58–60), host immune responses (60–62), disease manifestations (62–64), coincident infections with HIV-1 (65), and immunization and treatment modalities (59, 66–69). The evaluation of transgenic mice with an integrated HTLV-I genome or with specific HTLV-I genes (e.g. tax_1) have revealed other possible disease associations with HTLV-I. An inflammatory arthropathy resembling rheumatoid arthritis has been described in mice containing the HTLV-I genome (70). In addition, an exocrinopathy resembling Sjögren's syndrome (71) and a possible model for human neurofibromatosis has been identified in HTLV-I Tax_1 transgenic mice (30). However, evaluation of human neurofibromatosis tissue failed to reveal evidence of HTLV-I infection (32).

6. Tax structure and functional domains

The pX regions of HTLV-I and HTLV-II encode a positive *trans*-activator of viral and cellular transcription, Tax_1 and Tax_2, respectively (Fig. 1) (72–93). Tax_1 is a 40 kDa (353 amino acid) phosphoprotein which is predominantly localized to the nucleus of the host cell (Fig. 2) (73, 94–98). Tax_2 is a 331 amino acid protein, which is also located primarily in the nucleus of the infected cell (Fig. 2). Amino acid

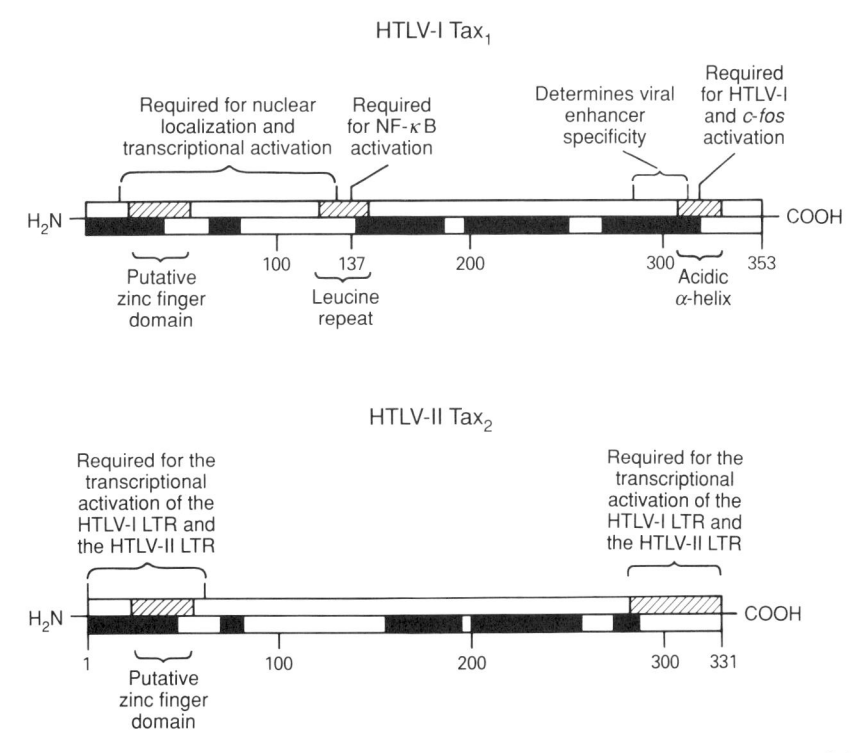

Fig. 2 Structural and functional domains of HTLV-I and HTLV-II Tax proteins. The lower portion of the schematic representation of these proteins indicate predicted hydrophobic (solid boxes) and hydrophilic/neutral (open boxes) domains. Functional domains are indicated by hatched boxes or brackets

sequence analysis of these two proteins reveal that there is approximately 77–85 per cent amino acid homology between these two proteins (72, 99). HTLV-II Tax$_2$ lacks the 22 carboxy-terminal amino acids of HTLV-I Tax$_1$ (99). HTLV-II Tax$_2$ can *trans*-activate both the HTLV-I and HTLV-II long terminal repeats (LTRs), but HTLV-I Tax$_1$ can only *trans*-activate the HTLV-I LTR (100).

The amino terminus of the HTLV-I Tax$_1$ protein has been analysed by deletion mutagenesis (98, 101) and by site-specific mutagenesis (97). These analyses have demonstrated that amino acids 2–59 of the Tax$_1$ protein contain a novel nuclear localization sequence. Select downstream amino acids, such as amino acids 62–63, 73–64, 82–83, 102–103, 123–124, 130–131, 189–190, 194–195, and 206–207 also appear to be important for the nuclear localization of Tax$_1$. All of the Tax$_1$ mutants which did not localize to the nucleus were unable to *trans*-activate the HTLV-1 LTR or heterologous LTRs (HTLV-II or HIV-1), suggesting that Tax$_1$ must be present in the nucleus in order to exert its transcriptional activation function. Substitution mutations of either amino acids 29, 39, 49, or 56 of Tax$_1$ resulted in loss of *trans*-activational activity on the HTLV-I LTR. These amino acids are conserved between HTLV-I, HTLV-II, and the related bovine leukemia virus Tax proteins. Additionally, substitution of amino acid 18 resulted in decreasing the ability of Tax$_1$ to *trans*-activate the HTLV-I LTR. Mutation of amino acid 47 had no significant effect on *trans*-activational activity. The mutation of amino acid 39 from arginine to glycine also resulted in a Tax$_1$ protein that exhibited *trans*-dominant activity in the presence of wild-type Tax$_1$ (102). A single amino acid substitution at position 5 of Tax$_1$ or Tax$_2$ altered the *trans*-activation phenotype of these proteins on various promoters. The mutated Tax$_2$ also exhibited *trans*-dominant behaviour in the presence of wild-type Tax$_2$ that was not exhibited by an identical mutation in Tax$_1$. Deletion of amino acids 2–17 from the Tax$_2$ protein eliminated *trans*-activation of the HTLV-I LTR, HTLV-II LTR, and the adenovirus EIII promoter (103). Deletion of amino acids 2–59 of HTLV-I Tax$_1$ resulted in a cytoplasmic Tax$_1$ protein that exhibited *trans*-dominant effects in the presence of wild-type Tax$_1$ (98).

The amino-terminal portion of the Tax$_1$ and Tax$_2$ proteins contain amino acid sequences which have some homology to known zinc-finger motifs (97, 98, 104). Tax$_1$ has been shown to bind to a zinc chelate affinity column (101, 104) and to interact with other divalent metals such as Cu^{2+} and Ni^{2+} with lesser affinity (105). Cysteines 23, 29, and 36 appear to be important for zinc binding (101). The exact significance of the zinc-finger motif and the ability of Tax$_1$ to bind to a zinc affinity column is uncertain at this time, but this domain may be involved with protein–protein interactions with other transcription factors or with dimerization of the Tax$_1$ protein. Tax$_1$ from *Escherichia coli* has been shown to form homomultimers that are dependent on the presence of functional sulphydryl groups, such as those present within the putative zinc-finger domain of the Tax$_1$ protein (98). As discussed above, deletion of the amino terminal (2–59) amino acids has also resulted in a truncated protein which acts as a *trans*-dominant in the presence of wild-type Tax$_1$ (98). Therefore, it is possible that the putative zinc-finger domain is required for the multimerization of Tax$_1$ and that the multimerized form of Tax$_1$ is the form which is transported to the nucleus. It is also possible that the zinc-finger domain allows Tax_1 to not only bind to other Tax_1 molecules, but also to cellular factors which are involved in the mediation of Tax_1 cellular and viral effects.

The middle portion of the Tax_1 protein is largely hydrophobic. The region from amino acid 124 to 146 resembles a leucine repeat domain similar to that found in the *jun/fos* transcription factor family (106). This region also contains a partial homology with the RB interaction domains of adenovirus E1a domain I, HPV 16 E7, and SV40 T-antigen (107). A two amino acid change at amino acids 137 and 138 of HTLV-I Tax_1 has been found to prevent Tax_1 from *trans*-activating NF-κB-responsive promoters and from inducing nuclear NF-κB expression and activity. This mutation did not affect the ability of Tax_1 to *trans*-activate CREB/ATF-responsive promoters (97).

The carboxyl terminus of Tax_1 is acidic in composition and is predicted to fold as an α-helix, suggesting that this domain could activate transcription by a mechanism similar to other *trans*-activators such as GAL4, GCN4, VP16, and NF-κB p65 (108–110). In contrast, the carboxyl terminus of Tax_2 is not predicted to fold as an α-helix. Deletion of the carboxy-terminal 22 amino acids from Tax_1 did not affect the ability of the resultant Tax_1 protein to *trans*-activate either CREB/ATF- or NF-κB-responsive promoters, indicating that the *trans*-activation functions of Tax_1 reside in sequences upstream of the carboxyl terminus and that there may be two distinct *trans*-activation domains. No transcriptional activation activity was reported when this carboxy-terminal region of Tax_1 was fused to the GAL4 DNA-binding domain (97). Of note, these mutagenesis studies also revealed that the c-*fos* promoter and the HTLV-I LTR are *trans*-activated by Tax_1 via a common mechanism (97). Deletion of the carboxy-terminal 69 amino acids of Tax_1 resulted in a Tax_1 protein which was unable to *trans*-activate the HTLV-I LTR in HeLa cells (102). These findings suggest that the region between amino acids 285 and 330 contain sequences that are necessary for the *trans*-activation function of Tax_1 on the HTLV-I LTR. Site-specific mutations at amino acids 315 to 322 resulted in Tax_1 proteins that are able to *trans*-activate the NF-κB-responsive HIV-1 LTR at levels comparable to wild type but that are impaired in their ability to *trans*-activate the HTLV-I LTR (97).

When deletion mutants of HTLV-I Tax_1 were fused downstream of the GAL4 DNA-binding domain, with and without the acidic activation domain of herpes simplex virus VP16, evidence for a transcriptional activation domain between amino acids 312 and 337 was reported. In addition, the region between amino acids 284 and 312, at the carboxy terminus of Tax_1, was found to determine specificity for the viral enhancer (111).

7. Rex

Through the use of an alternative AUG initiation codon, the doubly spliced mRNA from the *pX* region can support the synthesis of a 27 kDa protein (Rex_1) for HTLV-I and a 24 kDa/26 kDa protein (Rex_2) for HTLV-II (Fig. 3) (112, 113). The 26 kDa Rex_2 protein appears to be the phosphorylated form of p24 (114).

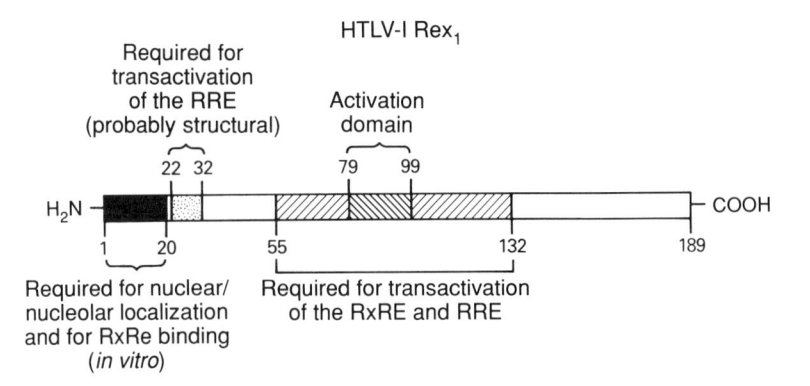

Fig. 3 Functional domains of HTLV-I and HTLV-II Rex proteins. The solid, stippled, and hatched boxes indicate identified functional domains within these proteins

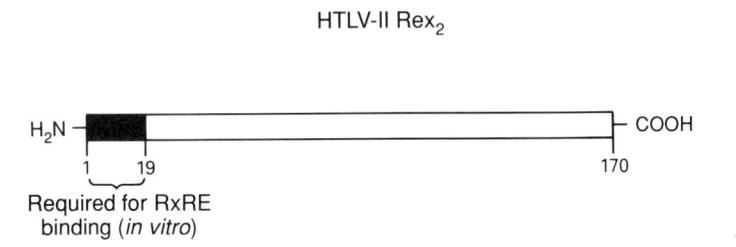

Rex$_1$ and Rex$_2$ are predominantly found in the nucleolus of the cell and are responsible for regulating the processing of the primary viral transcripts, leading to a balance of the three species of spliced and unspliced mRNAs that are necessary for virus replication (see Chapter 5). Rex is required for the expression of the unspliced *gag/pol* and the singly spliced *env* transcripts (115, 116). Rex$_1$ has been shown to be a positive regulator of expression of viral structural proteins and a negative regulator of viral gene transcription (115, 117). This effect is accomplished by a dose-dependent accumulation of unspliced *gag* and *env* transcripts in the nucleus and cytoplasm of the infected cell and a concomitant decrease of Tax/Rex-region mRNAs, suggesting that Rex$_1$ both mediates the nuclear transport of these transcripts and also regulates RNA processing at a post-transcriptional level (118). The basic, amino-terminal, approximately 20 amino acids are required for the nucleolar localization of Rex and for Rex binding to the Rex-response element (RxRE) (119, 120). The activation domain of Rex$_1$ has been mapped between amino acids 79 and 99 in the 189 amino acid Rex$_1$ protein (121).

One RxRE has been identified for HTLV-I in the 3′ LTR (R/U3 region) (Fig. 4). The HTLV-I RxRE$_1$ is a 255 nucleotide stem–loop structure which is required for Rex$_1$ responsiveness (122). The binding site for Rex$_1$ has been identified as a 10 bp, partially double-stranded sequence encompassing nucleotides 8761–8770/8792–8801 (506–515/537–546 relative to the 3′ LTR start site) (13, 123). The RxRe functions independent of 3′ mRNA cleavage and polyadenylation and is orientation

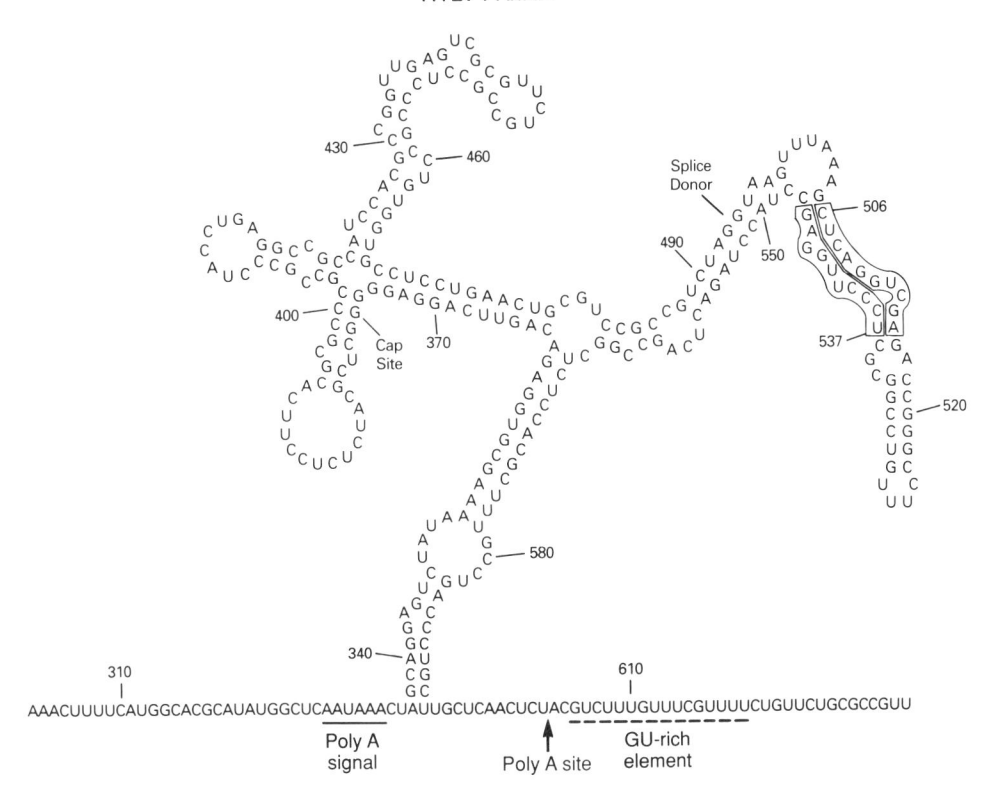

Fig. 4 Nucleotide sequence and predicted secondary structure of the HTLV-I RxRE. The sequences which have been shown to be required for Rex$_1$ responsiveness are shown (boxed sequences)

dependent but position independent. Independent of its importance for Rex$_1$ function, the stem–loop structure of the RxRE$_1$ plays a critical role in the poladenylation of HTLV-I transcripts. Unique RNA folding juxtaposes the polyadenylation (poly A) signal and the poly A site to the correct functional distance for polyadenylation to occur (117, 123).

Two RxREs have been identified for HTLV-II, one in the 5′ LTR (R/U5 region) (124) and one in the 3′ LTR (R/U3 region) (125). The HTLV-II 3′ LTR RxRE$_2$ is located between nucleotides 8454 and 8754 and contains a 281 nucleotide stem–loop structure that juxtaposes the poly A signal to the polyadenylation site (125), as is found for the HTLV-I 3′ LTR RxRE$_1$. The HTLV-II 5′ LTR RxRE$_2$ has been located between nucleotides 405 and 630 relative to the 5′ LTR start site (126). For efficient Rex binding, a stem–bulge–loop RNA structure (nucleotides 465–500) and the sequences between nucleotides 449–455 and 470–476 are required (127). The 5′ RxRE$_2$ includes a 'cis-acting repressive sequence' located between nucleotides 520 and 630, which has been reported to be involved in Rex-dependent regulation of

HTLV-II gene expression (126). The 5' RxRE$_2$ appears to be sufficient for the function of Rex$_2$ to accumulate unspliced HTLV-II RNAs (126).

Rex$_1$ and Rex$_2$ have been shown to be able to interact with the HIV-1 Rev-response element (RRE) and to be able to substitute for Rev in accumulating unspliced and singly spliced HIV transcript, thus having the ability to rescue Rev-deficient HIV-1 proviruses (122, 123, 125, 128). However, HIV-1 Rev cannot substitute for Rex$_1$ in HTLV-I RxRE$_1$-mediated post-transcriptional regulation (122). Interestingly, there is no significant sequence homology between Rex$_1$ and Rev, or between their respective response elements.

8. p21$^{x\text{-}III}$

With the use of antisera directed against a synthetic peptide corresponding to the carboxyl terminus of ORF III in an immunoblot assay, a 21 kDa and 27 kDa protein, p21$^{x\text{-}III}$ and p27$^{x\text{-}III}$ (Rex$_1$) respectively, were identified in the HTLV-I-infected human T-cell lines HUT-102, MT-2, and Hayai. p21$^{x\text{-}III}$ was also identified in the HTLV-I-infected cell line HOS/MT-2 (112). p21$^{x\text{-}III}$ was detected as a double band in these cell lines, suggesting that post-translational modification(5) of the protein occurs. Subcellular fractionation of HUT-102 cells resulted in the detection of p21$^{x\text{-}III}$ and pp21$^{x\text{-}III}$ in the soluble fraction, indicating that p21$^{x\text{-}III}$ is cytoplasmic or only loosely bound to the nucleus. Freshly isolated ATL leukemic cells do not express p21$^{x\text{-}III}$ or Rex$_1$. After 4 days in tissue culture, these cells express Rex$_1$ but not p21$^{x\text{-}III}$. No antibodies to p21$^{x\text{-}III}$ have been identified in the serum of HTLV-I-infected individuals (112). Recent experimental data suggest a possible antagonistic role of p21$^{x\text{-}III}$ against Rex$_1$ (129).

Both p21$^{x\text{-}III}$ and Rex$_1$ react with an antiserum to a synthetic peptide corresponding to the 3' end of ORF III. Therefore, both of these proteins share epitopes and both contain the carboxyl terminus encoded by ORF III (Fig. 1) (112). Evidence that these are two different proteins that contain similar sequences, and not modifications of one another, comes from *in vitro* translation experiments which show that these two proteins are translated independently of one another from different AUG initiation codons and from a single 2.1 kb mRNA (113). In fact, these proteins are translated from the same 2.1 kb *pX* mRNA that encodes the Tax$_1$ protein, indicating that the 2.1 kb *pX* mRNA is polycistronic (113).

9. Transcriptional regulation of HTLV gene expression

The HTLV LTRs are *cis*-acting transcriptional regulatory sequences which are divided into three regions, U3, R, and U5. The U3, R, and U5 regions of the HTLV-I LTR consist of 353, 221, and 120 bp, respectively (13, 14). The 5' U3 region contains several important elements for HTLV transcription regulation. The Tax$_1$ protein has not been shown to bind directly to Tax-responsive sequences (130, 131), suggesting that Tax$_1$ *trans*-activation occurs through indirect effects of Tax$_1$ on transcription factors which bind to the TREs (Fig. 5). Likely mechanisms for Tax$_1$

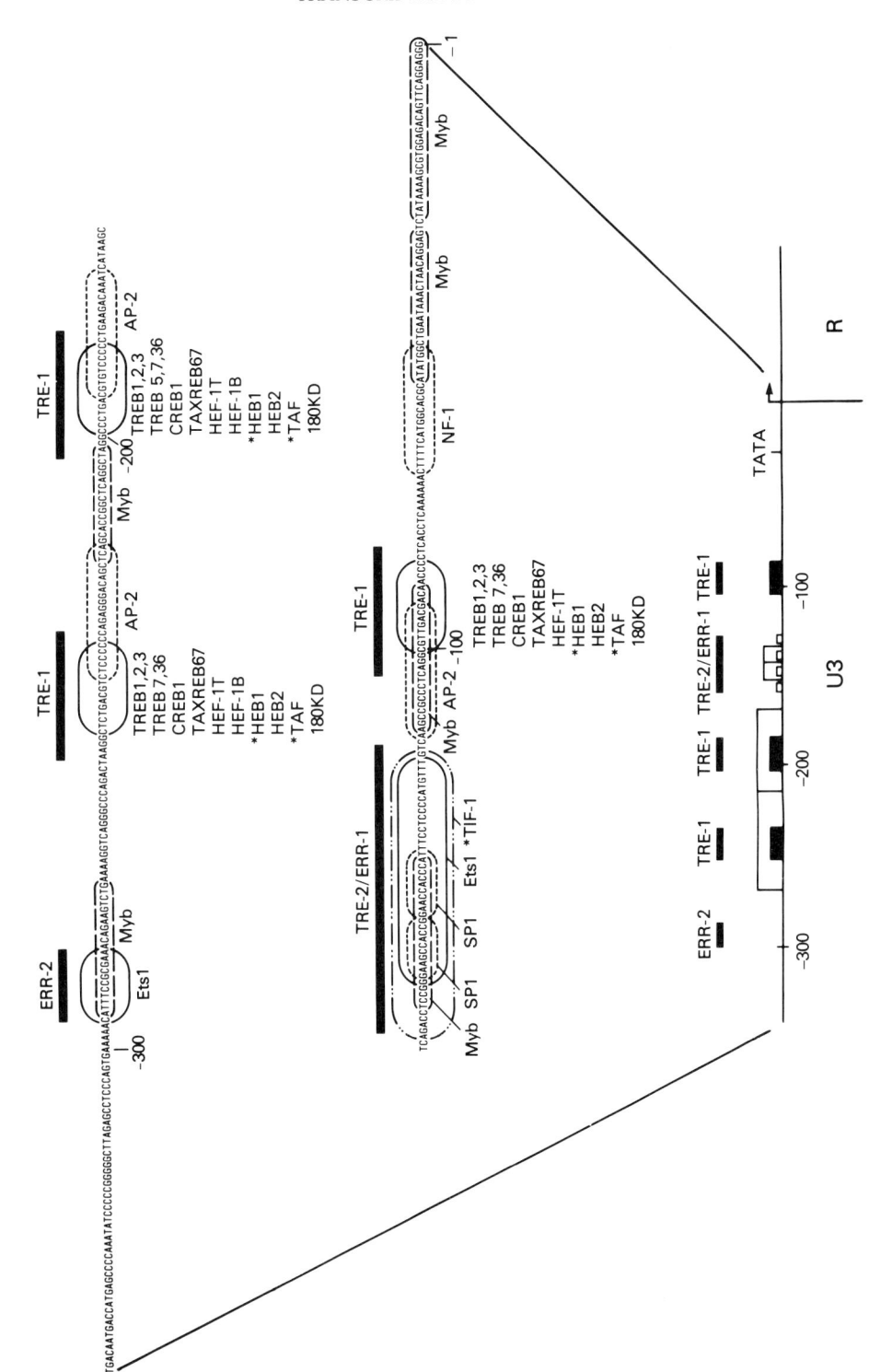

Fig. 5 Upstream control elements in the HTLV-I LTR. The HTLV-I sequence is that of the CR1 isolate (14). The transcriptional enhancer region (U3) contains binding sites for several cellular proteins, as shown. Asterisks (*) designate factors which have been shown to mediate an interaction between the Tax₁ protein and the HTLV-I LTR

trans-activation include: (1) transcriptional induction of TRE-binding transcription factors; (2) post-translational modification of TRE-binding factors; or (3) complex formation with transcription factors allowing indirect binding of the Tax₁ protein to the TRE(s).

9.1 Tax-responsive element 1 (TRE-1)

There are three 21 bp repeat elements (TRE-1) which have been shown to confer Tax₁ responsiveness and are highly conserved between HTLV-I and HTLV-II (Fig. 6) (132–138). The 21 bp repeats function in an orientation- and position-independent manner and confer Tax₁ responsiveness to a heterologous promoter (134). The TRE-1 element contains three conserved domains, designated from 5' to 3' as A, B, and C. Tax₁ responsiveness of the TRE-1 element requires regions A and B or regions B and C. Domain B consists of the first five of the 8-base cAMP-response element (CRE) *TGACGTCA*. Several techniques have been used to identify cellular proteins which interact with the TRE-1 elements, including the use of TRE-1 oligonucleotide affinity chromatography and the screening of cDNA expression libraries with ³²P-labelled TRE-1 probes (139–144). Several cellular proteins of the CREB/ATF family including CREB-1, TREB-1, -2, -3, -5, -36, HEF-1T, HEF-1B, HEB2, and TAXREB67 interact with TRE-1 (Fig. 5) (139, 141–144). The TREB-1, -2, and -3 proteins interact with the B domain of TRE-1 (139). Interestingly, the TREB-1 factor, which has been found to contain proteins of 35, 37, and 43 kDa, appears to be able to extend the DNase I footprint of the general transcription factor, TFIID, on the AdE4 promoter (139). TREB-7 and -36 are identical to ATF-2 and -1, respectively (145). TREB-5 (36 kDa), -7 (61 kDa) and -36 (34 kDa) contain leucine zipper domain homologies, a potential proline- and glutamine-rich *trans*-activation domain and a basic domain which may be involved in DNA binding (142). They also share homology with c-Fos, c-Jun, CREB-1, and C/EBP (142), and have been found to contain predicted phosphorylation sites for protein kinase A, protein kinase C, and casein kinase (142). TREB-36 shares 65 per cent homology

Fig. 6 The 21 bp repeat, TRE-1, domains of the HTLV-I LTR. The position of the TRE-1 regulatory sequences in the U3 region of the HTLV-I LTR are indicated (stippled boxes). Analysis of the TRE-1 enhancer elements has identified three functionally distinct conserved regulatory sequences, designated A, B, and C

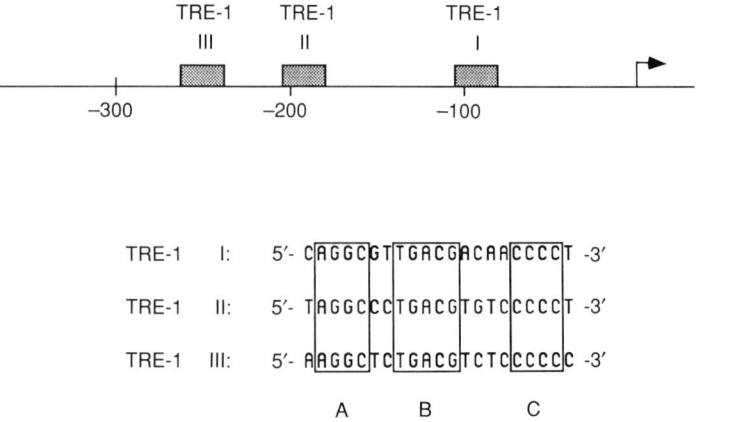

TRE-1 I: 5'- C|AGGCG|T|TGACGACAA|CCCC|T -3'

TRE-1 II: 5'- T|AGGCCC|TGACGTGTC|CCCC|T -3'

TRE-1 III: 5'- A|AGGCTC|TGACGTCTC|CCCC|C -3'

                    A   B   C

with the 45 kDa CREB-1. TAXREB67 is a 351 amino acid protein (52 kDa) whose RNA has been identified in several cell lines by Northern blot, and is identical to ATF4 (144, 145). TAXREB67 contains a leucine zipper near the C-terminus of its basic domain and has two acidic domains which could confer a *trans*-activation function (144). TAXREB67 expression in Jurkat cells was suppressed by dibutyryl-cAMP or by Vibrio cholera toxin, both of which increase cytoplasmic cAMP. TAXREB67 expression was increased by TPA or calcium ionophore A23187, inducers of T-cell proliferation (144). Additional proteins which interact with the TRE-1 include HEB1 (143), Tax activation factors (146), and a 180 kDa protein (137). Regulatory sequences including the second 21 bp repeat may also bind to cellular factors and confer TPA responsiveness to the LTR (147). The 21 bp repeats may also bind the Jun/Ap1 transcription factor (148) and Ap-2 binding sites are located adjacent to, and overlapping with, the TRE-1 domains (141).

Tax_1 has been shown to interact indirectly with the TRE-1 domain. Using an indirect binding assay, HEB1 was found to facilitate the indirect interaction of Tax_1 with TRE-1 DNA (143). In a separate study, Tax_1 was found to interact directly with the nuclear proteins, Tax_1 activation factors, that bind to TRE-1 (146). The interaction of Tax_1 with sequence-specific transcription factors and other cellular proteins (149) suggest that protein–protein interactions are important for HTLV transcription regulation.

9.2 Tax-responsive element 2 (TRE-2)

A second Tax-responsive element (TRE-2) is located at -117 to -163 (150), between the two proximal TRE-1 domains. TRE-2 contains binding sites for Sp1, TIF-1, Ets1 and Myb (Fig. 5) (141, 150–153). By use of an indirect binding assay, purified Tax_1 protein has also been shown to interact with TRE-2 DNA (131, 150). A 36 kDa HeLa nuclear protein, TIF-1, was purified and found to bind specifically to the TRE-2 domain and to mediate the binding of Tax_1 to the TRE-2 element in an indirect binding assay (131). TIF-1 has also been found to bind specifically and strongly to a single-stranded TRE-2 sense oligonucleotide (105).

The c-*ets*-1 proto-oncogene product, Ets1, is preferentially expressed in lymphoid tissues (154). Ets1 and the highly homologous Ets2 have been found to *trans*-activate the HTLV-I LTR in the absence of Tax_1. Evaluation of the Ets1 interaction with the HTLV-I LTR has revealed the presence of two distinct Ets1-responsive regions, ERR-1 located within TRE-2, and an upstream region designated ERR-2 (Fig. 5) (151, 153). The close proximity of the Ets binding sites to Tax-responsive elements suggests the possibility of an interaction between these cellular and viral factors. In fact, Tax_1 and Ets1 have been shown to cooperatively increase transcription from viral and cellular genes (155). The HTLV-I LTR also contains several binding sites for the Myb protein. *Trans*-activation of the HTLV-I LTR is greatest in the presence of the upstream Myb sites and little or no *trans*-activation resulted with only the downstream sites, suggesting that cooperation between the sites may occur (152).

9.3 R region of the HTLV-I LTR

Transcriptional control elements located downstream of the transcription start site, within the R region of the HTLV-I LTR, function in basal transcription and do not require Tax_1 for activity (156). These regulatory sequences enhance gene expression in an orientation-independent, but location-dependent manner (157). A 45-nucleotide *cis*-acting element which positively regulates basal transcription has been identified in the downstream R region of the HTLV-I LTR (105). This downstream element specifically binds a cellular factor which is important for *in vivo* and *in vitro* transcriptional activity (105). Further analysis of the R region of the HTLV-I LTR will determine if the observed increased transcriptional activity is at the level of transcriptional initiation or elongation.

10. Tax *trans*-activation of cellular genes

Tax_1 has been shown to *trans*-activate several cellular genes including IL-2, IL-2Rα, GM-CSF, and others (Table 1 and Fig. 7) (75–93). Tax_1 has also been shown to negatively regulate β-polymerase, a host DNA repair enzyme (36).

Tax_2, like Tax_1, is able to induce the expression of the IL-2 and IL-2Rα genes (158), and GM-CSF (79). Tax_2 is also able to *trans*-activate heterologous promoters, such as the adenovirus E1A and E3 promoters (159, 160).

Table 1 Cellular genes *trans*-activated by HTLV-I and HTLV-II Tax

HTLV-I TAX$_1$	HTLV-II TAX$_2$
IL-2	IL-2
IL-2Rα	IL-2Rα
GM-CSF	GM-CSF
IL-3	
Vimentin	
c-*fos*	
c-*sis*	
PTHrP	
TGF-β1	
Proenkephalin	
β globin	
ε globin	
Act-2 cytokine	
egr-1 (*Krox*-24)	
egr-2 (*Krox*-20)	
Class I MHC	
34 kDa membrane glycoprotein	
NF-κB	
Nerve growth factor	
β-polymerase (repressive effect)	

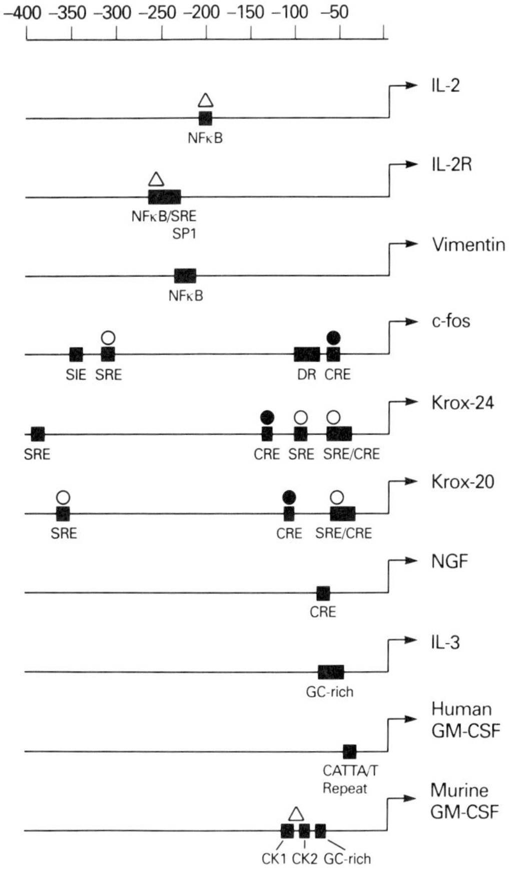

Fig. 7 Tax$_1$-responsive elements within the promoter/enhancer regions of cellular genes. The promoter/enhancer regions of several cellular genes which are *trans*-activated by HTLV-I Tax$_1$ have been defined (solid rectangles). These Tax$_1$-responsive sequences contain transcription factor consensus elements as indicated. Those regulatory sites in which transcription factor binding has been demonstrated are indicated by: NFκB (open triangle), SRF (open circle), or CREB (closed circle) transcription factors

Tax$_1$ induction of NF-κB is an important mechanism of transcriptional activation of some cellular genes. Tax$_1$ expression in virally infected and transfected cells induces IL-2Rα and IL-2 gene expression through induction of NF-κB proteins which bind to an 11 bp promoter element containing the NF-κB consensus sequence (75, 160–164). Other cellular genes which are stimulated by Tax$_1$ and contain NF-κB domains in the region of their promoters required for Tax$_1$ *trans*-activation include vimentin (a cytoskeletal growth-regulated gene) (81), TNF-β (165), and murine GM-CSF (166). It has been demonstrated that Tax$_1$-expressing cell lines show increased expression of the NF-κB DNA-binding proteins, p50, p55, p65, p75, p85, and p92 (167, 168). In addition, there is indirect evidence that Tax$_1$ activates NF-κB-dependent transcription mainly by dissociating the NF-κB–IκB complex (169).

It should be noted that Tax$_1$ activation of NF-κB-containing promoters may also require activation of other transcription factors. For example, complete IL-2Rα *trans*-activation by Tax$_1$ appears to require not only activation of NF-κB, but also appears to require a SRE/Sp1 consensus sequence which flanks the NF-κB element

(170). Likewise, elements in addition to the NF-κB sites are required for full activation of the IL-2 gene (164).

Although Tax$_1$ induction of NF-κB is an important mechanism of *trans*-activation of some cellular genes (Fig. 7), Tax$_1$ expression is not uniformly associated with increased levels of NF-κB or with the stimulation of genes which contain NF-κB promoter/enhancer elements. In some cell types, stable expression of Tax$_1$ has been associated with undetectable levels of NF-κB (171). In addition, some NF-κB element-containing genes are inactive in B-lymphocytes which constitutively express high levels of NF-κB (172). It is possible that these NF-κB elements may preferentially bind NF-κB factors which may repress transcription (173).

The role of the CREB/ATF site in Tax$_1$ *trans*-activation is also of interest because this motif is found in a wide variety of cellular genes as well as in the HTLV-I LTR (Figs 5 and 7). Not all CREB/ATF sites are Tax$_1$ responsive, suggesting that flanking promoter sequences are also important. Saturated mutagenesis of the HTLV-I 21 bp repeat has shown that the bases TGACGT are most important to Tax$_1$ *trans*-activation (130). Although there is considerable overlap in the Tax$_1$- and the CREB-responsive elements, promoter induction by these agents is independent and additive (174). As mentioned previously, the 21 bp repeat (TRE-1) has been divided into three 7 bp elements: domains A and C, which are important for Tax$_1$ responsiveness and which bind the HEB1 transcription factor which mediates the Tax$_1$ binding to the element, and region B, which is important for both Tax$_1$ and cAMP responsiveness (Fig. 6) (140, 143).

GM-CSF, a multipotent hematopoietic growth factor, which is expressed in normal T-lymphocytes after stimulation with mitogens such as PHA or phorbol esters, is expressed at high levels in Tax$_1$-producing Jurkat cells and in Tax$_1$-expressing transgenic mice (76, 168). Tax$_1$ and Tax$_2$ have been shown to mediate the activation of the GM-CSF promoter (Fig. 7) (77–79). The sequences responsible for Tax *trans*-activation appear to be species-specific. In the human GM-CSF promoter a sequence between base pairs –48 and –34 seems to be required and sufficient for both Tax$_1$/Tax$_2$ and PHA/TPA responsiveness (175). In the murine GM-CSF promoter three different elements (cytokine element 1 and 2 (CK1, CK2), and a GC-rich element) which respond to Tax$_1$ have been identified (78). The sequence which includes CK2 and the GC-rich region also shows TPA responsiveness. Another inducible growth factor, interleukin-3 (IL-3), can also be induced by Tax$_1$ through the CK2/GC-rich sequence (Fig. 7) (80).

The c-*fos* proto-oncogene is an immediate-early response gene that can be activated by several stimuli including lectins, phorbol ester, serum, growth factors, and cytokines (176). By forming a heterodimeric complex with the product of another early-response proto-oncogene, c-*jun*, c-Fos can activate the transcription of genes that respond to the phorbol ester TPA, an activator of protein kinase C (176). c-Fos can also repress the activity of genes, including its own gene (177). In non-transformed cells, c-Fos is only transiently expressed after stimulation due to the very short half-life of c-Fos mRNA and protein and to the negative autoregulation of c-*fos* gene expression (178). Tax$_1$ appears to interfere with c-Fos regulation.

A constitutively high level of c-Fos protein was observed in a Tax$_1$-producing HeLa cell line (179) and after induction of Tax$_1$ expression in JPX-9 cells (82). All four of the elements that confer responsiveness of *c-fos* to divergent stimuli, including the SRE, CRE and the sis/PDGF elements, can be used by Tax$_1$ to up-regulate *c-fos* transcription (Fig. 7) (179). In view of the fact that deregulated c-Fos expression can induce cellular transformation, this cellular protein could play an important role in the initial stages of HTLV-I transformation (180). Interestingly, several HTLV-I-infected cell lines were found to contain high Ap-1 binding activities (181), suggesting an increased expression of c-Jun. In addition, higher levels of the mRNA of c-Jun and other members of the Jun-family (JunB and JunD) have been found in Tax$_1$-expressing JPX-9 cells (181) and in HTLV-I- and HTLV-II-infected cell lines (182). This may be an indirect effect of *c-fos* activation. In addition, Ap-1 might be important for the Tax$_1$-mediated *trans*-activation of the TGF-β1 gene (93).

There are other early-response genes which bear SRE elements in their upstream flanking regions that can be induced by Tax$_1$ (Fig. 7). These include *egr-1* (also known as *Krox-24*, zif268, NGFI-A, or TIS-8) and *egr-2* (*Krox-20*) which belong to the Egr-family of zinc-finger proteins (89, 181). The *egr* genes also contain CREs (89). Like c-Fos, Egr-1 and Egr-2 are short-lived proteins whose expression is inducible by serum, growth hormones, and phorbol esters through a SRE-like sequence (177). This sequence is also responsible for the repression of these genes by c-Fos (177). Tax$_1$ can activate *egr*-1 and *egr*-2 expression through the SRE and CRE elements (89).

A *trans*-represssive effect of Tax$_1$ has been reported for the β-polymerase gene (36). This enzyme is important in the repair of DNA and could be involved in a secondary event required for HTLV-I transformation. Interestingly, c-Fos, whose gene is *trans*-activated by Tax$_1$, induces the transcription of the β-polymerase gene (183).

Obviously, Tax$_1$ activation and repression of cellular genes is complex and mediated through a variety of regulatory sequences. Although Tax$_1$ does not *trans*-activate all promoters, the use of a wide variety of promoter elements suggests that the number of cellular genes that Tax$_1$ *trans*-activates may be much larger than is presently known. This also suggests that Tax$_1$ may have a variety of indirect effects on cellular gene regulation.

11. Extracellular cytokine effects of Tax

Although much of the understanding of the functions of Tax$_1$ has come from studies with HTLV-I-infected or transfected cells, Tax$_1$, like HIV-1 Tat, has the additional properties of an extracellular cytokine, regulating cellular growth and gene expression in uninfected cells (Fig. 8) (184–188).

Tax$_1$ protein has been detected in the extracellular growth medium of HTLV-I-infected and transformed cells such as C81, MT4, and PX-1 (185, 186). In addition, recombinant extracellular Tax$_1$ protein introduced into cell culture media was taken

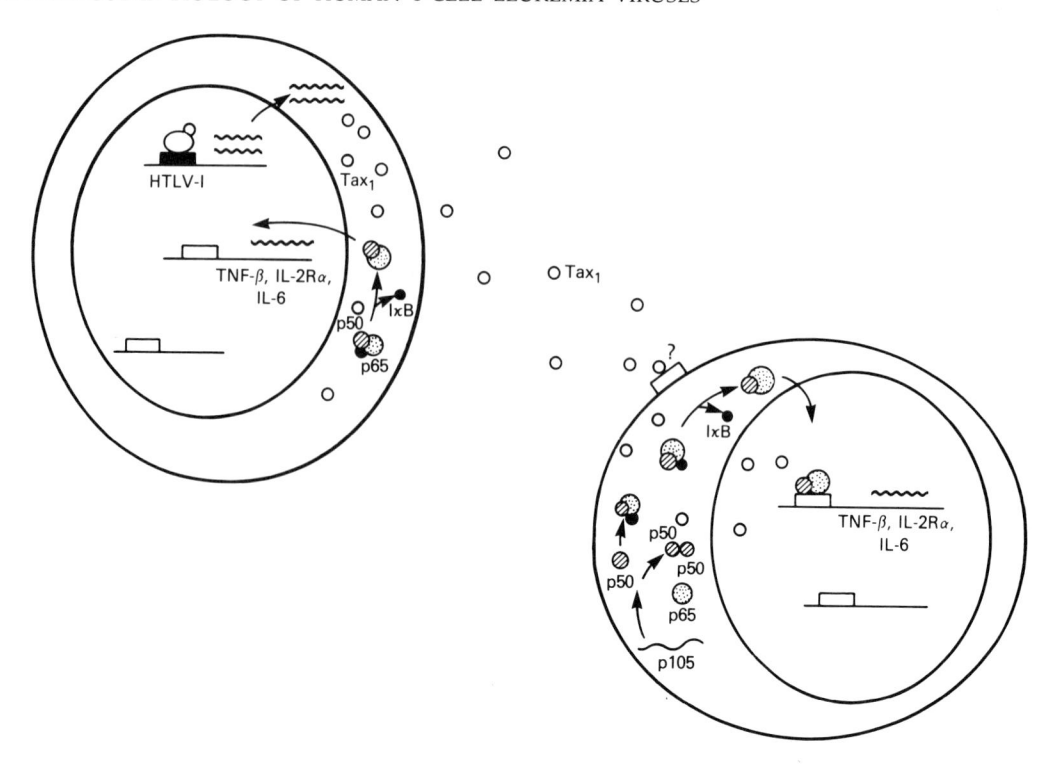

Fig. 8 Tax$_1$ regulation of cellular gene expression. In addition to *trans*-activation of the HTLV-I LTR in HTLV-I-infected cells (left), Tax$_1$ stimulates several cellular genes including IL-2Rα, TNF-β, and IL-6 via induction of NF-κB and other cellular transcription factors. Tax$_1$ may also gain access to uninfected cells (right), causing induction of NF-κB and *trans*-activation of cellular genes. The mechanism of Tax$_1$ induction of NF-κB is unknown, but may involve stimulation of NF-κB gene expression, stimulation of NF-κB–IκB dissociation by a membrane signalling pathway, processing or modification of NF-κB proteins, and/or indirect interactions with the NF-κB–IκB complex

up by cells and was found to induce the nuclear accumulation of NF-κB DNA-binding proteins in pre-B lymphocytes (185). The Tax$_1$ induction of NF-κB DNA-binding activity occurred in the presence of cycloheximide, indicating that new synthesis of NF-κB or related c-Rel proteins was not required (187). Tax$_1$ protein has not been shown to be able to directly dissociate NF-κB from the IκB inhibitor *in vitro*. It is, however, still possible that Tax$_1$ exerts its effects through post-translational modifications and/or through modulation of the composition of the NF-κB binding complex (105). The cellular uptake of Tax$_1$ and the induction of NF-κB was associated with the stimulation of cellular genes such as Igk light chain and TNF-β, which contain NF-κB binding sites in their promoters (Fig. 8) (187). More recently, Tax$_1$-expressing cell lines which have been cocultured with, but physically separated from, 70Z/3 pre-B cells has resulted in the induction of nuclear NF-κB DNA binding in the pre-B cells. This NF-κB induction was specifically blocked by anti-Tax$_1$ serum (D. Trinh and J. N. Brady, unpublished results).

Soluble Tax$_1$ protein was also able to stimulate cellular proliferation and expression of the IL-2Rα in PHA-primed competent human peripheral blood lymphocytes (PBLs) (186, 188). The stimulation of PBL proliferation was observed at extracellular Tax_1 concentrations as low as 25 pM, similar to the concentrations observed for growth factors such as nerve growth factor (189). This is in contrast to a concentration of 750 pM required for NF-κB induction in pre-B lymphocytes. The greater sensitivity of the PBL assay may be related to treatment of the PBLs with PHA to achieve a competent state. It is possible that Tax_1 may act synergistically with growth factors released from the PHA-stimulated or HTLV-I-infected lymphocytes. Potential candidates for this synergistic effect with Tax_1 include TNF, IL-6, and IL-2, which have been detected at increased levels in HTLV-I-infected lymphocytes (54, 165, 190–192) and granulocyte-macrophage colony stimulating factor (GM-CSF) which has been found to be released from Tax_1-expressing tumours of transgenic mice (76).

The mechanism by which intracellular Tax_1 gains access to the extracellular environment is unknown. Similar to several other cytokines, Tax_1 does not contain a classic signal peptide, although the amino-terminal region of Tax_1 is hydrophobic and contains limited homology to some cellular signal sequences (105). The mechanism by which Tax_1 is taken up by cells also remains to be established. The Tax_1 protein has been shown to associate with the cell membrane in a specific and saturable process, suggesting the existence of a membrane receptor (105).

HTLV-I-infected patients have circulating antibodies to the Tax_1 regulatory protein (193). Interestingly, the presence of Tax_1 serum antibodies has been associated with an increased risk of sexual transmission of HTLV-I infection (194). In addition, cytotoxic T-lymphocytes directed against Tax_1 have been demonstrated in TSP/HAM patients and have been suggested to correlate with disease progression (53). These findings suggest that the presence of Tax_1 may be related to productive viral infection or to viral pathogenesis and neurologic degeneration.

12. Conclusion

Much has been learned about the molecular biology of the HTLVs. Obviously, these retroviruses have a complex interaction with the host cell. Molecular studies of HTLV-I and HTLV-II have helped to clarify the epidemiology, prevalence of virus infection, and development of disease. In addition, we now have a much better understanding of the cellular and genetic effects which result from interaction of viral gene products with host genes and gene products. Understanding the virus and its relationship with the human host should provide important insight into disease prevention and treatment.

References

1. Poiesz, B. J., Ruscetti, F. W., Gazdar, A. F., Bunn, P. A., Minna, J. D., and Gallo, R. C. (1980) Detection and isolation of type C retrovirus particles from fresh and cultured lymphocytes of a patient with cutaneous T-cell lymphoma. *Proc. Natl Acad. Sci. USA*, **77**, 7415.
2. Yoshida, M., Miyoshi, I., and Hinuma, Y. (1982) Isolation and characterization of retrovirus from cell lines of human adult T-cell leukemia and its implication in the disease. *Proc. Natl Acad. Sci. USA*, **79**, 2031.
3. Kalyanaraman, V. S., Sarngadharan, M. G., Robert-Guroff, M., Miyoshi, I., Blayney, D., Golde, D., and Gallo, R. C. (1982) A new subtype of human T-cell leukemia virus (HTLV-II) associated with a T-cell variant of hairy cell leukemia. *Science*, **218**, 571.
4. Rosenblatt, J. D., Giorgi, J. V., Golde, D. W., Ezra, J. B., Wu, A., Winberg, C. D., Glaspy, J., Wachsman, W., and Chen, I. S. Y. (1988) Integrated human T-cell leukemia virus II genome in CD8$^+$ T cells from a patient with 'atypical' hairy cell leukemia: evidence for distinct T and B cell lymphoproliferative disorders. *Blood*, **71**, 363.
5. Lairmore, M. D., Jacobson, S., Gracia, F., De, B. K., Castillo, L., Larreategui, M., Roberts, B. D., Levine, P. H., Blattner, W. A., and Kaplan, J. E. (1990) Isolation of human T-cell lymphotropic virus type 2 from Guaymi Indians in Panama. *Proc. Natl Acad. Sci. USA*, **87**, 8840.
6. Manzari, V., Gismondi, A., Barillari, G., Morrone, S., Modesti, A., Albonici, L., De Marchis, L., Fazio, V., Gradilone, A., Zani, M., Frati, L., and Santoni, A. (1987) HTLV-V: a new human retrovirus isolated in a Tac-negative T cell lymphoma/leukemia. *Science*, **238**, 1581.
7. Wong-Staal, F. and Gallo, R. C. (1985) Human T-lymphotropic retroviruses. *Nature*, **317**, 395.
8. Tabor, E., Gerety, R. J., Cairns, J., and Bayley, A. C. (1990) Did HIV and HTLV originate in Africa? *J. Am. Med. Assoc.*, **264**, 691.
9. Sherman, M. P., Saksena, N. K., Dube, D. K., Yanagihara, R., and Poiesz, B. J. (1992) Evolutionary insights on the origin of human T-cell lymphoma/leukemia virus type I (HTLV-I) derived from sequence analysis of a new HTLV-I variant from Papua New Guinea. *J. Virol.*, **66**, 2556.
10. Larkin, J., Sinnott, J. T., Weiss, J., Holt, D. A. (1990) Human T-cell lymphotropic virus-type I. *Infect. Control Hosp. Epidemiol.*, **11**, 314.
11. Manns, A. and Blattner, W. A. (1991) The epidemiology of the human T-cell lymphotrophic virus type I and type II: etiologic role in human disease. *Transfusion*, **31**, 67.
12. Dixon, A. C., Dixon, P. S., and Nakamura, J. M. (1989) Infection with the human T-lymphotropic virus type I—a review for clinicians. *West. J. Med.*, **151**, 632.
13. Seiki, M., Hattori, S., Hirayama, Y., and Yoshida, M. (1983) Human adult T-cell leukemia virus: complete nucleotide sequence of the provirus genome integrated in leukemia cell DNA. *Proc. Natl Acad. Sci. USA*, **80**, 3618.
14. Josephs, S. F., Wong-Staal, F., Manzari, V., Gallo, R. C., Sodroski, J. G., Trus, M. D., Perkins, D., Patarca, R., and Haseltine, W. A. (1984). Long terminal repeat structure of an American isolate of type I human T-cell leukemia virus. *Virology*, **139**, 340.
15. Shimotohno, K., Takahashi, Y., Shimizu, N., Gojobori, T., Golde, D. W., Chen, I. S. Y., Miwa, M., and Sugimura, T. (1985) Complete nucleotide sequence of an infectious

clone of human T-cell leukemia virus type II: an open reading frame for the protease gene. *Proc. Natl Acad. Sci. USA,* **82,** 3101.
16. Ciminale, V., Pavlakis, G. N., Derse, D., Cunningham, C. P., and Felber, B. K. (1992) Complex splicing in the human T-cell leukemia virus (HTLV) family of retroviruses: novel mRNAs and proteins produced by HTLV type I. *J. Virol.,* **66,** 1737.
17. Berneman, Z. N., Gartenhaus, R. B., Reitz, Jr., M. S., Blattner, W. A., Manns, A., Hanchard, B., Ikehara, O., Gallo, R. C., and Klotman, M. E. (1992) Expression of alternatively spliced human T-lymphotropic virus type I pX mRNA in infected cell lines and in primary uncultured cells from patients with adult T-cell leukemia/lymphoma and healthy carriers. *Proc. Natl Acad. Sci. USA,* **89,** 3005.
18. Paine, E., Garcia, J., Philpott, T. C., Shaw, G., and Ratner, L. (1991) Limited sequence variation in human T-lymphotropic virus type 1 isolates from North American and African patients. *Virology,* **182,** 111.
19. Komurian, F., Pelloquin, F., and de The, G. (1991) In vivo genomic variability of human T-cell leukemia virus type I depends more upon geography than upon pathologies. *J. Virol.,* **65,** 3770.
20. Hjelle, B. and Chaney, R. (1992) Sequence variation of functional HTLV-II *Tax* alleles among isolates from an endemic population: lack of evidence for oncogenic determinant in *Tax. J. Med. Virol.,* **36,** 136.
21. Gessain, A., Saal, F., Morozov, V., Lasneret, J., Vilette, D., Gout, O., Emanoil-Ravier, R., Sigaux, F., de The, G., and Peries, J. (1989) Characterization of HTLV-I isolates and T lymphoid cell lines derived from French West Indian patients with tropical spastic paraparesis. *Int. J. Cancer,* **43,** 327.
22. Gazzolo, L. and Duc Dodon, M. (1987) Direct activation of resting T lymphocytes by human T-lymphotropic virus type I. *Nature,* **326,** 714.
23. Kimata, J. T. and Ratner, L. (1991) Temporal regulation of viral and cellular gene expression during human T-lymphotropic virus type I-mediated lymphocyte immortalization. *J. Virol.,* **65,** 4398.
24. Duc Dudon, M., Bernard, A., and Gazzolo, L. (1989) Peripheral T-lymphocyte activation by human T-cell leukemia virus type I interferes with the CD2 but not with the CD3/TCR pathway. *J. Virol.,* **63,** 5413.
25. Wucherpfennig, K. W., Höllsberg, P., Richardson, J. H., Benjamin, D., and Hafler, D. A. (1992) T-cell activation by autologous human T-cell leukemia virus type I-infected T-cell clones. *Proc. Natl Acad. Sci. USA,* **89,** 2110.
26. Gallo, R. C. (1991) Human retroviruses: a decade of discovery and link with human disease. *J. Infect. Dis.,* **164,** 235.
27. Maruyama, M., Shibuya, H., Harada, H., Hatakeyama, M., Seiki, M., Fujita, T., Inoue, J.-i., Yoshida, M., and Taniguchi, T. (1987) Evidence for aberrant activation of the interleukin-2 autocrine loop by HTLV-1-encoded p40^x and T3/Ti complex triggering. *Cell,* **48,** 343.
28. Grassmann, R., Dengler, C., Müller-Fleckenstein, I., Fleckenstein, B., McGuire, K., Dokhelar, M.-C., Sodroski, J. G., and Haseltine, W. A. (1989) Transformation to continuous growth of primary human T lymphocytes by human T-cell leukemia virus type I X-region genes transduced by Herpesvirus saimiri vector. *Proc. Natl Acad. Sci. USA,* **86,** 3351.
29. Pozzati, R., Vogel, J., and Jay, G. (1990) The human T-lymphotropic virus type I tax gene can cooperate with the ras oncogene to induce neoplastic transformation of cells. *Mol. Cell. Biol.,* **10,** 413.

30. Hinrichs, S. H., Nerenberg, M., Reynolds, R. K., Khoury, G., and Jay, G. (1987) A transgenic mouse model for human neurofibromatosis. *Science*, **237**, 1340.
31. Nerenberg, M., Hinrichs, S. H., Reynolds, R. K., Khoury, G., and Jay, G. (1987) The tat gene of human T-lymphotropic virus type 1 induces mesenchymal tumors in transgenic mice. *Science*, **237**, 1324.
32. Nerenberg, M. I., Minor, T., Price, J., Ernst, D. N., Shinohara, T., and Schwarz, H. (1991) Transgenic thymocytes are refractory to transformation by the human T-cell leukemia virus type I tax gene. *J. Virol.*, **65**, 3349.
33. Fujita, K., Yamasaki, Y., Sawada, H., Izumi, Y., Fukuhara, S., and Uchino, H. (1989) Cytogenetic studies on the adult T-cell leukemia in Japan. *Leuk. Res.*, **13**, 535.
34. Sadamori, N. (1991) Cytogenetic implication in adult T-cell leukemia. *Cancer Genet. Cytogenet.*, **51**, 131.
35. Maruyama, K., Fukushima, T., Kawamura, K., and Mochizuki, S. (1990) Chromosome and gene rearrangements in immortalized human lymphocytes infected with human T-lymphotropic virus type I. *Cancer Res.*, **50** (supplement), 5697s.
36. Jeang, K.-T., Widen, S. G., Seemes, O. J., and Wilson, S. H. (1990) HTLV-I trans-activator protein, Tax, is a transrepressor of the human β-polymerase gene. *Science*, **247**, 1082.
37. Kline, R. L., Brothers, T., Halsey, N., Boulos, R., Lairmore, M. D., and Quinn, T. C. (1991) Evaluation of enzyme immunoassays for antibody to human T-lymphotropic viruses type I/II. *Lancet*, **337**, 30.
38. Lipka, J. J., Bui, K., Reyes, G. R., Moeckli, R., Wiktor, S. Z., Blattner, W. A., Murphy, E. L., Shaw, G. M., Hanson, C. V., Sninsky, J. J., and Foung, S. K. H. (1990) Determination of a unique and immunodominant epitope of human T cell lymphotropic virus type I. *J. Infect. Dis.*, **162**, 353.
39. Khabbaz, R. F., Onorato, I. M., Cannon, R. O., Hartley, T. M., Roberts, B., Hosein, B., and Kaplan, J. E. (1992) Seroprevalence of HTLV-I and HTLV-II among intravenous drug users and persons in clinics for sexually transmitted diseases. *N. Engl. J. Med.*, **326**, 375.
40. Hjelle, B., Scalf, R., and Swenson, S. (1990) High frequency of human T-cell leukemia-lymphoma virus type II infection in New Mexico blood donors: determination by sequence-specific oligonucleotide hybridization. *Blood*, **76**, 450.
41. Hjelle, B., Cyrus, S., Swenson, S., and Mills, R. (1991) Serologic distinction between human T-lymphotropic virus (HTLV) type I and HTLV type II. *Transfusion*, **31**, 731.
42. Murphy, E. L., Figueroa, J. P., Gibbs, W. N., Brathwaite, A., Holding-Cobham, M., Waters, D., Cranston, B., Hanchard, B., and Blattner, W. A. (1989) Sexual transmission of human T-lymphotropic virus type I (HTLV-I). *Ann. Intern. Med.*, **111**, 555.
43. Chen, Y.-M. A., Okayama, A., Lee, T.-H., Tachibana, N., Mueller, N., and Essex, M. (1991) Sexual transmission of human T-cell leukemia virus type I associated with the presence of anti-Tax antibody. *Proc. Natl Acad. Sci. USA*, **88**, 1182.
44. Hino, S. (1989) Milk-borne transmission of HTLV-I as a major route in the endemic cycle. *Acta Paediatr. Jpn*, **31**, 428.
45. Ichimaru, M., Ikeda, S., Kinoshita, K., Hino, S., and Tsuji, Y. (1991) Mother-to-child transmission of HTLV-1. *Cancer Detect. Prev.*, **15**, 177.
46. Tsuji, Y., Doi, H., Yamabe, T., Ishimaru, T., Miyamoto, T., and Hino, S. (1990) Prevention of mother-to-child transmission of human T-lymphotropic virus type-I. *Pediatrics*, **86**, 11.
47. Matsumoto, R., Inaba, N., Shimizu, H., Yamaguchi, H., Cho, S., and Takamizawa, H.

(1991) A study of adult T-cell leukemia virus (ATLV) infection in the field of obstetrics: its epidemiology, vertical transmission and familial clustering. *Asia Oceania J. Obstet. Gynaecol.*, **17**, 57.

48. Murphy, E. L. and Blattner, W. A. (1988) HTLV-I-associated leukemia: a model for chronic retroviral diseases. *Ann. Neurol.*, **23** (suppl.), S174.

49. Waldmann, T. A., Pastan, I. H., Gansow, O. A., and Junghans, R. P. (1992) The multichain interleukin-2 receptor: a target for immunotherapy. *Ann. Intern. Med.*, **116**, 148.

50. Molgaard, C. A., Eisenman, P. A., Ryden, L. A., and Golbeck, A. L. (1989) Neuro-epidemiology of human T-lymphotropic virus type-I-associated tropical spastic paraparesis. *Neuroepidemiology*, **8**, 109.

51. Osame, M., Matsumoto, M., Usuku, K., Izumo, S., Ijichi, N., Amitani, H., Tara, M., and Igata, A. (1987) Chronic progressive myelopathy associated with elevated antibodies to human T-lymphotropic virus type I and adult T-cell leukemia-like cells. *Ann. Neurol.*, **21**, 117.

52. Janssen, R. S., Kaplan, J. E., Khabbaz, R. F., Hammond, R., Lechtenberg, R., Lairmore, M., Chiasson, M. A., Punsalang, A., Roberts, B., McKendall, R. R., Rosenblum, M., Brew, B., Farraye, J., Howley, D. J., Feraru, E., Sparr, S., Vecchio, J., Silverman, M., McHarg, M., Gorin, B., Rugg, D. R., Grenell, S., Trimble, B., Bruining, K., Guha, S., Amaraneni, P., and Price, R. W. (1991) HTLV-I-associated myelopathy/tropical spastic paraparesis in the United States. *Neurology*, **41**, 1355.

53. Jacobson, S., Shida, H., McFarlin, D. E., Fauci, A. S., and Koenig, S. (1990) Circulating CD8[+] cytotoxic T lymphocytes specific for HTLV-I pX in patients with HTLV-I associated neurological disease. *Nature*, **348**, 245.

54. Tendler, C. L., Greenberg, S. J., Blattner, W. A., Manns, A., Murphy, E., Fleisher, T., Hanchard, B., Morgan, O., Burton, J. D., Nelson, D. L., and Waldmann, T. A. (1990) Transactivation of interleukin 2 and its receptor induces immune activation in human T-cell lymphotropic virus type I-associated myelopathy: pathogenic implications and a rationale for immunotherapy. *Proc. Natl Acad. Sci. USA*, **87**, 5218.

55. Hjelle, B., Mills, R., Swenson, S., Mertz, G., Key, C., and Allen, S. (1991) Incidence of hairy cell leukemia, mycosis fungoides, and chronic lymphocytic leukemia in first known HTLV-II-endemic population. *J. Infect. Dis.*, **163**, 435.

56. Ogawa, K., Kawanishi, M., Matsuda, S., Isono, T., and Seto, A. (1989) Polyclonal increase of HTLV-I provirus-carrying lymphocytes in HTLV-I-carrier rabbits transplanted with Shope carcinoma cells. *Leuk. Res.*, **13**, 1009.

57. Seto, A., Isono, T., and Ogawa, K. (1991) Infection of inbred rabbits with cell-free HTLV-I. *Leuk. Res.*, **15**, 105.

58. Iwahara, Y., Takehara, N., Kataoka, R., Sawada, T., Ohtsuki, Y., Nakachi, H., Maehama, T., Okayama, T., and Miyoshi, I. (1990) Transmission of HTLV-I to rabbits via semen and breast milk from seropositive healthy persons. *Int. J. Cancer*, **45**, 980.

59. Kataoka, R., Takehara, N., Iwahara, Y., Sawada, T., Ohtsuki, Y., Dawei, Y., Hoshino, H., and Miyoshi, I. (1990) Transmission of HTLV-I by blood transfusion and its prevention by passive immunization in rabbits. *Blood*, **76**, 1657.

60. Yamade, I., Ishiguro, T., and Seto, A. (1991) Infection without antibody response in mother-to-child transmission of HTLV-I in rabbits. *J. Med. Virol.*, **33**, 268.

61. Cockerell, G. L., Lairmore, M., De, B., Rovnak, J., Hartley, T. M., and Miyoshi, I. (1990) Persistent infection of rabbits with HTLV-I: patterns of anti-viral antibody reactivity and detection of virus by gene amplification. *Int. J. Cancer*, **45**, 127.

62. Tseng, C.-T. K. and Sell, S. (1991) Protracted *Treponema pallidum*-induced cutaneous chancres in rabbits infected with human T-cell leukemia virus type I. *AIDS Res. Hum. Retroviruses*, **7**, 323.
63. Ogawa, K., Matsuda, S., and Seto, A. (1989) Induction of leukemic infiltration by allogeneic transfer of HTLV-I-transformed T cells in rabbits. *Leuk. Res.*, **13**, 399.
64. Lairmore, M. D., Roberts, B., Frank, D., Rovnak, J., Weiser, M. G., and Cockerell, G. L. (1992) Comparative biological responses of rabbits infected with human T-lymphotropic virus type I isolates from patients with lymphoproliferative and neurodegenerative disease. *Int. J. Cancer*, **50**, 124.
65. Truckenmiller, M. E., Kulaga, H., Gugel, E., Dickerson, D., and Kindt, T. J. (1989) Evidence for dual infection of rabbits with the human retroviruses HTLV-I and HIV-1. *Res. Immunol.*, **140**, 527.
66. Takehara, N., Iwahara, Y., Uemura, Y., Sawada, T., Ohtsuki, Y., Iwai, H., Hoshino, H., and Miyoshi, I. (1989) Effect of immunization on HTLV-I infection in rabbits. *Int. J. Cancer*, **44**, 332.
67. Isono, T., Ogawa, K., and Seto, A. (1990) Antiviral effect of zidovudine in the experimental model of adult T cell leukemia in rabbits. *Leuk. Res.*, **14**, 841.
68. Sawada, T., Iwahara, Y., Ishii, K., Taguchi, H., Hoshino, H., and Miyoshi, I. (1991) Immunoglobulin prophylaxis against milk-borne transmission of human T cell leukemia virus type I in rabbits. *J. Infect. Dis.*, **164**, 1193.
69. Shingu, M., Chinami, M., Taguchi, T., and Shingu, Jr., M. Therapeutic effects of bovine enterovirus infection on rabbits with experimentally induced adult T cell leukaemia. *J. Gen. Virol.*, **72**, 2031.
70. Iwakura, Y., Tosu, M., Yoshida, E., Takiguchi, M., Sato, K., Kitajima, I., Nishioka, K., Yamamoto, K., Takeda, T., Hatanaka, M., Yamamoto, H., and Sekiguchi, T. (1991) Induction of inflammatory arthropathy resembling rheumatoid arthritis in mice transgenic for HTLV-I. *Science*, **253**, 1026.
71. Green, J. E., Hinrichs, S. H., Vogel, J., and Jay, G. (1989) Exocrinopathy resembling Sjögren's syndrome in HTLV-1 *tax* transgenic mice. *Nature*, **341**, 72.
72. Cann, A. J., Rosenblatt, J. D., Wachsman, W., Shah, N. P., and Chen, I. S. Y. (1985) Identification of the gene responsible for human T-cell leukaemia virus transcriptional regulation. *Nature*, **318**, 571.
73. Felber, B. K., Paskalis, H., Kleinman-Ewing, C., Wong-Staal, F., and Pavlakis, G. N. (1985) The pX protein of HTLV-I is a transcriptional activator of its long terminal repeats. *Science*, **229**, 675.
74. Sodroski, J. G., Rosen, C. A., and Haseltine, W. A. (1985) *Trans*-acting transcriptional activation of the long terminal repeat of human T lymphotropic viruses in infected cells. *Science*, **225**, 381.
75. Leung, K. and Nabel, G. J. (1988) HTLV-1 transactivator induces interleukin-2 receptor expression through an NF-κB-like factor. *Nature*, **333**, 776.
76. Green, J. E., Begley, C. G., Wagner, D. K., Waldmann, T. A., and Jay, G. (1989) *trans*-activation of granulocyte-macrophage colony-stimulating factor and the interleukin-2 receptor in transgenic mice carrying the human T-lymphotropic virus type 1 *tax* gene. *Mol. Cell. Biol.*, **9**, 4731.
77. Wano, Y., Feinberg, M., Hosking, J. B., Bogerd, H., and Greene, W. C. (1988) Stable expression of the *tax* gene of type I human T-cell leukemia virus in human T cells activates specific cellular genes involved in growth. *Proc. Natl Acad. Sci. USA*, **85**, 9733.
78. Miyatake, S., Seiki, M., Yoshida, M., and Arai, K.-I. (1988) T-cell activation signals

and human T-cell leukemia virus type I-encoded p40^x protein activate the mouse granulocyte-macrophage colony-stimulating factor gene through a common DNA element. *Mol. Cell. Biol.*, **8**, 5581.

79. Nimer, S. D., Gasson, J. C., Hu, K., Smalberg, I., Williams, J. L., Chen, I. S. Y., and Rosenblatt, J. D. (1989) Activation of the GM-CSF promoter by HTLV-I and -II *tax* proteins. *Oncogene*, **4**, 671.
80. Nishida, J., Yoshida, M., Arai, K.-i., and Yokota, T. (1991) Definition of a GC-rich motif as regulatory sequence of the human IL-3 gene: coordinate regulation of the IL-3 gene by CLE2/GC box of the GM-CSF gene in T cell activation. *Int. Immunol.*, **3**, 245.
81. Lilienbaum, A., Duc Dodon, M., Alexandre, C., Gazzolo, L., and Paulin, D. (1990) Effect of human T-cell leukemia virus type I Tax protein on activation of the human vimentin gene. *J. Virol.*, **64**, 256.
82. Nagata, K., Ohtani, K., Nakamura, M., and Sugamura, K. (1989) Activation of endogenous c-*fos* proto-oncogene expression by human T-cell leukemia virus type I-encoded p40^{tax} protein in the human T-cell line, Jurkat. *J. Virol.*, **63**, 3220.
83. Ratner, L. (1989) Regulation of expression of the c-sis proto-oncogene. *Nucleic Acids Res.*, **17**, 4101.
84. Watanabe, T., Yamaguchi, K., Takatsuki, K., Osami, M. and Yoshida, M. (1990) Constitutive expression of parathyroid hormone-related protein gene in human T cell leukemia virus type 1 (HTLV-1) carriers and adult T cell leukemia patients that can be *trans*-activated by HTLV-1 tax gene. *J. Exp. Med.*, **172**, 759.
85. Dittmer, J., Gitlin, S. D., and Brady, J. N. Manuscript in preparation.
86. Joshi, J. B. and Dave, H. P. G. (1992) Transactivation of the proenkephalin gene promoter by the Tax$_1$ protein of human T-cell lymphotropic virus type I. *Proc. Natl Acad. Sci. USA*, **89**, 1006.
87. Fox, H. B., Gutman, P. D., Dave, H. P. G., Cao, S. X., Mittelman, M., Berg, P. E., and Schechter, A. N. (1989) *Trans*-activation of human globin genes by HTLV-I Tax$_1$. *Blood*, **74**, 2749.
88. Napolitano, M., Modi, W. S., Cevario, S. J., Gnarra, J. R., Seuanez, H. N., and Leonard, W. J. (1991) The gene encoding the Act-2 cytokine. *J. Biol. Chem.*, **266**, 17531.
89. Alexandre, C., Charnay, P., and Verrier, B. (1991) Transactivation of *Krox-20* and *Krox-24* promoters by the HTLV-1 Tax protein through common regulatory elements. *Oncogene*, **6**, 1851.
90. Sawada, M., Suzumura, A., Yoshida, M., and Marunouchi, T. (1990) Human T-cell leukemia virus type I *trans* activator induces class I major histocompatibility complex antigen expression in glial cells. *J. Virol.*, **64**, 4002.
91. Miura, S., Ohtani, K., Numata, N., Niki, M., Ohbo, K., Ina, Y., Gojobori, T., Tanaka, Y., Tozawa, H., Nakamura, M., and Sugamura, K. (1991) Molecular cloning and characterization of a novel glycoprotein, gp34, that is specifically induced by the human T-cell leukemia virus type I transactivator p40^{tax}. *Mol. Cell. Biol.*, **11**, 1313.
92. Green, J. E. (1991) *Trans* activation of nerve growth factor in transgenic mice containing the human T-cell lymphotropic virus type I *tax* gene. *Mol. Cell. Biol.*, **11**, 4635.
93. Kim, S.-J., Kehrl, J. H., Burton, J., Tendler, C. L., Jeang, K.-T., Danielpour, D., Thevenin, C., Kim, K. Y., Sporn, M. B., and Roberts, A. B. (1990) Transactivation of the transforming growth factor β1 (TGF-β1) gene by human T lymphotropic virus type 1 Tax: a potential mechanism for the increased production of TGF-β1 in adult T cell leukemia. *J. Exp. Med.*, **172**, 121.
94. Goh, W. C., Sodroski, J., Rosen, C., Essex, M., and Haseltine, W. A. (1985) Subcellular

localization of the product of the long open reading frame of human T-cell leukemia virus type I. *Science,* **227,** 1227.
95. Jeang, K.-T., Brady, J., Radonovich, M., Duvall, J., and Khoury, G. (1988) p40^x *trans*-activation of the HTLV-I LTR promoter. In *Mechanisms of control of gene expression.* B. Cullen, L. P. Gage, M. A. Q. Siddiqui, A. M. Skalka, and H. Weissbach (eds). Alan R. Liss, Inc., New York, p. 181.
96. Slamon, D. J., Boyle, W. J., Keith, D. E., Press, M. F., Golde, D. W., and Souza, L. M. (1988) Subnuclear localization of the *trans*-activating protein of human T-cell leukemia virus type I. *J. Virol.,* **62,** 680.
97. Smith, M. R. and Greene, W. C. (1990) Identification of HTLV-I *tax trans*-activator mutants exhibiting novel transcriptional phenotypes. *Genes Dev.,* **4,** 1875.
98. Gitlin, S. D., Lindholm, P. F., Marriott, S. J., and Brady, J. N. (1991) Transdominant human T-cell lymphotropic virus type I Tax_1 mutant that fails to localize to the nucleus. *J. Virol.,* **65,** 2612.
99. Haseltine, W. A., Sodroski, J., Patarca, R., Briggs, D., and Perkins, D. (1984) Structure of 3' terminal region of type II human T lymphotropic virus: evidence for new coding region. *Science,* **225,** 419.
100. Cann, A., Rosenblatt, J. D., Wachsman, W., and Chen, I. S. Y. (1989) In vitro mutagenesis of the human T-cell leukemia virus types I and II *tax* genes. *J. Virol.,* **63,** 1474.
101. Smith, M. R. and Greene, W. C. (1992) Characterization of a novel nuclear localization signal in the HTLV-I tax transactivator protein. *Virology,* **187,** 316.
102. Heder, A., Paca-Uccaralertkun, S., and Boros, I. (1991) Mutational analysis of the HTLV-I *trans*-activator, Tax. *FEBS Lett.,* **292,** 210.
103. Wachsman, W., Cann, A. J., Williams, J. L., Slamon, D. J., Souza, L., Shah, N. P., and Chen, I. S. Y. (1987) HTLV *x* gene mutants exhibit novel transcriptional regulatory phenotypes. *Science,* **235,** 674.
104. Lindholm, P. F., Marriott, S. J., Gitlin, S. D., and Brady, J. N. (1991) Differential precipitation and zinc chelate chromatography purification of biologically active HTLV-I Tax_1 expressed in *E. coli. J. Biochem. Biophys. Methods,* **22,** 233.
105. Lindholm, P. F., Kashanchi, F., and Brady, J. N. (1992) Transcriptional regulation in the human retrovirus HTLV-I. *Semin. Virol.,* in press.
106. Landschulz, W. H., Johnson, P. F., and McKnight, S. L. (1988) The leucine zipper: a hypothetical structure common to a new class of DNA binding proteins. *Science,* **240,** 1759.
107. Dyson, N., Howley, P. M., Münger, K., and Harlow, E. (1989) The human papilloma virus-16 E7 oncoprotein is able to bind to the retinoblastoma gene product. *Science,* **243,** 934.
108. Ptashne, M. (1988) How eukaryotic transcriptional activators work. *Nature,* **335,** 683.
109. Hope, I. A. and Struhl, K. (1986) Functional dissection of a eukaryotic transcriptional activator protein, GCN4 of yeast. *Cell,* **46,** 885.
110. Schmitz, M. L. and Baeuerle, P. A. (1991) The p65 subunit is responsible for the strong transcription activating potential of NF-κB. *EMBO J.,* **10,** 3805.
111. Fujii, M., Tsuchiya, H., and Seiki, M. (1991) HTLV-1 Tax has distinct but overlapping domains for transcriptional activation and for enhancer specificity. *Oncogene,* **6,** 2349.
112. Kiyokawa, T., Seiki, M., Iwashita, S., Imagawa, K., Shimizu, F., and Yoshida, M. (1985) p27^{x-III} and p21^{x-III}, proteins encoded by the *pX* sequence of human T-cell leukemia virus type I. *Proc. Natl Acad. Sci. USA,* **82,** 8359.

113. Nagashima, K., Yoshida, M., and Seiki, M. (1986) A single species of *pX* mRNA of human T-cell leukemia virus type I encodes *trans*-activator p40^x and two other phosphoproteins. *J. Virol.*, **60**, 394.
114. Green, P. L., Xie, Y., and Chen, I. S. Y. (1991) The rex proteins of human T-cell leukemia virus type II differ by serine phosphorylation. *J. Virol.*, **65**, 546.
115. Hidaka, M., Inoue, J., Yoshida, M., and Seiki, M. (1988) Post-transcriptional regulator (*rex*) of HTLV-1 initiates expression of viral structural proteins but suppresses expression of regulatory proteins. *EMBO J.*, **7**, 519.
116. Ohta, M., Nyunoya, H., Tanaka, H., Okamoto, T., Akagi, T., and Shimotohno, K. (1988) Identification of a cis-regulatory element involved in accumulation of human T-cell leukemia virus type II genomic mRNA. *J. Virol.*, **62**, 4445.
117. Ahmed, Y. F., Gilmartin, G. M., Hanly, S. M., Nevins, J. R., and Greene, W. C. (1991) The HTLV-I Rex response element mediates a novel form of mRNA polyadenylation. *Cell*, **64**, 727.
118. Inoue, J.-I., Itoh, M., Akizawa, T., Toyoshima, H., and Yoshida, M. (1991) HTLV-1 Rex protein accumulates unspliced RNA in the nucleus as well as in cytoplasm. *Oncogene*, **6**, 1753.
119. Nosaka, T., Siomi, H., Adachi, Y., Ishibashi, M., Kubota, S., Maki, M., and Hatanaka, M. (1989) Nucleolar targeting signal of human T-cell leukemia virus type I rex-encoded protein is essential for cytoplasmic accumulation of unspliced viral mRNA. *Proc. Natl Acad. Sci. USA*, **86**, 9798.
120. Grassmann, R., Berchtold, S., Aepinus, C., Ballaun, C., Boehnlein, E., and Fleckenstein, B. (1991) In vitro binding of human T-cell leukemia virus rex proteins to the rex-response element of viral transcripts. *J. Virol.*, **65**, 3721.
121. Weichselbraun, I., Farrington, G. K., Rusche, J. R., Böhnlein, E., and Hauber, J. (1992) Definition of the human immunodeficiency virus type 1 Rev and human T-cell leukemia virus type I Rex protein activation domain by functional exchange. *J. Virol.*, **66**, 2583.
122. Hanly, S. M., Rimsky, L. T., Malim, M. H., Kim, J. H., Hauber, J., Duc Dodon, M., Le, S.-Y., Maizel, J. V., Cullen, B. R., and Greene, W. C. (1989) Comparative analysis of the HTLV-I rex and HIV-1 rev trans-regulatory proteins and their RNA response elements. *Genes Dev.*, **3**, 1534.
123. Bogerd, H. P., Huckaby, G. L., Ahmed, Y. F., Hanly, S. M., and Greene, W. C. (1991) The type I human T-cell leukemia virus (HTLV-I) rex trans-activator binds directly to the HTLV-I rex and the type 1 human immunodeficiency virus rev RNA response elements. *Proc. Natl Acad. Sci. USA*, **88**, 5704.
124. Bar-Shira, A., Panet, A., and Honigman, A. (1991) An RNA secondary structure juxtaposes two remote genetic signals for human T-cell leukemia virus type I RNA 3'-end processing. *J. Virol.*, **65**, 5165.
125. Kim, J. H., Kaufman, P. A., Hanly, S. M., Rimsky, L. T., and Greene, W. C. (1991) Rex transregulation of human T-cell leukemia virus type II gene expression. *J. Virol.*, **65**, 405.
126. Black, A. C., Chen, I. S. Y., Arrigo, S., Ruland, C. T., Allogiamento, T., Chin, E., and Rosenblatt, J. D. (1991) Regulation of HTLV-II gene expression by rex involves positive and negative cis-acting elements in the 5' long terminal repeat. *Virology*, **181**, 433.
127. Black, A. C., Ruland, C. T., Yip, M. T., Luo, J., Tran, B., Kalsi, A., Quan, E., Aboud, M., Chen, I. S. Y., and Rosenblatt, J. D. (1991) Human T-cell leukemia virus II rex binding and activity require an intact splice donor site and a specific RNA secondary structure. *J. Virol.*, **65**, 6645.

128. Yip, M. T., Dynan, W. S., Green, P. L., Black, A. C., Arrigo, S. J., Torbati, A., Heaphy, S., Ruland, C., Rosenblatt, J. D., and Chen, I. S. Y. (1991) Human T-cell leukemia virus (HTLV) type II rex protein binds specifically to RNA sequences of the HTLV long terminal repeat but poorly to the human immunodeficiency virus type 1 rev-responsive element. *J. Virol.*, **65**, 2261.

129. Furukawa, K., Furukawa, K., and Shiku, H. (1991) Alternatively spliced mRNA of the *pX* region of human T lymphotropic virus type I proviral genome. *FEBS Lett.*, **295**, 141.

130. Giam, C.-Z. and Xu, Y.-L. (1989) HTLV-I *tax* gene product activates transcription via pre-existing cellular factors and cAMP responsive element. *J. Biol. Chem.*, **264**, 15236.

131. Marriott, S. J., Lindholm, P. F., Brown, K. M., Gitlin, S. D., Duvall, J. F., Radonovich, M. F., and Brady, J. N. (1990) A 36-kilodalton cellular transcription factor mediates an indirect interaction of human T-cell leukemia/lymphoma virus type I TAX$_1$ with a responsive element in the viral long terminal repeat. *Mol. Cell. Biol.*, **10**, 4192.

132. Shimotohno, K., Golde, D. W., Miwa, M., Sugimura, T., and Chen, I. S. Y. (1984) Nucleotide sequence analysis of the long terminal repeat of human T-cell leukemia virus type II. *Proc. Natl Acad. Sci. USA*, **81**, 1079.

133. Sodroski, J., Trus, M., Perkins, D., Patarca, R., Wong-Staal, F., Gelmann, E., Gallo, R., and Haseltine, W. A. (1984) Repetitive structure in the long-terminal-repeat element of a type II human T-cell leukemia virus. *Proc. Natl Acad. Sci. USA*, **81**, 4617.

134. Paskalis, H., Felber, B. K., and Pavlakis, G. N. (1986) Cis-acting sequences responsible for the transcriptional activation of human T-cell leukemia virus type I constitute a conditional enhancer. *Proc. Natl Acad. Sci. USA*, **83**, 6558.

135. Shimotohno, K., Takano, M., Teruuchi, T., and Miwa, M. (1986) Requirement of multiple copies of a 21-nucleotide sequence in the *U3* regions of human T-cell leukemia virus type I and type II long terminal repeats for trans-acting activation of transcription. *Proc. Natl Acad. Sci. USA*, **83**, 8112.

136. Brady, J., Jeang, K.-T., Duvall, J., and Khoury, G. (1987) Identification of p40x-responsive regulatory sequences within the human T-cell leukemia virus type I long terminal repeat. *J. Virol.*, **61**, 2175.

137. Jeang, K.-T., Boros, I., Brady, J., Radonovich, M., and Khoury, G. (1988) Characterization of cellular factors that interact with the human T-cell leukemia virus type I p40[x]-responsive 21-base-pair sequence. *J. Virol.*, **62**, 4499.

138. Seiki, M., Inoue, J.-i., Hidaka, M., and Yoshida, M. (1988) Two cis-acting elements responsible for posttranscriptional trans-regulation of gene expression of human T-cell leukemia virus type I. *Proc. Natl Acad. Sci. USA*, **85**, 7124.

139. Tan, T.-H., Horikoshi, M., and Roeder, R. G. (1989) Purification and characterization of multiple nuclear factors that bind to the TAX-inducible enhancer within the human T-cell leukemia virus type 1 long terminal repeat. *Mol. Cell. Biol.*, **9**, 1733.

140. Montagne, J., Béraud, C., Crenon, I., Lombard-Platet, G., Gazzolo, L., Sergeant, A., and Jalinot, P. (1990) Tax1 induction of the HTLV-I 21 bp enhancer requires cooperation between two cellular DNA-binding proteins. *EMBO J.*, **9**, 957.

141. Nyborg, J. K., Matthews, M.-A. H., Yucel, J., Walls, L., Golde, W. T., Dynan, W. S., and Wachsman, W. (1990) Interaction of host cell proteins with the human T-cell leukemia virus type I transcriptional control region. *J. Biol. Chem.*, **265**, 8237.

142. Yoshimura, T., Fujisawa, J.-i., and Yoshida, M. (1990) Multiple cDNA clones encoding nuclear proteins that bind to the *tax*-dependent enhancer of HTLV-1: all contain a leucine zipper structure and basic amino acid domain. *EMBO J.*, **9**, 2537.

143. Béraud, C., Lombard-Platet, G., Michal, Y., and Jalinot, P. (1991) Binding of the

HTLV-I Tax1 transactivator to the inducible 21 bp enhancer is mediated by the cellular factor HEB1. *EMBO J.*, **10**, 3795.

144. Tsujimoto, A., Nyunoya, H., Morita, T., Sato, T., and Shimotohno, K. (1991) Isolation of cDNAs for DNA-binding proteins which specifically bind to a *tax*-responsive enhancer element in the long terminal repeat of human T-cell leukemia virus type I. *J. Virol.*, **65**, 1420.

145. Hai, T., Liu, F., Coukos, W., and Green, M. (1989) Transcription factor ATF cDNA clones: an extensive family of leucine zipper proteins able to selectively form DNA-binding heterodimers. *Genes Dev.*, **2**, 1216.

146. Zhao, L.-J. and Giam, C.-Z. (1991) Interaction of the human T-cell lymphotrophic virus type I (HTLV-I) transcriptional activator Tax with cellular factors that bind specifically to the 21-base-pair repeats in the HTLV-I enhancer. *Proc. Natl Acad. Sci. USA*, **88**, 11445.

147. Radonovich, M. and Jeang, K.-T. (1989) Activation of the human T-cell leukemia virus type I long terminal repeat by 12-O-tetradecanoylphorbol-13-acetate and by Tax (p40x) occurs through similar but functionally distinct target sequences. *J. Virol.*, **63**, 2987.

148. Jeang, K.-T., Chiu, R., Santos, E., and Kim, S.-J. (1991) Induction of the HTLV-I LTR by Jun occurs through the Tax-responsive 21-bp elements. *Virology*, **181**, 218.

149. Nagata, K., Ide, Y., Takagi, T., Ohtani, K., Aoshima, M., Tozawa, H., Nakamura, M., and Sugamura, K. (1992) Complex formation of human T-cell leukemia virus type I p40^{tax} transactivator with cellular polypeptides. *J. Virol.*, **66**, 1040.

150. Marriott, S. J., Boros, I., Duvall, J. F., and Brady, J. N. (1989) Indirect binding of human T-cell leukemia virus type I tax$_1$ to a responsive element in the viral long terminal repeat. *Mol. Cell. Biol.*, **9**, 4152.

151. Bosselut, R., Duvall, J. F., Gégonne, A., Bailly, M., Hémar, A., Brady, J., and Ghysdael, J. (1990) The product of the c-*ets*-1 proto-oncogene and the related Ets2 protein act as transcriptional activators of the long terminal repeat of human T cell leukemia virus HTLV-1. *EMBO J.*, **9**, 3137.

152. Bosselut, R., Lim, F., Romond, P.-C., Frampton, J., Brady, J., and Ghysdael, J. (1992) Myb protein binds to multiple sites in the human T cell lymphotropic virus type 1 long terminal repeat and transactivates LTR-mediated expression. *Virology*, **186**, 764.

153. Gitlin, S. D., Bosselut, R., Gégonne, A., Ghysdael, J., and Brady, J. N. (1991) Sequence-specific interaction of the Ets1 protein with the long terminal repeat of the human T-lymphotropic virus type I. *J. Virol.*, **65**, 5513.

154. Bhat, N. K., Thompson, C. B., Lindsten, T., June, C. H., Fujiwara, S., Koizumi, S., Fisher, R. J., and Papas, T. S. (1990) Reciprocal expression of human *ETS1* and *ETS2* genes during T-cell activation: regulatory role for the protooncogene *ETS1*. *Proc. Natl Acad. Sci. USA*, **87**, 3723.

155. Gitlin, S. D., Dittmer, J., and Brady, J. N. Manuscript submitted.

156. Ohtani, K., Nakamura, M., Saito, S., Noda, T., Ito, Y., Sugamura, K., and Hinuma, Y. (1987) Identification of two distinct elements in the long terminal repeat of HTLV-I responsible for maximum gene expression. *EMBO J.*, **6**, 389.

157. Nakamura, M., Ohtani, K., Hinuma, Y., and Sugamura, K. (1988) Functional mapping of the activity of the R region in the human T-cell leukemia virus type I long terminal repeat to increase gene expression. *Virus Genes 2*, **2**, 147.

158. Greene, W. C., Leonard, W. J., Wano, Y., Svetlik, P. B., Peffer, N. J., Sodroski, J. G., Rosen, C. A., Goh, W. C., and Haseltine, W. A. (1986) *Trans*-activator gene of HTLV-II induces IL-2 receptor and IL-2 cellular gene expression. *Science*, **232**, 877.

159. Chen, I. S. Y., Cann, A. J., Shah, N. P., and Gaynor, R. B. (1985) Functional relation between HTLV-II x and Adenovirus E1A proteins in transcriptional activation. *Science*, **230**, 570.
160. Williams, J. L., Cann, A. J., Leff, T., Sassone-Corsi, P., and Chen, I. S. Y. (1989) Studies of heterologous promoter trans-activation by the HTLV-II tax protein. *Nucleic Acids Res.*, **17**, 5737.
161. Böhnlein, E., Lowenthal, J. W., Siekevitz, M., Ballard, D. W., Franza, B. R., and Greene, W. C. (1988) The same inducible nuclear proteins regulates mitogen activation of both the interleukin-2 receptor-alpha gene and type I HIV. *Cell*, **53**, 827.
162. Ballard, D. W., Böhnlein, E., Lowenthal, J. W., Wano, Y., Franza, B. R., and Greene, W. C. (1988) HTLV-I Tax induces cellular proteins that activate the κB element in the IL-2 receptor α gene. *Science*, **241**, 1652.
163. Ruben, S., Poteat, H., Tan, T.-H., Kawakami, K., Roeder, R., Haseltine, W., and Rosen, C. A. (1988) Cellular transcription factors and regulation of IL-2 receptor gene expression by HTLV-I *tax* gene product. *Science*, **241**, 89.
164. Hoyos, B., Ballard, D. W., Böhnlein, E., Siekevitz, M., and Greene, W. C. (1989) Kappa B-specific DNA binding proteins: role in the regulation of human interleukin-2 gene expression. *Science*, **244**, 457.
165. Paul, N. L., Lenardo, M. J., Novak, K. D., Sarr, T., Tang, W.-L., and Ruddle, N. H. (1990) Lymphotoxin activation by human T-cell leukemia virus type I-infected cell lines: role for NF-κB. *J. Virol.*, **64**, 5412.
166. Schreck, R. and Baeuerle, P. A. (1990) NF-κB as inducible transcriptional activator of the granulocyte-macrophage colony-stimulating factor gene. *Mol. Cell. Biol.*, **10**, 1281.
167. Arima, N., Molitor, J. A., Smith, M. R., Kim, J. H., Daitoku, Y., and Greene, W. C. (1991) Human T-cell leukemia virus type I Tax induces expression of the Rel-related family of κB enhancer-binding proteins: evidence for a pretranslational component of regulation. *J. Virol.*, **65**, 6892.
168. Lacoste, J., Cohen, L., and Hiscott, J. (1991) NF-κB activity in T cells stably expressing the Tax protein of human T cell lymphotropic virus type I. *Virology*, **184**, 553.
169. Ruben, S., Perkins, A., and Rosen, C. A. (1989) Activation of NF-κB by the HTLV-I trans-activator protein tax requires an additional factor present in lymphoid cells. *New Biol.*, **1**, 275.
170. Ballard, D. W., Böhnlein, E., Hoffman, J. A., Bogerd, H. P., Dixon, E. P., Franza, B. R., and Greene, W. C. (1989) Activation of the interleukin-2 receptor α gene: regulatory role for DNA-protein interactions flanking the κB enhancer. *New Biol.*, **1**, 83.
171. Ruben, S. M. and Rosen, C. A. (1990) Suppression of signals required for activation of transcription factor NF-κB in cells constitutively expressing the HTLV-I Tax protein. *New Biol.*, **2**, 894.
172. Mauxion, F., Jamieson, C., Yoshida, M., Arai, K.-i., and Sen, R. (1991) Comparison of constitutive and inducible transcriptional enhancement mediated by κB-related sequences: modulation of activity in B cells by human T-cell leukemia virus type I *tax* gene. *Proc. Natl Acad. Sci. USA*, **88**, 2141.
173. Inoue, J.-i., Kerr, L. D., Ransone, L. J., Bengal, E., Hunter, T., and Verma, I. M. (1991) c-rel activates but v-rel suppresses transcription from κB sites. *Proc. Natl Acad. Sci. USA*, **88**, 3715.
174. Poteat, H. T., Kadison, P., McGuire, K., Park, L., Park, R. E., Sodroski, J. G., and

Haseltine, W. A. (1989) Response of the human T-cell leukemia virus type 1 long terminal repeat to cyclic AMP. *J. Virol.*, **63**, 1604.

175. Nimer, S. (1991) Tax responsiveness of the GM-CSF promoter is mediated by mitogen-inducible sequences other than κB. *New Biol.*, **3**, 997.

176. Angel, P. and Karin, M. (1991) The role of Jun, Fos and the AP-1 complex in cell-proliferation and transformation. *Biochim. Biophys. Acta*, **1072**, 129.

177. Gius, D., Cao, X., Rauscher III, F. J., Cohen, D. R., Curran, T., and Sukhatme, V. P. (1990) Transcriptional activation and repression by Fos are independent functions: The C terminus represses immediate-early gene expression via CArG elements. *Mol. Cell. Biol.*, **10**, 4243.

178. Rivera, V. M. and Greenberg, M. E. (1990) Growth factor-induced gene expression: the ups and downs of c-*fos* regulation. *New Biol.*, **2**, 751.

179. Alexandre, C. and Verrier, B. (1991) Four regulatory elements in the human c-fos promoter mediate transactivation by HTLV-1 tax protein. *Oncogene*, **6**, 543.

180. Miller, A. D., Curran, T., and Verma, I. M. (1984) c-*fos* Protein can induce cellular transformation: a novel mechanism of activation of a cellular oncogene. *Cell*, **36**, 51.

181. Fujii, M., Niki, T., Mori, T., Matsuda, T., Matsui, M., Nomura, N., and Seiki, M. (1991) HTLV-1 Tax induces expression of various immediate early serum responsive genes. *Oncogene*, **6**, 1023.

182. Hooper, W. C., Rudolph, D. L., Lairmore, M. D., and Lal, R. B. (1991) Constitutive expression of c-jun and jun-B in cell lines infected with human T-lymphotropic virus types I and II. *Biochem. Biophys. Res. Commun.*, **181**, 976.

183. Scanlon, R. J., Jiao, L., Funato, T., Wang, W., Tone, T., Rossi, J. J., Kashani-Sabet, M. (1991) Ribozyme-mediated cleavage of c-fos mRNA reduces gene expression of DNA synthesis enzymes and metallothionein. *Proc. Natl Acad. Sci. USA*, **88**, 10591.

184. Ensoli, B., Barillari, G., Salahuddin, S. Z., Gallo, R. C., and Wong-Staal, F. (1990) Tat protein of HIV-1 stimulates growth of cells derived from Kaposi's sarcoma lesions of AIDS patients. *Nature*, **345**, 84.

185. Lindholm, P. F., Marriott, S. J., Gitlin, S. D., Bohan, C. A., and Brady, J. N. (1990) Induction of nuclear NF-κB DNA binding activity after exposure of lymphoid cells to soluble Tax$_1$ protein. *New Biol.*, **2**, 1034.

186. Marriott, S. J., Lindholm, P. F., Reid, R. L., and Brady, J. N. (1991) Soluble HTLV-I Tax$_1$ protein stimulates proliferation of human peripheral blood lymphocytes. *New Biol.*, **3**, 678.

187. Lindholm, P. F., Reid, R. L., and Brady, J. N. (1992) Extracellular Tax$_1$ protein stimulates tumor necrosis factor-β and immunoglobulin kappa light chain expression in lymphoid cells. *J. Virol.*, **66**, 1294.

188. Marriott, S. J., Trinh, D., and Brady, J. N. (1992) Activation of interleukin 2 receptor alpha expression by extracellular HTLV-I Tax$_1$ protein: a potential role in HTLV-I pathogenesis. *Oncogene*, **7**, 1749.

189. Rudkin, B. B., Lazarovici, P., Levi, B.-Z., Abe, Y., Fujita, K., and Guroff, G. (1989) Cell cycle-specific action of nerve growth factor in PC12 cells: differentiation without proliferation. *EMBO J.*, **8**, 3319.

190. Hirano, T., Taga, T., Nakano, N., Yasukawa, K., Kashiwamura, S., Shimizu, K., Nakajima, K., Pyun, K. H., and Kishimoto, T. (1985) Purification to homogeneity and characterization of human B-cell differentiation factor (BCDF or BSFp-2). *Proc. Natl Acad. Sci. USA*, **82**, 5490.

191. Tschachler, E., Robert-Guroff, M., Gallo, R. C., and Reitz Jr., M. S. (1989) Human

T-lymphotropic virus I-infected T cells constitutively express lymphotoxin in vitro. *Blood*, **73**, 194.
192. Libermann, T. A. and Baltimore, D. (1990) Activation of interleukin-6 gene expression through the NF-κB transcription factor. *Mol. Cell. Biol.*, **10**, 2327.
193. Gessain, A., Barin, F., Vernant, J. C., Gout, O., Maurs, L., Calander, A., and de The, G. (1985) Antibodies to human T-lymphotropic Virus Type-I in patients with tropical spastic paraparesis. *Lancet*, **ii**, 407.
194. Chen, Y.-M. A., Okayama, A., Lee, T.-H., Tachibana, N., Mueller, N., and Essex, M. (1991) Sexual transmission of human T-cell leukemia virus type I associated with the presence of anti-Tax antibody. *Proc. Natl Acad. Sci. USA*, **88**, 1182.

8 | The molecular biology of the human spumavirus

ROLF M. FLÜGEL

1. Introduction

Twenty years after the first isolation and morphological characterization of a human spumaretrovirus (HSRV) from the lymphoblastoid cells of a nasopharyngeal carcinoma patient (1) and 5 years after the first molecular-biological characterization of this isolate (2, 3), spumaviruses have begun to attract significant research interest. The spumaviruses are of interest for several reasons:

1. Spumaviruses, also called foamy viruses, occupy a biological niche that sets them significantly apart from other retroviruses and, in particular, from the intensely studied oncoviruses (4, 5).
2. Foamy viruses, as complex retroviruses (4), encode a number of regulatory genes, the *bel* genes (derived from between env and long terminal repeat (LTR)), that are unique in structure and function.
3. Sero-epidemiological surveys show that antibodies against HSRV occur in sera of patients with a wide spectrum of diseases, including neurologic disease in particular (6). These surveys also showed that HSRV antibody prevalence is higher in African countries than in Europe. Certain tumour patients from Tanzania and Kenya had a particularly high HSRV antibody prevalence.
4. HSRV seems to be a neuropathogen, at least in the animal model of transgenic mice (7, 8).

Taken together, these insights have rekindled efforts to identify the human disease(s) associated with HSRV infection and to gain insights into regulation of gene expression in HSRV.

This review will cover recent advances on HSRV genome structure, viral mRNA expression patterns, and the function of the HSRV *bel* genes, including *bel 1* in particular. The review will not deal with the older literature before the advent of molecular biology and the classification of spumaviruses within the family Retroviridae, since other reviews have been published that focus on spumavirinae as a genus, on the evolution of spumaviruses, and on the primate foamy viruses (4, 9). For a review on foamy virus research before 1975, the reader is referred to refs 10 and 11.

2. Morphological and virological properties of spumaviruses

At first glance, spumaviruses appear similar to C-type viruses in that they possess a similar core and envelope structure. Closer examination of electron-microscopic morphology reveals that foamy viruses are characterized by long surface projections and cores (nucleocapsids) that rarely condense like those of other retroviruses (12). HSRV particles are slightly larger than C-type viruses, with diameters ranging from 100 to 140 nm. Virions can bud either from vacuoles or from the cytoplasmic membrane, but have a relatively strong tendency to remain cell associated. These properties are shared by foamy viruses independent of the source and history of a given isolate.

Spumaviruses have been isolated repeatedly as passengers from many species, including pets, captive and wild animals, and man. One reason that members of this subfamily of retroviruses were isolated so often is that spumaviruses induce characteristic cytopathic effects in tissue culture, leading to the lysis of host cells. The appearance of multinucleated giant and vacuolated cells is fairly characteristic of spumavirus-infected cell cultures. It is from this 'foamy' cytopathic effect that the name spumaviruses was derived. Foamy viruses have been isolated from apparently healthy animals as well as from humans and animals with a broad spectrum of diseases (4, 9, and references cited therein). Spumaviruses have been obtained from virtually every organ and body fluid of infected animals, including the brain. In addition, spumaviruses seem to occur ubiquitously, since isolates and HSRV-antibody positives have been reported from many different countries and geographic locations (6). It therefore appears that spumaviruses have a broad host range with respect to species and to cell types. Nevertheless, virus replication *in vitro* proceeds relatively slowly and seldom reaches high yields. *In vivo*, spumaviral infections have been reported to lead to a persistent low-level infection, as has also been observed for other complex retroviruses. It is of great interest that experimentally infected rabbits show a transient immunosuppression with concomitant opportunistic infections (13).

Primate foamy viruses have frequently been isolated from Old World monkeys and apes (11). The different isolates have been classified into approximately a dozen stereotypes. So far, nucleotide sequences from only one human (HSRV), and two simian (SSRV types 1 and 3, also called SFV-1 and SFV-3) isolates are known (2, 3, 14–16). A bovine spumavirus (BSRV) has been biochemically characterized recently (17). The overall genetic organization is similar, but distinct features in their coding capacities are clearly recognizable, as will be discussed below.

3. An overview of the replication cycle of HSRV in comparison with other retroviruses

In general, the replication cycle of spumaviruses is similar to that of other known human and animal retroviruses as shown in Fig. 1. After an infectious retrovirus

Fig. 1 Schematic view of a retrovirus life cycle. The infection process starts with the attachment to a cellular receptor (upper right corner). After virus entry, reverse transcription takes place inside the cytoplasm in a ribonucleoprotein complex (brackets). The viral RNA genome (thin line) is copied into a double-stranded DNA (thick line with filled boxes at both ends that stand for the 5' and 3' LTR). The linear viral DNA moves to the cell nucleus where it integrates into chromosomal DNA (double line) and forms the provirus. Transcription and splicing provide viral genomic and subgenomic mRNAs that are transported to the cytoplasm to be translated into viral proteins. Self assembly (small circles) leads to new virus particles that finally mature by budding into infectious progeny virions

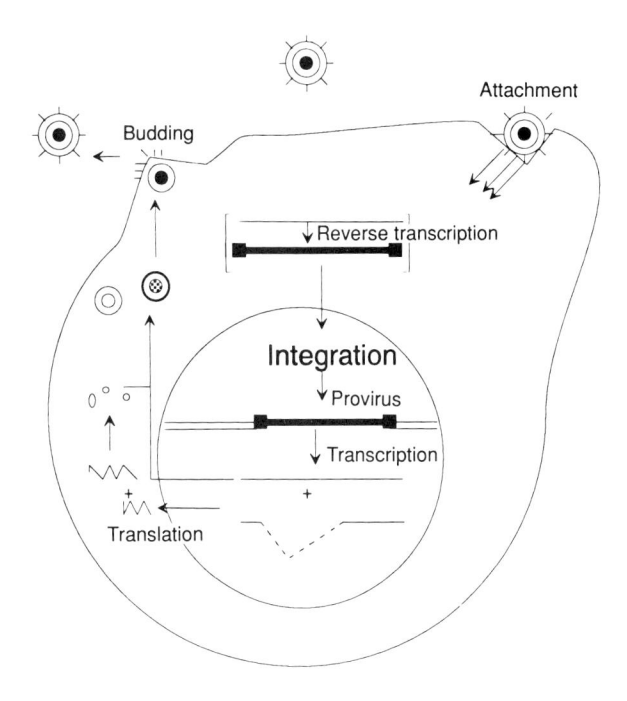

particle has attached to a susceptible cell that carries a receptor on its surface, virus entry into the cytoplasm takes place. The known cellular receptors of retroviruses bind to a specific domain of the viral surface glycoprotein (SU). In uninfected cells, the receptors seem to function as transporters of amino acids or phosphate or as markers of cell recognition during development and immune response (18, 19). The details of virus penetration, the fusion events and entry, the precise cytoplasmic site, and the degree of uncoating of the retrovirus ribonucleoprotein complex have not been clarified. However, retrovirus entry does not require receptor-mediated endocytosis or pH-dependent entrance via endosomes (20, 21).

One mature and infectious retrovirus particle contains two molecules of single-stranded RNA as genome. The dimeric nature of the genome, in conjunction with a mode of replication that includes template switches and strand transfer reactions, enables retroviruses to achieve a high rate of homologous recombination (22).

Retroviruses are the only RNA viruses that replicate via a DNA molecule as intermediate. The virus-encoded enzyme reverse transcriptase, a DNA polymerase that catalyzes the characteristic replication reactions, together with a second virus-encoded enzyme ribonuclease H, converts genomic RNA into a double-stranded DNA copy (Fig. 1). HSRV also possesses two essentially identical molecules of genomic RNA packaged into virus particles together with the enzyme reverse transcriptase (23). In infected cells, reverse transcription apparently takes place in the cytoplasm in a viral ribonucleoprotein complex that at least contains the capsid antigen, CA (24). The newly synthesized, double-stranded viral DNA is then transported into the cell nucleus. In the replication cycle of most retroviruses, the major part of the linear viral DNA is covalently and stably inserted into host cell DNA, a process called integration (5, 25). The viral enzyme Int (integrase) has a DNA ligase and nuclease activities, hence its second name Endo derived from endonuclease. These enzymatic activities reside in the carboxy-terminal domain of the Pol gene product (Fig. 2). Whereas the Endo activity trims nucleotides from the ends of the LTRs and is responsible for opening host cell DNA, the integrase catalyses the ligation of the ends of the linear viral DNA to the termini of the cellular DNA (25–27). This reaction results in the covalent integration of retroviral DNA into host DNA and leads to the formation of the provirus (5) (Fig. 1).

Apparently, proviral integration sites are not specific with respect to host cellular DNA sequences, although there are clear indications from related systems, such as retrotransposons, for preferred integration into transcription initiation regions of tRNA and rRNa genes (28). The integration sites, however, are specific with respect to the retroviral genomes, i.e. the points of integration are the 5' and 3' ends of the LTRs. Evidence for integration of HSRV DNA is indirect, but HSRV DNA fragments larger than genome size, and therefore likely due to covalent integration, were detectable by Southern blot hybridization (R. M. Flügel, unpublished results). However, the amount of integrated HSRV DNA in infected cells is significantly less than that of unintegrated viral DNA.

The integrated provirus enjoys all the advantages of a normal cellular gene, i.e. it can exploit the cellular transcription and translation machinery for the synthesis of viral gene products. Reverse transcriptase and newly synthesized genomic RNA molecules are packaged into virions in a complex with several components that are necessary for virus replication, e.g. cellular tRNA molecules of which the 3' end serves as primer for the synthesis of minus strand DNA (5). In the case of HSRV and other spumaviruses, a cellular tRNA specific for lysine, tRNA-$(lys_{1,2})$ is the primer for viral DNA synthesis (3). Spumaviruses bud inefficiently from the cytoplasm into culture medium. Apparently, the process of budding does not kill the cells (Fig. 1). It remains to be determined whether the accumulation of linear unintegrated HSRV DNA molecules observed in infected cells is responsible for the highly cytopathic effect of HSRV infection in culture. This correlation has been proposed for other retroviral systems (29, 30).

4. The genetic structure of HSRV is complex

The primary structure of the genome of HSRV was determined from an isolate that was obtained from a Kenyan nasopharyngeal carcinoma patient (1–3, 31). Genetic and molecular-biological analysis of this prototypic HSRV have shown that the genome structure encodes several regulatory and accessory gene products. Fig. 2 (centre line) depicts the structure of the HSRV genome, which is almost 12 kb in size.

The HSRV Gag precursor migrates as double bands of 78 and 74 kDa under denaturing conditions (32). The values are consistent with the calculated one of 72.6 kDa for the 648 amino acid Gag gene product. The viral proteinase (PR) generates the individual Gag and Pol proteins by proteolytic cleavage of distinct

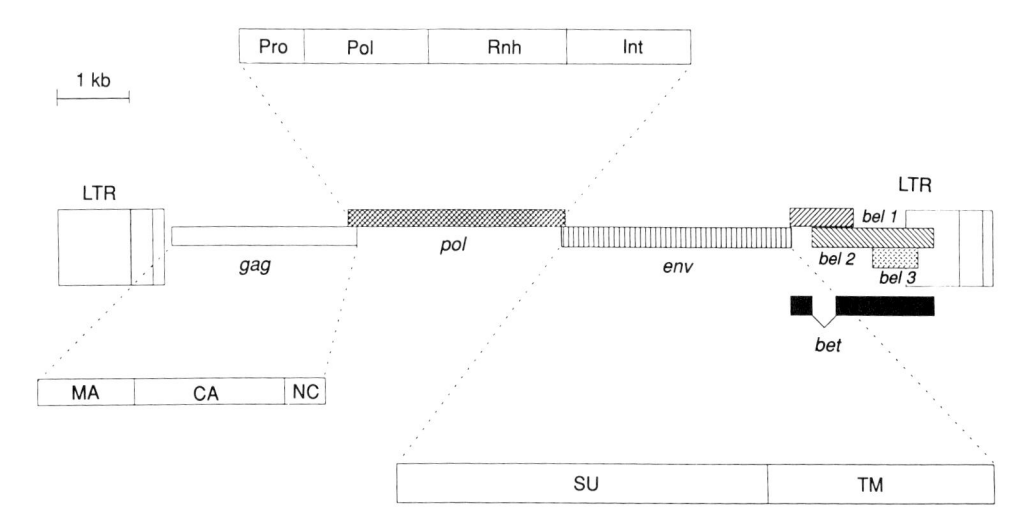

Fig. 2 Organization of the HSRV DNA genome (central part). The LTRs are symbolized by boxes, the genes by different shading. The bipartite structure of *bet* (filled boxes) indicates that it is formed by splicing from genomic RNA. The stippled lines indicate the corresponding gene products

peptide bonds in the HSRV Gag precursor, as shown schematically in Fig. 3. These include the mature MA (matrix protein), CA (capsid antigen), and NC (nucleocapsid antigen) Gag proteins. The viral PR is encoded by the amino-terminal region of the *pol* gene, as has also been shown for simian foamy viruses (4–16) and lentiviruses (33). Antibodies against the HSRV CA domain specifically immunoprecipitated a 33 kDa protein from HSRV-infected human cells and, in addition, two double bands of 78 and 74 (Gag precursors) and of 60 and 58 kDA (processing intermediates MA–CA). The double bands of the HSRV Gag precursor and of the processing intermediates probably result from post-translational modification. Recombinant CA-Gag antigen can be used for the detection of HSRV-specific antibodies in human sera (see Section 9 of this chapter).

Fig. 3 Schematic drawing of the HSRV Gag precursors of 78 and 74 kDa (34) and their proteolytic processing into mature HSRV Gag proteins. Arrowheads mark the cleavage sites

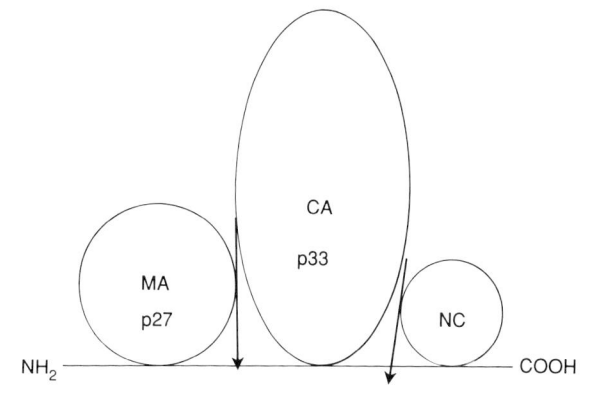

There are other features of the HSRV Gag proteins that appear common to all spumaviruses but are not seen in other retroviruses. The most obvious and distinct difference is that the NC sequence has a strongly basic glycine- and arginine-rich repeat structural motif instead of the cysteine-rich zinc finger found in most retroviruses and also in the pararetrovirus cauliflower mosaic virus (3). In other retroviruses, the Gag zinc-finger motif is believed to mediate the specific interaction of Gag with viral genomic RNA. The highly basic nature of this Gly-Arg-rich Gag domain, and its very high conservation in the three spumavirus genomes sequenced to date, suggests that this novel motif may serve the same purpose in the foamy virus replication cycle. It is of phylogenetic interest that hepadnaviruses and retrotransposons that also lack the cysteine structural motifs also possess strongly basic sequences in their core proteins that appear similar to those of spumaviruses.

The Gag–Pol polyprotein of about 190 kDa (6) is postulated to be synthesized via translational frameshifting (34). It is worth mentioning that spumaviruses, like some retrotransposons, seem to use +1 frameshifting. A frameshifting sequence has been postulated for SFV-1 Gag–Pol that lies just downstream of the end of the *gag* open reading frame (14). However, a purine-rich heptanucleotide that ends three nucleotides upstream of the *gag* stop codon in all three known spumaviral sequences seems to be the better candidate, since it conforms to the general rules for frameshifting sequences (34) and is none the less unique to foamy viruses.

The protein sequence of HSRV reverse transcriptase is distinct from those of other retrovirus groups. Surprisingly, the degree of homology of HSRV Pol is closest to that of Moloney murine leukemia virus (Mo-MLV) (2, 3, 35). The structural motifs essential for the functional activities of the polymerase, the ribonuclease H, and the integrase activities that have been defined in other retroviral reverse transcriptases (36), are also well conserved at the corresponding sites of the HSRV Pol protein sequence. The long and characteristic tether or hinge region between the reverse transcriptase domain and the *rnh* region is similar to that of Mo-MLV, both in protein homology and predicted secondary structure. The carboxy-terminal *pol–int* domain overlaps the HSRV *env* coding region. A *pol/env* overlap region is common to all genomes of retroviruses, except for those of lentiviruses. HSRV accessory proteins are not present in this region, in contrast to the genomes of lentiviruses. There are, however, a number of non-coding exons located in this part of the HSRV genome.

The HSRV Env protein has the characteristic features of a retroviral glycoprotein: an amino-terminal signal peptide sequence that is preceded by a hydrophilic leader sequence, an internally located, basic processing site (Arg-Lys-Arg-Arg) followed by an internal fusion peptide sequence, and a long, strongly hydrophobic transmembrane (TM) peptide sequence that is located proximal to the carboxy-terminus of the TM protein (2). The HSRV TM sequence ends with a short hydrophilic tail. The HSRV Env sequence contains fifteen potential N-linked glycosylation signals and 21 cysteine residues. The glycosylated Env precursors of approximately 160 and 130 kDa are assumed to be processed by a cellular signal peptidase between

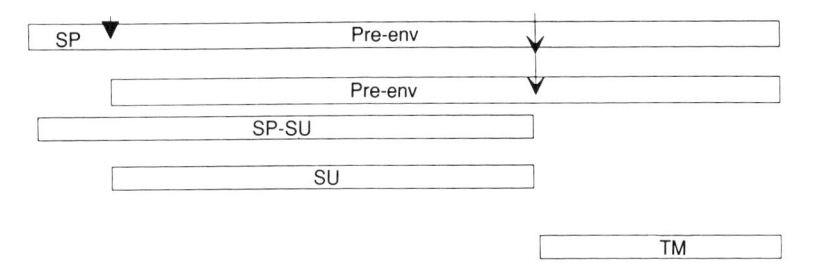

Fig. 4 Schematic drawing of HSRV Env gene products. The longest rectangle (top line) symbolizes the Env precursor glycoprotein with the uncleaved signal peptide (SP) at the amino terminus. The arrowhead marks the cleavage site at the SP/SU boundary. The centrally located arrows indicate the basic proteolytic cleavage sites that define the boundaries of the mature SU and TM proteins. An alternative form of the Env glycoprotein SP–SU is also shown (third line from top) that has the SP still attached to the SU protein

residues 91 and 92, Ser-Arg (Fig. 4). The proteasome complex is responsible for the cleavage of the Env precursor into an SU protein that migrates as a broad band, of 70 to 80 kDa, and into a TM protein of 45 kDa (Fig. 4) (6, 37–39). The 130 kDa Env glycoprotein precursor lacks the signal peptide as well as the relatively long leader signal sequence with the SU–TM domains still covalently linked as shown in Fig. 4 (second line from top).

Further downstream, the *bel 1* open reading frame overlaps the *env* COOH-terminus by nine residues (Fig. 2). It encodes a nuclear phosphoprotein of 36 kDa (300 residues) in size (40). The 89 amino acid amino-terminal part of the chimeric Bet protein consists of *bel 1* sequences while the carboxy-terminal part of Bet contains the entire *bel 2* sequence (41). Bet seems to be unique to spumaviruses, since it does not have any homologies to other retroviral or known cellular proteins. It is the largest accessory HSRV protein with a size of 56 kDa (480 amino acids) (40). Bet is abundantly expressed in HSRV-infected cells. The Bel 2 protein, with a measured molecular mass of 43 kDa (40), initiates downstream of the *bet* splice acceptor in the same reading frame as *bet* and is fully contained within the *bet* sequence (Fig. 2). Both Bel 2 and Bet are predominantly localized in the cytoplasm late after infection; their functions are unknown.

The 1120 bp HSRV LTR is unusually large in comparison with those of other retroviruses. A detailed discussion of the HSRV LTR will be given below, see Section 7 and Table 1. The overall genetic organization of HSRV indicates that spumaviruses take an intermediate position between the more complex lentiviruses and the less complex human T-cell leukemia virus/bovine leukemia virus (HTLV/BLV) group of retroviruses (4).

Recently, infectious HSRV DNA clones were constructed that upon transfection into susceptible cells produced viral transcripts and proteins showed cytopathic effects typical for HSRV after repeated passaging in cell cultures, and yielded infectious virions (40, 42). The infectious HSRV molecular clones have been used for the analysis of functions of the various accessory gene products.

The proviral DNA intermediates of spumaviruses and lentiviruses seem to possess gaps of up to several hundred nucleotides in length that can be detected by S1 nuclease analysis (43–45). The significance for replication of these DNA gaps, which are located in the central part of the viral genome, remains to be determined.

5. The transcriptional pattern of the HSRV genome is complex

The cap site or start site of transcription begins with a G as determined by nucleotide sequencing (3). The major cap site is located at position 778 of the HSRV LTR (see Fig. 7). Per definition the cap site occupies the first nucleotide position of the HSRV RNA genome, and this is the start site of the R (repeat) region of the LTR of the HSRV DNA genome. A minor cap site at position +3 also coincides with a G base (3). The R region extends over 190 nucleotides to the poly(A) addition site (41). There is, however, still an inherent ambiguity with respect to the exact nucleotide portion of the major poly(A) addition site, since alternative, closely nested sites are utilized during addition of the poly(A) tail, as also observed in some other retroviruses (11).

5.1 HSRV *env* transcripts

The splicing pattern of HSRV is complex (Figs 5 and 6). All subgenomic HSRV transcripts share a short RNA leader of 51 nucleotides derived from the 5' end of the R region of the LTR (41). There are at least four *env* mRNAs that have different splice acceptors resulting in slightly variable 5' ends. The longest *env*-specific

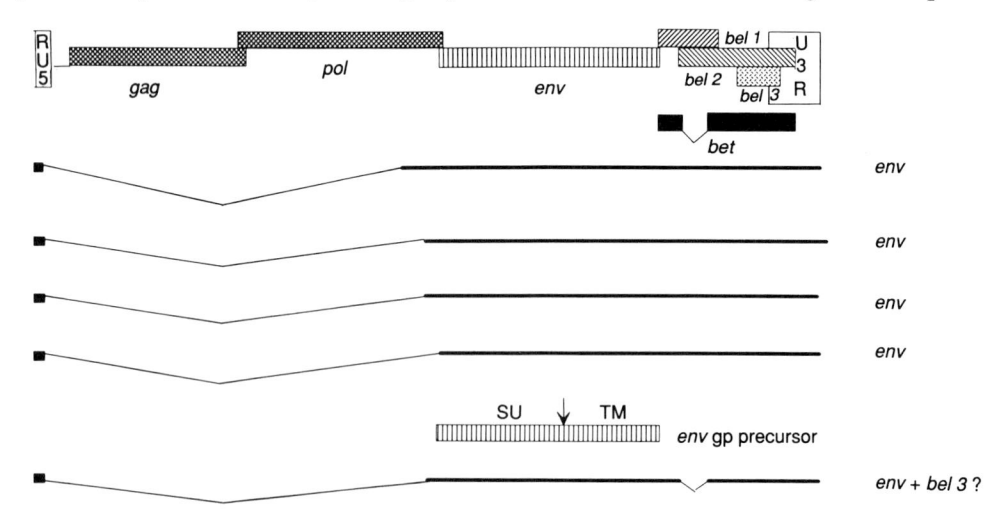

Fig. 5 The HSRV *env* transcripts identified by PCR. The top line shows the HSRV RNA genome with the R–U5 region at the 5' end and the U3–R region at the 3' end. Black thick lines and boxes show exons thin lines symbolize introns. The question mark in the last line indicates that it is unknown which viral proteins are translated from this doubly spliced *env* mRNA. For further explanations, see legends to Figs 2 and 3

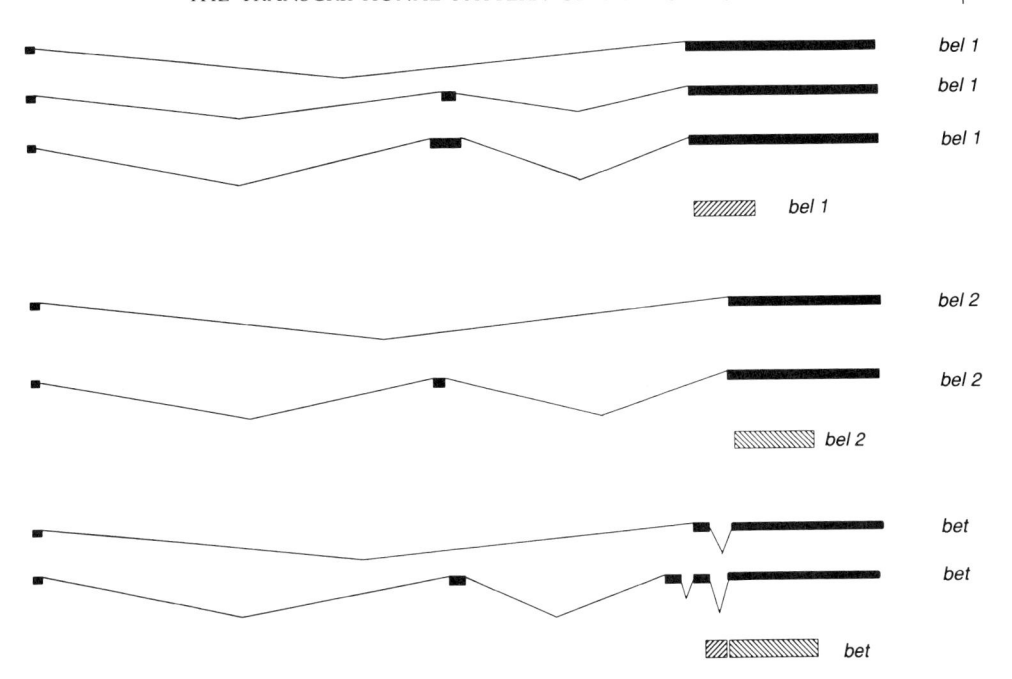

Fig. 6 The seven HSRV *bel 1*, *bel 2*, and *bet* transcripts (41). Black boxes show exons, thin lines introns. Shaded boxes indicate gene coding sequences

transcripts start in the *int* domain of the *pol* gene (Fig. 5). The splice acceptor of the shortest *env* mRNA is located just one nucleotide from the consensus initiation codon of *env*. This *env* transcript was also detectable by S1 nuclease analysis, whereas all other *env* mRNAs could only be identified by polymerase chain reaction (PCR). Figure 5 (bottom line) shows yet another *env* transcript that formally might also encode the HSRV Env protein. It lacks an excised intron in the *bel 1* region (41) that is also lacking in the *bet* mRNAs (Fig. 6). While up to four or even five Env-encoding subgenomic mRNAs were found, the Env protein does not seem to be as abundantly expressed as Bet in infected cells.

5.2 *bel* and *bet* transcripts

Three species of *bel 1* transcripts were identified (41). One is singly spliced, while the remaining two are multiply spliced mRNA species containing non-coding exons from the middle of the HSRV genome (Fig. 6). Further analysis of the expression capacity of the individually cloned transcripts will reveal whether or not there are differences in the concentration and regulation of these mRNAs. However, there is no apparent difference in the coding capacities of the *bel 1* transcripts. These *bel 1*-specific mRNAs are not abundantly expressed, particularly when compared with *bet* transcripts (41). The splice acceptor that is located just 5' to the *bel 1* initiator codon has also been verified by S1 nuclease analysis. The initiator codon of the *bel 1* (and *bet*) mRNAs obeys the consensus sequence established for initiation of eukaryotic translation (46).

There are, furthermore, two spliced *bel 2* mRNAs (41). Again, the first is singly spliced, while the second contains a non-coding exon from the pre-*env* region (Fig. 6). The *bel 2* coding region is completely contained within *bet*. Thus, it is possible that differential splicing contributes to the regulation of *bel 2* expression, since a *bel 2*-specific protein with the expected size was found by protein blotting (40). The assumed *bel 2* protein starts with a non-consensus AUG initiator codon, and, in addition, most of the downstream AUG codons of *bel 2* also appear weak. Therefore, by analogy to the translation mechanisms used for HIV-1 *nef*, the weak AUGs of *bel 2* might allow for the translation of the downstream *bel 3* gene product by a leaky scanning mechanism of translation (47). A *bel 3*-specific transcript has not been identified.

Two multiply spliced *bet* mRNAs were detected by PCR. It is interesting that the *bel 2/bet* splice sites were also found with S1 nuclease protection analysis (41). This result indicates that the overall concentration of *bet* mRNA was greater than that of other subgenomic transcripts, e.g. *bel 2* mRNA was not detectable by S1 nuclease analysis.

Overall, the HSRV transcriptional pattern resembles that of HIV-1 in complexity, although certain features, such as the apparent absence of a *rev*-like gene, are striking. The centrally located, short and non-coding exons of the HSRV *bel/bet* transcripts seem to be characteristic for foamy viruses. The locations of the major splice donor and acceptor sites as determined for the HSRV genome (41) is conserved in the three sequenced spumaviruses, so that the splicing pattern of the SSRV will undoubtedly be similarly complex.

6. Bel 1 is the HSRV transcriptional *trans*-activator

The Bel 1 gene product has been identified in HSRV-infected cells as a nuclear phosphoprotein (40, 48). It migrates with an apparent molecular mass of 36 kDa, a value consistent with that predicted from the deduced protein sequence (300 amino acid residues). The *bel 1* gene is absolutely required for virus replication, as shown by site-specific mutation of an infectious HSRV DNA clone (40).

It has been demonstrated that Bel 1 acts as a specific *trans*-activator of the HSRV LTR in a variety of different cell lines (48). The *trans*-activation activity is confined to the *bel 1* coding region, thus establishing that it is the Bel 1 protein that is responsible. In fact, *bel* regions further downstream of the *bel 1* gene do not show any *trans*-activator activity (48, 49). Limited mutagenesis of the *bel 1* gene indicated that most of its coding region is essential for the activation function except for the 27 carboxy-terminal amino acid residues (50).

To test whether or not Bel 1 is capable of cross *trans*-activating any of the other human retroviral LTRs, the Bel 1 *trans*-activation activity was directly compared with the activities of the HIV-1 Tat and HTLV-1 Tax in similar indicator gene constructs that contained either the homologous HSRV LTR or the heterologous HIV-1 and HTLV-I LTRs. It was found that Bel 1 can *trans*-activate the HIV-1 LTR but not the HTLV-I LTR (48). This remarkable result indicates that in patients infected with both HIV-1 and HSRV Bel 1 might activate HIV-1 transcription and thereby potentially might contribute to HIV-1 pathogenesis. This hypothesis is supported by the identification of a novel *bel 1* DNA response sequence in the unique 3' (U3) region of the HIV-1 LTR. The enhancement of HIV-specific transcription was found not only in fibroblast cells but also in human T-cells (51). The DNA target sequence is close to, and may overlap, the two NF-κB sites present in the HIV LTR (51). It is interesting that upon inactivation of these sites, *trans*-activation by Bel 1 is enhanced significantly, again strengthening the view that HSRV might be a cofactor in the pathogenesis of dually infected individuals. The Bel 1 *trans*-activator functions effectively in various mammalian cells, such as those derived from human, simian, murine, and avian sources (48). This result is consistent with the wide host range of spumaviruses.

Recently, the corresponding *trans*-activator of the simian spumavirus type 1 (Taf) was also identified (52). Taf has no effect on the HSRV LTR nor does Bel 1 *trans*-activate the SFV-1 LTR (52–54). These results are consistent with the absence of any significant homology in the U3 region of both spumaviruses that otherwise are closely related (54). However, the degree of homology of the Bel 1/Taf protein sequences is 39 per cent. These findings seem to indicate that the assumed interaction of cellular transcription factors with both the Bel 1 protein and the corresponding U3 DNA response elements might be species- and/or cell-specific. Further experiments are required to settle this question.

7. The HSRV LTR is unusually long and contains different DNA response elements for the *trans*-activator Bel 1

The HSRV LTR sequence was determined from linear viral DNA. Starting from the 5' end, the HSRV DNA genome is subdivided into the U3 domain of 777 bp, the R region of 190 bp in size with the cap site as the first nucleotide, and the U5 unique 5' (U5) region that is 155 bp long, the first nucleotide of U5 being the poly(A) addition site (Table 1 and Fig. 7). Thus, it is mainly the U3 region that contributes to the overall length of the HSRV LTR. Some of the simian spumaviruses possess even longer U3 regions (14–16, 52). The only other non-foamy retroviruses that are comparable to HSRV with respect to the lengths of the U3 domains are mouse mammary tumour virus (MMTV) and human immunodeficiency virus type 1 (HIV-1) (11). This is at least in part due to coding regions of *sag* (superantigen), *nef*, and *bel 3* that all overlap into the U3 regions. The HSRV *bet/bel 2* coding sequences overlap into the U3 region, too (Fig. 2). In addition, retroviruses use parts of their U3 domains as target sequences for regulatory factors that directly or indirectly control viral gene expression. Viral regulatory proteins are assumed to be com-

Table 1 Comparison of HSRV with primate spumaviruses genomes

Virus type	Provirus length (bp)	LTR length (bp)	Presence of *bel 3*
HSRV	11 955	1120	Present
SFV-1	12 972	1621	Absent
SFV-3	13 111	1708	Absent

Homologies between LTRs of HSRV and those of SFV types 1 and 3 are insignificant for the U3 regions, and 81 to 86% for the R–U5 regions.

plexed to diverse cellular transcription factors and can function either as positive activators or alternatively, as negative regulators (55). An impressive spectrum of different cellular transcriptional activators and coactivators have been observed and identified that can bind to their cognate DNA response elements in retroviral LTRs (21, 56). This is particularly valid for HIV-1 and murine leukemia virus (MLV) (5); for HSRV, however, corresponding studies are just beginning to reveal some of the complexities of the regulation of transcription.

Two distinct Bel 1 response elements (BREs) were mapped to the U3 region of HSRV (Fig. 7). The major BRE site is located upstream of the HSRV cap site, between −94 and −162, and contributes most of the *trans*-activator activity of the LTR (48). Another group has mapped the primary BRE between −94 to −136 that overlaps with the first region by 44 nucleotides (50). It is likely that the minimal BRE DNA sequence is contained within this domain. A second weaker BRE site was found to be located further upstream in the U3 region between positions −389 to −500 (Fig. 7). However, the removal of two apparent Ap-1 sites in this region resulted in only a slight decrease of LTR-induced chloramphenicol acetyl transferase (CAT) expression (50, 57). It remains to be proven whether Ap-1 or other cellular transcription factors in fact bind to these sites in the HSRV LTR, although it was claimed that treatment with phorbol ester induced Ap-1-mediated CAT gene expression (57). In any case, the BRE site closer to the TATA box is the essential and

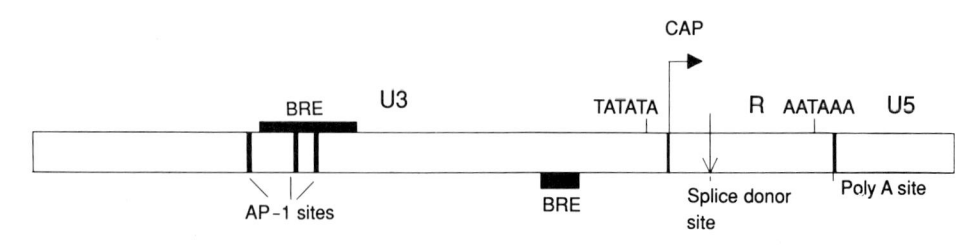

Fig. 7 Schematic drawing of the HSRV LTR. BREs are the Bel 1 *trans*-activator response elements (48, 50). Three Ap-1 sites are marked. The cap site and the splice donor (vertical arrow) at nucleotide position 51 are indicated

dominant LTR DNA target site for Bel 1 *trans*-activation compared with that located further upstream.

Preliminary results from different laboratories indicate that there are *cis*-acting negative regulatory elements (NRE) in the HSRV LTR (50). Upon deletion of these NREs a moderate increase in CAT gene expression was observed. The HSRV NREs were localized to the R–U5 regions of the LTRs (50). An inhibitory effect of Bel 1-mediated *trans*-activation was also observed in the closely related SFV-1 LTR (54). Again, the NREs were mapped to the R–U5 regions. Finally, an intragenic enhancer within the *env* coding region of HSRV has been detected (M. Löchelt, personal communication). This region *trans*-activated CAT and secreted alkaline phosphatase (SEAP) expression when *bel 1* was cotransfected in either COS or human embryonic lung cells. A novel intragenic BRE site has to be postulated for this observation. The results of the studies on the Bel 1 *trans*-activator and its target sites, BREs, are schematically shown in Fig. 8.

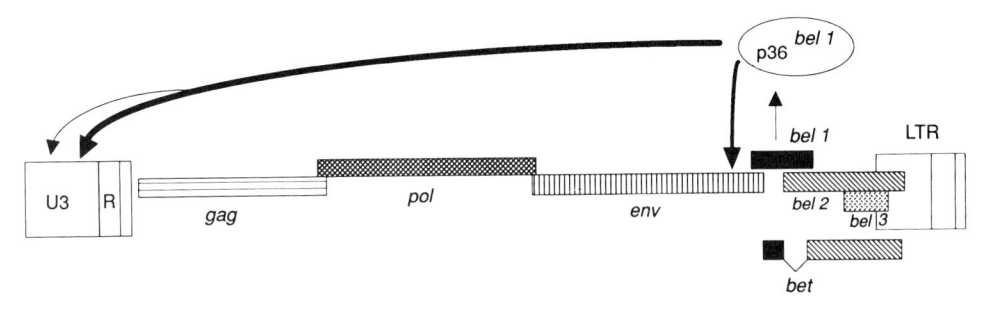

Fig. 8 Schematic drawing of the HSRV DNA genome as *trans*-activated by Bel 1. The ellipse shows the Bel 1 *trans*-activator protein of 36 kDa. The strong arrows mark the extra- and intragenic DNA targets of Bel 1, the thin line indicates the weaker BRE located in the upstream region of the U3 domain

8. Is Bel 3 a spumaviral superantigen?

Recent advances in the understanding of the function of the long open reading frame (ORF) of MMTV has solved one of the long-standing enigmas in retrovirology. It was demonstrated *in vitro* and *in vivo* that the Orf protein of MMTV functions as a superantigen (for a review on superantigens, see 58, 59). Hence Orf of MMTV was renamed Sag, an acronym for superantigen, see Fig. 9. Sag leads to a strong and rapid stimulation of T-cells *in vitro* (60). It was demonstrated *in vivo* that expression of MMTV Sag causes a specific deletion of one of the members of the Vβ subunits of the T-cell receptor, namely that of Vβ 14 (61).

Sag of MMTV has peculiar properties that sets it apart from most known retroviral gene products. Its coding region is largely located in the 3' LTR (11). Subgenomic *sag*-specific mRNA was detectable in MMTV-infected cells, but the corresponding protein was not found for unknown reasons. It was, however, identified in phorbol ester-treated murine T-lymphoma cell lines in a cytoplasmic localization (62). Sag can be produced in *in vitro* translation and baculovirus expression systems

(63). Apparently, it occurs in various glycosylated and carbohydrate-free forms ranging in size from 20 to 36 kDa (64). Sag is not required for MMTV replication *in vitro*, since there are isolates of MMTV that do not encode an intact reading frame for Sag, but the virus strains are capable of replication. Deletion of the *Sag* region that coincides with parts of U3 can lead to T-cell lymphomas in mice (65).

There are a number of curious resemblances between Bel 3 of HSRV and Sag of MMTV. Bel 3 was not detectable in HSRV-infected cells; however, Bel 3 expressed in transfected COS cells was found to localize to the cytoplasm (J. Weissenberger, personal communication). The *bel 3* coding region is located in the 3' LTR comparable to the *sag* and *nef* genes of HIV-1. The *bel 3* gene seems not to be required for virus replication *in vitro*, since there are simian spumaviruses like SFV types 1 and 3 that do not encode any *bel 3* gene, but replicate as efficiently as HSRV (9). In a series of pilot experiments, purified Bel 3 protein was found to act like a HSRV superantigen, since it led to a strong stimulation of human T-cells *in vitro* (J. Weissenberger and A. Altmann, personal communication). In contrast, HSRV Bel 2 protein did not stimulate T-cells. The results seem to indicate that Bel 3 of HSRV might function as a spumaviral superantigen. It is fascinating that Bel 3 has a segmental homology to the HIV-2 Nef protein sequence (66), and to that of Sag (J. Weissenberger and R. M. Flügel, unpublished). In view of the proven *in vivo* requirement for HIV-1 Nef for high virus loads, full pathogenesis, and, last but not least, decreases in CD4$^+$ cells (67, 68), this homology is of interest. Of potential importance, a recent report claimed to have detected deletions of specific Vβ elements of the T-cell receptor lymphocytes of acquired immunodeficiency syndrome (AIDS) patients but not in normal individuals (69).

9. HSRV-specific detection systems

There has been some recent interest in the possible involvement of HSRV in chronic fatigue syndrome (CFS) (70, for a review on CFS, see ref. 71). However, using HSRV-specific PCRs and ELISA techniques, the Heidelberg laboratory has found no evidence for consistent association of HSRV with CFS (72).

The question of whether HSRV is a human pathogen remains open, but certain clues have been found. To address this interesting problem, it was necessary to develop a systematic approach that relies on HSRV-specific sequences and corresponding antibodies suitable and specific for the identification of HSRV DNA, RNA, and antigens. Straightforward methods and sensitive detection systems are being developed in several laboratories. Southern blot hybridization with labelled HSRV-specific probes is still the method of choice, if sufficient sample DNA is at hand.

Recently, very sensitive PCR techniques in various versions have been employed (73). However, this technique suffers from its sensitivity to false positives. In laboratories where work on HSRV and/or closely related retroviruses is done, extreme caution is required to avoid any laboratory cross-contamination. Another assumption on which PCR is also based is that the viral nucleic acid potentially present in the patient biopsy material contains sequences that are at least in part homologous to the oligonucleotides used as primers. It is, however, unknown how far the present-day HSRV sequence in an acutely infected individual might have diverged from that of the cloned HSRV that was passaged for about 20 years in human fibroblastic cells. Hence it is not straightforward to take at random any spumavirus-specific DNA sequence to design primers for PCR. Although the rate of variability is probably lower for spumaviruses than for HIV, this assumption might lead to difficulties. One way out of this dilemma is to use alternative and specific detection methods specific for HSRV, e.g. sensitive, but HSRV-specific immunological tests and Southern hybridization. Since in natural infection by retroviruses, antibodies against Gag and Env antigens have been identified in most cases, HSRV-specific ELISA and protein blots have been developed, established and employed (6, 32, 37).

10. Sero-epidemiological studies on prevalence of HSRV antibodies

A recombinant HSRV DNA clone containing the HSRV Gag capsid antigen domain was constructed and expressed in *E. coli* (32). The purified CA-Gag protein was employed to set up HSRV Gag-specific ELISAs as screening tests and a corresponding Western blotting test system was developed as confirmatory test (32). This HSRV Gag-based protein blot is particularly suited for testing a large number of human patients' sera. Some patients' sera contained antibodies against the two characteristic double bands of 78 and 74 kDa that are derived from two versions of the HSRV Gag precursor. Smaller HSRV antigens from processed forms of the HSRV Gag proteins can also be detected. Similar tests have been developed for the detection of antibodies against HSRV Env antigens (37). The results of a sero-epidemiological survey were based on these assays (6) and confirm the results of older studies that used different methods (74, 75). It was found that HSRV antibodies are more prevalent in African patients with a broad spectrum of diseases than in European patients. Certain patient groups, including those with multiple sclerosis, CFS, and Graves disease were invariantly seronegative (6). There was a certain number of tumour and neurologic disease patients that had a relatively high percentage of antibody positives. The percentage of HSRV Gag- and Env-antibody seropositives was highest among patients suffering from cervical myelopathies and amyotrophic lateral sclerosis (76). These results are of interest in view of the recently reported brain-specific HSRV gene expression in transgenic mice (7).

11. Neuropathogenicity of HSRV genes in transgenic mice

Recently, HSRV genes were introduced into the germ line of mice to study the tissue tropism and the pathogenic potential of spumaviruses in an *in vivo* animal model system (7, 8). HSRV DNA constructs were introduced into fertilized mouse eggs by microinjection techniques. One HSRV wild-type construct contained a frameshift mutation that resulted in a non-functional integrase protein; the other construct was deleted in most of *gag* and *env* and completely lacked the *pol* gene. The second plasmid construct coded for the *bel* genes. At 6 to 8 weeks of age the HSRV transgenic mice developed severe neurological symptoms, including ataxia, spastic tetraparesis, and to a lesser degree blindness (8). All mice showing neurologic disease died within 4 to 6 weeks. Stable transgenic mice containing multiple copies of the second construct could be bred, since disease progression was slow, whereas transgenic mice harbouring the first construct died earlier. In all animals that were examined histochemically and by *in situ* hybridization, the histopathological alterations were confined to the central nervous system and the striated muscles. Nerve cell lesions were prominent in the forebrain. Foci of degeneration were observed in the striated muscles, particularly in the autochthonous spinal musculature (8). HSRV RNA was identified by blot hybridization (7). Bel 1 protein expression was detected by using antisera against Bel 1 in the affected tissues (A. Aguzzi, personal communication). The *bel 1* gene is postulated to be the most likely candidate that is responsible for the observed neuromyodegenerative disease without ruling out the remaining aminoterminal HSRV *gag* sequences that might contribute to the pathological symptoms. In view of the proven strong *trans*-activating activity of Bel 1, the increase in HSRV transcription in single cells could be responsible for the cytopathic effects in the transgenic mouse model system. It is of interest that the expression of HSRV genes during development in transgenic mice was shown to occur in two distinct phases (77). There was low HSRV gene expression in mid-gestation that later increased to moderate levels in tissue originating from embryonic mesoderm, neuroectoderm, and neural crest. Later, however, and after birth expression of HSRV transgenes was suppressed. After 8 weeks HSRV gene transcription increased in the central nervous system and in skeletal muscle. The results imply that there exists a developmental control of HSRV gene expression in transgenic mice (77). Doubtless, the potential of HSRV genes to function in nerve cells will be a challenge for further work directed to the analysis of the underlying mechanisms.

It has long been established that several retroviruses from different groups can induce neurologic disease or lesions of nerve cells (11). Well known are the demyelinating effects of visna virus on brain tissues of sheep (78), HIV-1 AIDS dementia (79), and tropical spastic paraparesis (TSP), also called HTLV-associated myelopathy (HAM), found in certain patients infected with HTLV-I (80). It is less well known that certain oncoviruses, e.g. including some mutants of MLV and ecotropic strains of wild mouse MLV, can induce a spongiform neurodegenerative disease that is manifested by an encephalopathy, hind-leg paralysis, gliosis, and chronic non-inflammatory neuronal degeneration (for a review, see ref. 81). The disease occurs naturally in old animals but it can be induced by experimental infection of new-born mice. The neurovirulent determinants were mapped to the *env* gene and to the R–U5 5' leader sequence of Cas-BR-E MLV that controls the onset of disease (82, 83). These observations and results indicate that neurogenerative diseases that are caused by different retroviruses are not exceptions but seem rather to be the rule.

12. Concluding remarks

Several results and observations that were discussed in this review will determine future research efforts in this area. The brain-specific effects of the complex retrovirus HSRV (84) in transgenic mice are of potentially great interest. In this context, it is intriguing that foamy viruses were often isolated from the brain of monkeys and apes. In one case, another human spumavirus isolate that had originally been obtained from a kidney dialysis patient with concomitant encephalopathy was recently rescued and is currently being characterized in detail (R. A. Weiss, personal communication). Studies on the Bel 1 *trans*-activator will undoubtedly focus on interactions with cellular molecules required for viral transcription in different cell types. The recently sequenced genome of SFV-3 is of interest, since in contrast to the other spumaviruses, it has been reported to grow in the lymphoblastoid cell line MOLT-4 (85). It is also of interest to compare the gene expression of foamy viruses with that of other complex retroviruses. The main distinguishing feature is the presence of Rev and Rex proteins and their cognate response elements in lentiviruses and the HTLV/BLV group of retroviruses (56, 84). The HIV-1 Rev protein acts post-transcriptionally and is absolutely required for the expression of viral mRNAs that encode the viral structural proteins (56). In marked contrast, HSRV appears to lack an RRE or a *rev*-like gene. The role and mechanism of action of the HSRV Bet, Bel 2, and Bel 3 proteins, and their potential mechanistic relationship to accessory genes, such as *nef*, encoded by other complex retroviruses, is also of considerable interest. However, the most important issue remains the degree to which HSRV plays a role in the development and pathology of human diseases, including those of a neurodegenerative nature in particular.

Acknowledgements

I thank Bryan Cullen, Martin Löchelt, and Walter Muranyi for critically reading the manuscript and Harald zur Hausen for support. I wish to thank Adriano Aguzzi, Vienna for communicating manuscripts prior to publication. The work described in this review was financed by grants (DFLR KI8906/1) from the Bundesministerium fur Forschung und Technologie and in part by a grant (TS2 0186) from the Commission of European Community in Brussels.

References

1. Achong, G., Mansell, P. W. A., Epstein, M. A., and Clifford, P. (1971) An unusual virus in cultures from a human nasopharyngeal carcinoma. *J. Natl Cancer Inst.*, **42,** 299.

2. Flügel, R. M., Rethwilm, A., Maurer, B., and Darai, G. (1987) Nucleotide sequence analysis of the *env* gene and its flanking regions of the human spumaretrovirus reveals two novel genes. *EMBO J.*, **6**, 2077.
3. Maurer, B., Bannert, H., Darai, G., and Flügel, R. M. (1988) Analysis of the primary structure of the long terminal repeat and the *gag* and *pol* genes of the human spumaretrovirus. *J. Virol.*, **62**, 1590.
4. Flügel, R. M. (1991) Spumaviruses: a group of complex retroviruses. *J. Acquired Immun. Deficiency Syndr.*, **4**, 739.
5. Coffin, J. M. (1990) Replication of retroviridae. In *Virology*. B. N Fields (ed.). Raven Press, New York, Chapter 51, p. 1437.
6. Mahnke, C., Kashaiya, P., Rössler, J., Bannert, H., Levin, A., Blattner, W. A., Dietrich, M., Luande, J., Löchelt, M., Friedman-Kien, A. E., Komaroff, A. L., Loh, P. C., Westarp, M. E., and Flügel, R. M. (1992) Human spumavirus antibodies in sera from African patients. *Arch. Virol.*, **123**, 243.
7. Bothe, K., Aguzzi, A., Lassmann, H., Rethwilm, A., and Horak, I. (1991) Progressive encephalopathy and myopathy in transgenic mice expressing human foamy virus genes. *Science*, **253**, 555.
8. Aguzzi, A., Bothe, K., Wagner, E. F., Rethwilm, A., and Horak, I. (1992) Human foamy virus: an underestimated neuropathogen? *Brain Pathol.*, **2**, 61.
9. Mergia, A. and Luciw, P. A. (1991) Replication and regulation of primate foamy viruses. *Virology*, **184**, 475.
10. Hooks, J. J. and Gibbs, C. J. (1975) The foamy viruses. *Bacteriol. Rev.*, **39**, 169.
11. Weiss, R. A., Teich, N., Varmus, H. E., and Coffin, J. M. (ed.) (1985) In *RNA Tumor Viruses*. 2nd edn. Cold Spring Harbor, Cold Spring Harbor Laboratory, NY, pp. 25–207.
12. Gelderblom, H., Özel, M., Gheysen, D., Reupke, H., Winkel, T., Grund, C., and Pauli, G. (1990) Morphogenesis and fine structure of lentiviruses. In *Animal Models in AIDS*. H. Schellekens, and M. C. Horzinek (eds). Elsevier Science Publishers, Amsterdam, p. 1.
13. Hooks, J. J. and Detrick-Hooks, B. (1981) Simian foamy virus-induced immunosuppression in rabbits. *J. Gen. Virol.*, **44**, 383.
14. Kupiec, J.-J., Kay, A., Hayat, M., Ravier, R., Peries, J., and Galibert, F. (1991) Sequence analysis of the simian foamy virus type 1 genome. *Gene*, **101**, 185.
15. Mergia, A., Shaw, K. E. S., Lackner, J. E., and Luciw, P. A. (1990) Relationship of the *env* genes and the endonuclease domain of the *pol* genes of simian foamy virus type 1 and human foamy virus. *J. Virol.*, **64**, 406.
16. Renne, R., Friedl, E., Schweizer, M., Fleps, U., Turek, R., and Neumann-Haefelin, D. (1992) Genomic organization and expression of simian foamy virus type 3 (SFV-3). *Virology*, **186**, 597.
17. Renshaw, R. W., Gonda, M. A., and Casey, J. W. (1991) Viral DNA structure and transcriptional status of bovine syncytial virus in cytopathic infections. *Gene*, **105**, 179.
18. Vile, R. G. and Weiss, R. A. (1991) Virus receptors as permeases. *Nature*, **352**, 666.
19. Capon, D. J. and Ward, R. H. R. (1991) The CD4–gp120 interaction and AIDS pathogenesis. *Ann. Rev. Immunol.*, **9**, 649.
20. McClure, M. O., Sommerfelt, M. A., Marsh, M., and Weiss, R. A. (1990) The pH independence of mammalian retrovirus infection. *J. Gen. Virol.*, **71**, 767.
21. Camerini, D. and Seed, B. (1990). A CD4 domain important for HIV-medicated syncytium formation lies outside the virus binding site. *Cell*, **60**, 747.
22. Hu, W.-S. and Temin, H. M. (1990) Retroviral recombination and reverse transcription. *Science*, **250**, 1227.

23. Loh, P. C. and Matsuura, F. (1981) The RNA of the human syncytium-forming (foamy) virus. *Arch. Virol.*, **68**, 53.
24. Bowerman, B., Brown, P. O., Bishop, J. M., and Varmus, H. E. (1989) A nucleoprotein complex mediates the integration of retroviral DNA. *Genes Dev.*, **3**, 469.
25. Brown, P. O. (1991) Integration of retroviral DNA. *Curr. Topics Microbiol. Immunol.*, **157**, 21.
26. Engleman, A., Mizuuchi, K., and Craigie, R. (1991) HIV-1 DNA integration: mechanism of viral DNA cleavage and DNA strand transfer. *Cell*, **67**, 1211.
27. Chow, S. A., Vincent, K. A., Ellison, V., and Brown, P. O. (1992) Reversal of integration and DNA splicing mediated by integrase of human immunodeficiency virus. *Science*, **255**, 723.
28. Chalker, D. L. and Sandmeyer, S. B. (1992) Ty3 integrates within the region of RNA polymerase III transcription initiation. *Genes Dev.*, **6**, 117.
29. Weller, S. K. and Temin, H. M. (1981) Cell killing by avian leukosis viruses. *J. Virol.*, **39**, 713.
30. Harris, J. D., Blum, H., Scott, J., Traynor, B., Ventura, P., and Haase, A. (1984) Slow virus visna: reproduction *in vitro* of virus from extrachromosal DNA. *Proc. Natl Acad. Sci. USA*, **81**, 7212.
31. Flügel, R. M., Rethwilm, A., Maurer, B., and Darai, G. (1990) Nucleotide sequence analysis of the *env* gene and its flanking regions of the human spumaretrovirus reveals two novel genes. Corrigenda. *EMBO J.*, **9**, 3806.
32. Bartholomä, A., Muranyi, W., and Flügel, R. M. (1992) Bacterial expression of the capsid antigen domain and identification of native *gag* proteins in spumavirus-infected cells. *Virus Res.*, **23**, 27.
33. Orozlan, S. and Luftig, R. B. (1990) Retroviral proteinases. *Curr. Topics Microbiol. Immunol.*, **158**, 153.
34. Jacks, T., Madhani, H. D., Masiarz, F. R., and Varmus, H. E. (1988) Signals for ribosomal frameshifting in the Rous sarcoma virus *gag–pol* region. *Cell*, **55**, 447.
35. Maurer, B. and Flügel, R. M. (1988) Genomic organization of the human spumaretrovirus and its relatedness to AIDS and other retroviruses. *AIDS Res. Hum. Retrovirol.*, **4**, 467.
36. Jacobo-Molina, A. and Arnold, E. (1991) HIV reverse transcriptase structure–function relationships. *Biochemistry*, **30**, 6351.
37. Mahnke, C., Löchelt, M., Bannert, H., and Flügel, R. M. (1990) Specific enzyme-linked immunosorbent assay for the detection of antibodies to the human spumavirus. *J. Virol. Methods*, **29**, 13.
38. Netzer, K. O., Rethwilm, A., Maurer, B., and ter Meulen, V. (1990) Identification of major immunogenic structural proteins of human foamy virus. *J. Gen. Virol.*, **71**, 1237.
39. Tobaly-Tapiero, J., Santillana-Hayat, M., Giron, M. L., Guillemin, M. C., Rozain, F., Peries, J., and Emanoil-Ravier, R. (1990) Molecular differences between two immunologically related spumaretroviruses: the human prototype HSRV and the chimpanzee isolate SFV 6. *AIDS Res. Human Retroviruses*, **6**, 951.
40. Löchelt, M., Zentgraf, H., and Flügel, R. M. (1991) Construction of an infectious DNA clone of the full-length human spuma retrovirus genome and mutagenesis of the *bel 1* gene. *Virology*, **184**, 43.
41. Muranyi, W. and Flügel, R. M. (1991) Analysis of splicing patterns of human spumaretrovirus by polymerase chain reaction reveals complex RNA structures. *J. Virol.*, **65**, 727.
42. Rethwilm, A., Baunach, G., Netzer, K. O., Maurer, B., Borisch, B., and ter Meulen, V. (1990) Infectious DNA of the human spumaretrovirus. *Nucleic Acids Res.*, **18**, 773.

43. Kupiec, J. J., Tobaly-Tapiero, J., Canivet, M., Santillana-Hayat, M., Flügel, R. M., Peries, J., and Emanoil-Ravier, R. (1988) Evidence for a gapped linear duplex DNA intermediate in the replicative cycle of human and simian spumaviruses. *Nucleic Acids Res.*, **16,** 9557.
44. Tobaly-Tapiero, J., Kupiec, J. J., Santillana-Hayat, M., Canivet, M., Peries, J., and Emanoil-Ravier, R. (1991) Further characterization of the gapped DNA intermediates of human spumaretrovirus: evidence for a dual initiation of plus-strand DNA synthesis. *J. Gen. Virol.*, **72,** 605.
45. Harris, J. D., Scott, J. V., Traynor, B., Brahic, M., Stowring, L., Ventura, P., Haase, A. T., and Peluso, R. (1981) Visna virus DNA: discovery of a novel gapped structure. *Virology*, **113,** 573.
46. Kozak, M. (1989) The scanning model of translation: an update. *J. Cell. Biol.*, **108,** 229.
47. Schwartz, S., Felber, B., and Pavlakis, G. N. (1992) Mechanism of translation of monocistronic and multicistronic human immunodeficiency virus type 1 mRNAs. *Mol. Cell. Biol.*, **12,** 207.
48. Keller, A., Partin, K. M., Löchelt, M., Bannert, H., Flügel, R. M., and Cullen, B. R. (1991) Characterization of the transcriptional *trans*-activator of human foamy retrovirus. *J. Virol.*, **65,** 2589.
49. Rethwilm, A., Erlwein, O., Baunach, G., and Maurer, B. (1991) The transcriptional transactivator of human foamy virus maps to the *bel 1* genomic region. *Proc. Natl Acad. Sci. USA*, **88,** 941.
50. Venkatesh, L. K., Theodorakis, P. A., and Chinnadurai, G. (1991) Distinct *cis*-acting regions in U3 regulate *trans*-activation of the human spumaretrovirus long terminal repeat by the viral *bel 1* gene product. *Nucleic Acids Res.*, **19,** 3661.
51. Keller, A., Garrett, E. D., and Cullen, B. R. (1992) The *bel 1* protein of human foamy virus activates human immunodeficiency virus type-1 gene expression via a novel DNA target site. *J. Virol.*, **66,** 3946.
52. Mergia, A., Shaw, K. E. S., Pratt-Lowe, E., Barry, P. A., and Luciw, P. A. (1990) Simian foamy virus type 1 is a retrovirus which encodes a transcriptional activator. *J. Virol.*, **64,** 3598.
53. Mergia, A., Shaw, K. E. S., Pratt-Lowe, E., Barry, P. A., and Luciw, P. A. (1991) Identification of the simian foamy virus transcriptional transactivator gene (*taf*). *J. Virol.*, **65,** 2903.
54. Mergia, A., Pratt-Lowe, E., Shaw, K. E. S., Renshaw-Gegg, L. W., and Luciw, P. A. (1992) *Cis*-acting regulatory regions in the long terminal repeat of simian foamy virus types 1. *J. Virol.*, **66,** 251.
55. Pugh, B. F. and Tijan, R. (1992) Diverse transcriptional functions of the multisubunit eukaryotic TFIID complex. *J. Biol. Chem.*, **267,** 679.
56. Cullen, B. R. (1991) Regulation of human immunodeficiency virus replication. *Ann. Rev. Microbiol.*, **45,** 219.
57. Maurer, B., Serfling, E., ter Meulen, V., and Rethwilm, A. (1991) Transcriptional factor AP-1 modulates the activity of the human foamy virus long terminal repeat. *J. Virol.*, **65,** 6353.
58. Herman, A., Kappler, J. W., Marrack, P., and Pullen, A. M. (1991) Superantigens: mechanism of T-cell stimulation and role in immune responses. *Ann. Rev. Immunol.*, **9,** 745.
59. Coffin, J. M. (1992) Superantigens and endogenous retroviruses: a confluence of puzzles. *Science*, **255,** 411.

60. Choi, Y., Kappler, J. W., and Marrack, P. (1991) A superantigen encoded in the open reading frame of the 3' long terminal repeat of mouse mammary tumour virus. *Nature*, **350**, 203.
61. Acha-Orbea, H., Shakov, A. N., Scarpellino, L., Kolb, E., Müller, V., Vessaz-Shaw, A., Fuchs, R., Blöchinger, K., Rollini, P., Billotte, J., Sarafidou, M., Robson MacDonald, H., and Diggelmann, H. (1991) Clonal deletion of V b14-bearing T cells in mice transgenic for mammary tumour virus. *Nature*, **350**, 207.
62. Racevskis, J. (1986) Expression of the protein product of the mouse mammary tumor virus long terminal repeat gene in phorbol ester-treated mouse T-cell-leukemia cells. *J. Virol.*, **58**, 441.
63. Racevskis, J. and Prakash, O. (1984) Proteins encoded by the long terminal repeat region of mouse mammary tumor virus: identification by hybrid-selected translation. *J. Virol.*, **51**, 604.
64. Brandt-Carlson, C. and Butel, J. S. (1991) Detection and characterization of a glycoprotein encoded by the mouse mammary tumor virus long terminal repeat gene. *J. Virol.*, **65**, 6051.
65. Michalides, R., Wagenaar, E., and Weijers, P. (1985) Rearrangements in the long terminal repeat of extra mouse mammary tumor proviruses in T-cell leukemias of mouse strain GR result in a novel enhancer-like structure. *Mol. Cell. Biol.*, **5**, 823.
66. Maurer, B. and Flügel, R. M. (1987) The 3' orf protein of human immunodeficiency virus 2 shows sequence homology with the *bel 3* gene of the human spumaretrovirus. *FEBS Lett.*, **222**, 286.
67. Kestler III, H. W., Ringler, D. J., Mori, K., Panicalli, D. L., Sehgal, P. K., Daniel, M. D., and Desrosiers, R. C. (1991) Importance of the *nef* gene for maintenance of high virus loads and for development of AIDS. *Cell*, **65**, 651.
68. Cullen, B. R. (1991) The positive effect of the negative factor. *Nature*, **351**, 698.
69. Imberti, L., Sottini, A., Bettinardi, A., Puoti, M., and Primi, D. (1991) Selective depletion in HIV infection of T cells that bear specific T cell receptor Vb sequences. *Science*, **254**, 860.
70. Palca, J. (1991) On the track of an elusive disease. *Science*, **254**, 1726.
71. Komaroff, A. L. (1988) Chronic fatigue syndromes: relationship to chronic viral infections. *J. Virol. Methods*, **21**, 3.
72. Flügel, R. M., Mahnke, C., Geiger, Komaroff, A. L. (1992) Absence of antibody to human spumaretrovirus in patients with chronic fatigue syndrome. *Clin. Infect. Dis.*, **14**, 623.
73. Muranyi, W. and Flügel, R. M. (1992) Detection of human spumaviruses by PCR. In *Frontiers of Virology*, Y. Becker and G. Darai (eds). Springer-Verlag, Heidelberg, Vol. 1, Chapter 5, pp. 46–53.
74. Loh, P. C., Matsuura, F., and Mizumoto, C. (1980) Seroepidemiology of human syncytial virus: antibody prevalence in the Pacific. *Intervirology*, **13**, 87.
75. Muller, K. H., Ball, G., Epstein, M. A., Achong, B. G., Lenoir, G., and Levin, A. (1980) The prevalence of naturally occurring antibodies to human syncytial virus in East African populations. *J. Gen. Virol.*, **47**, 399.
76. Westarp, M. E., Kornhuber, H. H., Rössler, J., and Flügel, R. M. (1992) Human spumaretrovirus in amyotrophic lateral sclerosis. *Neurol., Psychiatry Brain Res.* **1**, 1–4.
77. Aguzzi, A., Bothe, K., Anhauser, I., Horak, I., Rethwilm, A., and Wagner, E. F. (1992) Expression of human foamy virus is differentially regulated during development in transgenic mice. *New Biol.*, **4**, 1.

78. Narayan, O. and Clements, J. E. (1989) Biology and pathogenesis of lentiviruses. *J. Gen. Virol.*, **70**, 1617.
79. Price, R. W., Brew, B., Sidtis, J., Rosenblum, M., Scheck, A. C., and Cleary, P. (1988) The brain in AIDS: central nervous system HIV-1 infection and AIDS dementia complex. *Science,* **239,** 586.
80. Gessain, A., Barin, F., Vernan, J., Gout, O., Maurs, L., Calendar, A., and de The, G. (1985) Antibodies to human T-lymphotropic virus type I in patients with tropical spastic parasis. *Lancet,* **ii,** 407.
81. Gardner, M. B. (1985) Retroviral spongiform polioencephalomyelopathy. *Rev. Infect. Dis.*, **7,** 99.
82. Rassart, E., Nelbach, L., and Jolicoeur, P. (1986) Cas-Br-E murine leukemia virus: sequencing of the paralytogenic region of its genome and derivation of specific probes to study its origin and the structure of its recombinant genomes in leukemic tissues. *J. Virol.*, **60,** 910.
83. Portis, J. L., Perryman, S., and McAtee, F. J. (1991) The R-U5-5' leader sequence of neurovirulent wild mouse retrovirus contains an element controlling the incubation period of neurodegenerative disease. *J. Virol.*, **65,** 1877.
84. Cullen, B. R. (1991) Human immunodeficiency virus as a prototypic complex retrovirus. *J. Virol.*, **65,** 1053.
85. Neumann-Hefelin, D., Rethwilm, A., Bauer, G., Gudat, F., and zur Hausen, H. (1983) Characterization of a foamy virus isolated from *Cercopithecus aethiops* lymphoblastoid cells. *Med. Microbiol. Immunol.*, **172,** 75.

Index

N-acetyl cysteine, HIV transcriptional regulation and 60
adult T-cell leukemia/lymphoma 163
adult T-cell leukemia retrovirus, see human T-cell leukemia virus type I
AIDS
 discovery/historical aspects 2
 Vpu and 146
alimentary tract, HIV infection 20–1
cAMP response element (CRE) 172, 177
cAMP response element binding protein (CREB), Tax$_1$ and 167, 167, 172, 173, 176
animal models
 of HSRV-associated disease 207–9
 of HTLV-associated disease 164–5
antibodies
 HIV infection enhanced by 31–2
 HSRV, sero-epidemiological studies on prevalence of 207
anti-oxidants, HIV transcriptional regulation and 60
antiretroviral therapy 125–6
antisense DNA/oligonucleotides as antiretroviral agents 125
assembly, virion 8–9
 Vpu and 146
ATF, Tax$_1$ and 167, 172, 173, 174
ATLV, see human T-cell leukemia virus type I
ATP, Tat and the effects of 85
auxiliary proteins 137–58
 of non-primate lentiviruses/ immunodeficiency viruses 140–1
 of primate lentiviruses/ immunodeficiency viruses 137–58

bel 1 (gene) 199, 201–2
 neurodegeneration in transgenic mice and the 208
 transcription 201–2

Bel 1 (protein) 202–5
 neurodegeneration in transgenic mice and the 208
 as transcriptional trans-activator 202–3
Bel 1–response element 204, 205
bel 3 (gene) 206
Bel 3 (protein) 205–6
β-polymerase, Tax$_1$ and the 177
bet gene, transcription 201–2
Bet protein 199
blood cells, HIV infection 20
BRE 204, 205
BT cells, HIV tropism for 31

CA, HSRV 196
cAMP response element, see AMP
capsid antigen, HSRV 196
CD4 glycoprotein receptor for HIV 2, 3, 24, 26–7, 32, 146, 147
 down-regulation 149–50
  Nef protein and 149
  significance 150
 gp120/gp160 interactions with 22, 23, 32, 33, 146, 147
  Vpu and 146, 147
 soluble recombinant version (sCD4) 35
  HIV neutralized by 26
 tropism and the envelope region binding to 27–8
CD4$^+$ lymphocytes, see T-cell
central nervous system, HIV infection 20
c-fos proto-oncogene, Tax$_1$ and the 176–7
Chinese hamster ovary (CHO) cells, TAT and TAR in 82–3
chronic fatigue syndrome, HSRV and 206
CK1 and CK2, Tax responsiveness 176
classification, retroviral 5, 7, 9–10
complement receptors, HIV infection via 31–2

complex retroviruses 5–6, 7–8, 103–20
 foamy viruses/spumaviruses as 193
 gene expression 5–6, 7–8, 103–20
 post-transcriptional regulation 103–20
CRE and CREB, see cAMP response element; cAMP response element binding protein
c-rel proto-oncogene 57
cutaneous lesions, HIV-associated, Tat and 89
cutaneous T-cell leukemia/ lymphoma 164
cyclic AMP, see AMP
cysteine protease inhibitor, Vif-dependent processing of gp41 and effects of 142
cytokines
 extracellular, Tax as 177–9
 HIV gene expression and 51, 53
 HTLV Tax inducing expression of 174, 176
 see also specific cytokines
cytoplasm, transport of RNA from nucleus to, Rev-like proteins involved in 119

DNA
 antisense, as antiretroviral agents 125
 transcription, see transcription
DNA-binding proteins, Tat trans-activation and 85–7; see also specific proteins
Dorsal gene (Drosophila) 57

E1A 53
E2 promoter 53
E64 (cysteine protease inhibitor), Vif-dependent processing of gp41 and effects of 142

INDEX

effector domain of Rev/Rev-like protein 114–15
  antiretroviral agents acting at 126
EIAV, *see* equine infectious anaemia virus
Elf-1 protein 62
enhancer
  HIV-1 53–7, 60
    negative regulation 57
  HIV-2 61–2
env gene 6, 18
  of HSRV 198, 200–1
  of SIV, *see* simian immunodeficiency virus
Env glycoprotein, human spumavirus 198–9
Env glycoprotein, lentiviral (gp160) 123–5
  CD4 and, *see* CD4
  external surface domain, *see* gp120 HIV
    structure 22–3
    translation 123–5
  SIV, species specificity and the 28
  transmembrane region/cytoplasmic tail, *see* gp41
equine infectious anemia virus (EIAV)
  TAR 78
  TAT 75, 77–8, 80
Ets-like binding proteins 62
expression/induction/activation of genes, *see* gene expression

fatigue, syndrome of chronic, HSRV and 206
Fc receptors, HIV infection via 31–2
fibroblasts, HIV tropism for 31
foamy viruses (spumaretroviruses) 1, 193–214
  antibodies, sero-epidemiological studies on prevalence of 207
  detection system 206–7
  genetic structure 196–200
  human, *see* human spumaretrovirus
  morphological and virological properties 194
  pathogenicity 206, 207–9
  replication cycle 194–6
  simian 194
  superantigen 205–6
  taxonomy 9, 10
  transcription 200–5

foot-and-mouth disease virus protein p3A, Vpu functional similarities with 147
c-fos proto-oncogene, Tax_1 and the 176–7
frameshifting, ribosomal, *see* ribosomal frameshifting

gag (gene) 6, 18
Gag (polyprotein) 8, 9, 121–2, 196–7, 198
  HSRV 196–7, 198, 207
  synthesis/assembly 8, 121–2, 196–7
Gag–Pol polyproteins 8, 9
  synthesis/assembly 8, 121
Gal-4 87
galactosyl ceramide, HIV infection and 32
GALV, receptor 2
gastrointestinal tract, HIV infection 20–1
gene expression
  cellular, Tax regulation of 174–7
  retroviral, regulation 5–8, 49–73, 75–136; *see also specific viruses see also specific genes*
genome (retroviral)
  reverse transcription 4–5
  structure/organization 6
    HSRV 196–200
    HTLV 160–1
    lentivirus 138, 139
gibbon ape leukemia virus, receptor 2
gp41 (transmembrane protein)
  HIV 22–3
    Vif-dependent proteolysis 142
  SIV 28
  synthesis 22–3, 146, 147
gp120 (Env glycoprotein) 18, 22, 22–3
  CD4 interactions with, *see* CD4
  structure 22–3
  synthesis 22–3, 146, 147
  V3 region, *see* V3
gp160, *see* Env glycoprotein
granulocyte–macrophage colony stimulating factor, HTLV Tax inducing expression of 175
GTPase activity of Nef protein 149

GTP-binding activity of Nef protein 149
guanine nucleotide binding by Nef protein 149
gut, HIV infection 20–1

HEB1 173
HeLa cells
  gp160–CD4 interaction in, VPU and 146
  HIV tropism for 31
HIP 53
HIV, *see* human immunodeficiency virus
HSRV, *see* human spumaretrovirus
HTLV, *see* human T-cell leukemia virus
human immunodeficiency virus(es) 17–73, 75–100 *passim*
  CD4 glycoprotein receptor, *see* CD4
  gene expression (transcription etc.) 49–73, 83–7, 103–4
    anti-oxidants and 60
    HSVR Bel 1 and 203
    inducible/induced/activated 53–6, 203
    suppression (negative regulation) 56–7
  genetic variation/heterogeneity 17–18, 24
  pathogenesis
    HSRV infection contributing to 203
    tropism and, relationship between 21–2
  species specificity 19
  Tat transactivator 75–100
  tissue distribution 19–22
  tropism 17–48
    for fibroblasts 31
    historical perspectives 23–4
    at level of entry, important events controlling 32–5
    for macrophages, *see* macrophages
    non-envelope regions involved in 29–31
    non-V3 envelope regions involved in 27–8
    pathogenesis and, relationship between 21–2
    soluble CD4 neutralization correlating with 26

INDEX | 217

for T-cell lines 26, 27
V3 loop as determinant of 24–7, 33–5, 35
human immunodeficiency virus initiator protein (HIP) 53
human immunodeficiency virus type 1 (HIV-1) 17
discovery/historical aspects 2
gag/pol region of, ribosomal frameshifting and the 121
replication, *see* replication
Rev protein 105–8, 110, 117
translation 123–5
tropism
for macrophages, *see* macrophages
non-V3 envelope regions involved 27, 28
for T-cell lines 26, 27
vif gene mutants 142
vpr gene 142–3
human immunodeficiency virus type 2 (HIV-2) 17, 61–2
nef gene and gene mutant 148, 149, 150
Bel 1 (of HSRV) homology with 206
transcriptional regulation 61–2
vpr gene mutants 144
vpx gene mutants 145
human spumaretrovirus (HSRV; human foamy virus) 193–214
antibodies, sero-epidemiological studies on prevalence of 207
detection systems 206–7
genetic structure 196–200
morphological and virological properties 194
pathogenicity 206, 207–9
replication cycle 194–6
superantigen 205–6
transcription 200–5
human T-cell leukemia virus(es) (HTLV) 159–92
disease associated with 163–5
epidemiology 162–3
genome structure 160–1
transcriptional regulation 170–7
transformation 161–2
human T-cell leukemia virus type I (HTLV-I; ATLV) 1–2
discovery/isolation 159
discriminating assays 162
diseases associated with 163–4, 164–5

Rex 7, 113, 114, 115, 117, 160, 167, 168, 170
Rex response elements (XRE; RxRE$_2$) 115, 116, 117, 118, 168–9, 169
Tax$_1$, *see* Tax$_1$
transformation by 161–2
transmission 162–3
human T-cell leukemia virus type II (HTLV-II)
discovery/isolation 159
discriminating assays 162
disease associated with 164
Rex 115, 167, 170
Rex response elements (XRE; RxRE$_2$) 115, 169–70
Tax$_2$, *see* Tax$_2$
human T-cell leukemia virus type V (HTLV-V) 157
disease caused by 174

IκB 59
immunodeficiency viruses, *see* human immunodeficiency virus; simian immunodeficiency virus
influenza A virus M2 protein, Vpu functional similarities with 147
initiator-binding protein 53
int domain of *pol* gene, HSRV 198, 201
integration, proviral, and integrated proviruses, *see* provirus
interleukin-1
HIV gene expression and 51, 53
mRNA, Nef products blocking induction of 150
interleukin-2, HTLV Tax inducing expression of 175
interleukin-2Rα (interleukin-2 receptor), HTLV Tax inducing expression of 175
intestine, HIV infection 20–1

Kaposi's sarcoma, Tat and 89
κB site
in HIV-1 57
in HIV-2 61

latency, Rev-like *trans*-activators and 110–11

LBP-1 53
Tat and 86
lentiviruses
auxiliary proteins, *see* auxiliary proteins
promoters, TAR RNA element in, *see trans*-acting responsive region
taxonomy 9, 10
leukemia/lymphoma, T-cell, *see* T-cell leukemia/lymphoma
life cycle, retroviral, *see* replication
long terminal repeat elements (LTRs) 4, 5, 30–1
HIV 203
HSRV 199, 202–3
HTLV-1 166, 167, 174
R region 174
SIV tropism and pathogenesis and 30–1
lung, HIV infection 20–1

M2 protein, influenza A virus, Vpu functional similarities with 147
MA, HSRV 196
macrophages 25–6, 28, 60–1
HIV expression in 60–1
HIV tropism for 25–6, 28
region 5' of V3 modulating infection efficacy 28
MAD-3 gene 59
matrix protein, HSRV 196
messenger RNA, *see* mRNA
mice, transgenic, HSRV gene pathogenicity in 207–9; *see also entries under* murine
MLV, *see* murine leukemia virus
MMTV, *see* murine (mouse) mammary tumour virus
models, animal, *see* animal models
Moloney murine leukemia virus, genetic structure 198
Mo-MLV, genetic structure 198
monocytes, HIV expression in 60–1
mouse, transgenic, HSRV gene pathogenicity in 207–9; *see also entries under* murine
murine (mouse) leukemia virus
ecotropic (MLV)
Gag/Pol production 121
genome organization 6

murine (mouse) leukemia virus (*cont.*)
  ecotropic (MLV) (*cont.*)
    neurodegenerative disorders associated with 208–9
    receptor 2, 3
  Moloney (Mo-MLV), genetic structure 198
murine (mouse) mammary tumour virus (MMTV)
  Orf protein/Sag 205–6
  ribosomal frameshifting 121
myelopathy, HTLV-1 associated 163–4
myristylation of Gag and Gag–Pol polyproteins 8

nef gene and Nef protein (negative factor) 124, 138, 148–51
  Bel 3 (of HSRV) homology with 206
  HIV tropism and 29–30, 56
negative factor gene, *see nef*
negative regulatory elements in HSRV LTR 205
nervous system, *see* central nervous system; neurological disorders
neurological disorders
  HIV-associated 20
  HSRV gene-associated, in transgenic mice 308
  MLV-associated 208–9
NFAT-1 complex/site/element 55–6, 56, 176–7
  Tax₁ activation of promoters containing 175–6
NF-κB 55, 57–9, 178
  regulation 59
NRE in HSRV LTR 205
nuclear factor κB, *see* NF-κB
nucleocytoplasmic transport of RNA, Rev-like proteins involved in 119
nucleus
  proteins in, TAR-binding 82
  Rev/Rev-like protein localization in 113

oligomerization, Rev protein 113–14
  agents interfering with 126
oligonucleotides, antisense, as antiretroviral agents 125

oncogenes, *see* specific genes/gene products
Orf protein (Sag) of MMTV 205–6

p3A, foot-and-mouth disease virus protein, Vpu functional similarities with 147
p21^{x-III} 170
p36 protein, TAT and 79
p49 protein 57, 58
p50 protein 58
p65 protein 58, 59
p68 protein, TAR-binding 82
p100 protein 57, 59
p105 protein 57, 59
paraparesis, tropical spastic 163–4
phosphorylation, Tat 85
pol gene 4, 6, 18, 120–3
  expression 120–3
  HSRV 198, 201
  see also Gag–Pol polyprotein
Pol protein, HSRV 198
polymerase chain reaction, HSRV detection via 206–7
post-transcriptional processing, *see* mRNA post-transcriptional regulation (of gene expression)
  antiretroviral therapy aimed at 125
  Tat and 87–8
post-translational modification 8
post-viral (chronic) fatigue syndrome, HSRV and 206
processing (splicing etc.), mRNA, *see* mRNA
promoters
  HIV 52–3
  RNA element in lentiviral, *see trans*-acting responsive region
  auxiliary, *see* auxiliary proteins
  DNA-binding, Tat *trans*-activation and 85–7
  post-translational modification 8
  synthesis, *see* translation
  TAR-associating 80–2
  TAT-associating 79–80
  see also specific proteins
protein kinases, Tat activity and 85
proto-oncogenes 10; *see also* specific genes and gene products

provirus
  integrated HSRV 199–200
  transcription, HSRV 196
  integrated lentivirus 199–200
  transcription 6–8
  integration, HSRV 196
pulmonary tissue, HIV infection 20–1
pX region of HTLV 160, 161

receptors for retroviruses 22–4, 31–2; *see also* specific receptors
regulator of expression of virion proteins (HIV), *see rev*; Rev
regulator of expression of virion proteins (HTLV), *see* Rex
Rel B gene product/protein 58
release, virion 8–9
  Vpu and 146
rel oncogene and *c-rel* proto-oncogene 57
Rel protein 58
replication cycle/life cycle 2–9, 88, 110–11, 194–6
  biphasic, of complex retroviruses, Rev-like *trans*-activators and 110–11
  HIV-1 10–11
  Tat and the 88
  HSRV 194–6
  see also specific stages of replication
retroposons 10
rev (gene) 124
Rev (protein) 7, 105–8, 110
Rev-like transactivators/proteins 108–15
  experimental systems for studying 111–12
  functional architecture 112–15
  latency and biphasic life cycle and 110–11
  mutants 115
Rev response element (RRE) 107–8, 113, 115, 116, 117, 118, 119
  analogues, as antiretroviral agents 125–6
reverse transcriptase
  HIV tropism and 30
  HSRV 195, 196
reverse transcription 4–5
Rex 167–70
  HTLV-I (Rex₂) 7, 113, 114, 115, 117, 160, 167, 168, 170
  HTLV-II (Rex₂) 115, 167, 170

INDEX | 219

Rex response element of HTLVs (RxRE) 115, 116, 117, 118, 168–70
ribosomal frameshifting 120–3
  inhibitors 125
RNA
  nuclear, transport to cytoplasm, Rev-like proteins involved in 119
  Rev binding to 113
  viral genome of, see genome
  mRNA
    interleukin-1, Nef products blocking induction of 150
    multiply spliced 105
    processing (splicing etc.) 7–8, 100–5
      alternative 100–5
      HSRV 201–2
      inhibition 118–19
    single spliced 104
    synthesis 6–8
    translation, see translation
    unspliced 104
  tRNA, as reverse transcription primer 4
RNA element in lentiviral promoters, see trans-acting responsive region
RNA polymerase I 51
  HIV gene expression and 52
RNA polymerase II 51
  HIV gene expression and 52
RNA polymerase III 51
rnh, Mo-MLV 198
Rof 160
Rous sarcoma virus gene expression, post-transcriptional regulation 102
RRE, see Rev response element
R region of HTLV-1 LTR 174
RSV gene expression, post-transcriptional regulation 102
RxRE (Rex response element) of HTLVs 115, 116, 117, 118, 168–70

Sag, MMTV 205–6
sarcoma, Kaposi's, Tat and 89
Schistosoma mansoni surface antigen, vif and 141
simian immunodeficiency virus (SIV) 17, 137, 139

species specificity, env gene product (transmembrane protein) associated with 28
tropism
  env mutations determining 28
  LTRs and 30–1
  virulence/pathogenicity 22, 30–1
  LTRs and 30–1
  nef and 150–1
simian spumaretrovirus (SSRV) 194
simple retroviruses (in general), gene expression 5, 7, 102–3
  post-transcriptional regulation 102–3
skin, see entries under cutaneous
SLII (stem–loop II) of RRE 117
Sp1
  HIV-1 and 52
  HIV-2 and 61
spastic paraparesis, tropical 163–4
species specificity, HIV 19
splicing of mRNA 7, 101–5
  alternative 101–5
  in HSRV 201–2
  inhibition 118–19
spongiform neurodegenerative disease, MLV-associated 208
spuma viruses, see foamy viruses
SSRV 194
stem–loop II of RRE 117
superantigen, Bel 3 as 205–6

TAR (and TAR RNA-binding protein and TAR decoys), see trans-activation response element
tat (gene) 124
Tat (trans-activator) 29, 75–100, 105, 124–5
  domains 75–6, 77
  HIV tropism and 29
  mechanisms of action 83–8, 110
  proteins associating with 79–80
  synthesis 124–5
  TAR and, interactions 80–3, 84, 87, 87–8, 88
TAT-binding protein (trans-activator-binding protein) 79
TATA(A) box/element 52
  Tat and 86
Tax₁ 160, 161, 165–7
  as extracellular cytokine 177–9
  structure and function/functional domains 165, 170–2
  trans-activation 174, 175–7

Tax₂ 165–6, 174
  trans-activation 174, 176
taxonomy, retroviral 5, 7, 9–10
TAXREB67 173
Tax-responsive element 1 172–3
Tax-responsive element 2 173
Tax-responsive element binding protein (TREB-1) 172, 173
TBP-1 79
T-cell (T lymphocyte), CD4⁺/helper subset 22
  as HIV target 19–20, 22, 24, 26
  immortalized lines/cell lines
    HIV tropism for 26, 27
    HTLV infection resulting in 161
    vif gene HIV mutant growth in 142
    vpr gene HIV mutant growth in 143–4
    vpx gene HIV mutant growth in 145
T-cell leukemia/lymphoma
  adult 163
  cutaneous 164
T-cell leukemia viruses 157–92
  human, see human T-cell leukemia virus
  taxonomy 9, 10
Tev protein (=Tnv) 75, 76
TFIID 52
therapy, antiretroviral 125–6
thyroid hormone receptor element, HIV enhancer site related to 56
Tnv protein (=Tev) 75, 76
Tof 160
trans-activation response element (TAR RNA) 75–100
  life cycle of HIV and 88
  structure in different lentiviruses 78
  Tat and, interactions 80–3, 84, 87, 87–8, 88
trans-activation response element-binding protein-1 (TAR RNA-binding protein-1; TRP1; TRP185) 82
trans-activation response element-binding protein-2 (TAR RNA-binding protein-2; TRP2) 82
trans-activation response element-binding protein-185 (TAR RNA-binding protein-1; TRP185; TRP1) 82

trans-activation response element decoys (TAR decoys) 83
trans-activation 83–8, 105–20
  antiretroviral therapy aimed at 125
  in complex retroviruses 105–20
  by Rev and Rev-like proteins 105–20
  by Tat 83–8, 105
  by Tax 174–7
trans-activator
  Bel 1 202–3
  Tat, *see* Tat; TAT-binding protein
  Tax, *see* Tax
transcription
  events following, *see entries under* post-transcriptional *and specific events*
  of HIV, *see* human immunodeficiency virus
  of HSRV 200–5
  of HTLVs 170–7
  inhibition by antiretroviral agents 125
  of integrated proviruses, *see* provirus
  reverse, *see* reverse transcriptase; reverse transcription
  see also mRNA
transcription factors, cellular, HIV gene expression and the role of 49–73
transfer RNA as reverse transcription primer 4
transgenic mice, HSRV gene pathogenicity in 207–9; *see also entries under* murine
translation of mRNA 120–5
  protein modification following 8
  Tat-controlled 87–8
transmembrane protein
  lentiviral, *see* gp41
  spumaviral 198–9
transport of RNA, nucleocytoplasmic, Rev-like proteins involved in 119
TRE-1 172–3
TRE-2 173
TREB-1 172, 173
tropical spastic paraparesis 163–4
tropism
  HIV, *see* human immunodeficiency virus
  SIV, LTRs and 30–1

TRP, *see trans*-activation response element-binding protein
tumour necrosis factor-α, HIV gene expression and 51, 53

UBP-1 53

V3 loop/region of gp120 23, 24–7, 33–5
  as tropism determinant 24–7, 33–5, 35
vif gene and gene product (Vif) 138, 141–2
virion, assembly and release, *see* assembly; release
virion infectivity factor gene (*vif*) and gene product (Vif) 138, 141–2
VP-16 87
vpr gene and gene product (Vpr) 138–9, 142–4
vpu 124, 139, 145–7
vpx gene and gene product (Vpx) 138–9, 143, 144–5